The Forecasting Accuracy of Major
Time Series Methods

The Forecasting Accuracy of Major Time Series Methods

S. Makridakis
INSEAD, Fontainebleau, France

A. Andersen
University of Sydney, Australia

R. Carbone
Université Laval, Quebec, Canada

R. Fildes
Manchester Business School, Manchester, England

M. Hibon
INSEAD, Fontainebleau, France

R. Lewandowski
Marketing Systems, Essen, Germany

J. Newton
E. Parzen
Texas A & M University, Texas, USA

R. Winkler
Indiana University, Bloomington, USA

JOHN WILEY & SONS
Chichester · New York · Brisbane · Toronto · Singapore

Library of Congress Cataloging in Publication Data:

Main entry under title:

The Forecasting accuracy of major time series methods.
  Includes index.
    1. Time-series analysis. 2. Economic forecasting—
Statistical methods. 3. Business forecasting—
Statistical methods. I. Makridakis, Spyros G.
HA30.3.F67 1984   338.5′442′0151955   83-17055

ISBN 0 471 90327 2

British Library Cataloguing in Publication Data:

The Forecasting accuracy of major time series methods.
  1. Forecasting
  I. Makridakis, Spyros
  658.4′0355   HD30.27

ISBN 0 471 90327 2

Printed in Great Britain by Galliard (Printers) Ltd, Great Yarmouth

Contents

Preface

Interest in forecasting has been increasing exponentially during the last twenty years. Many books and hundreds of articles are written each year on the topic. The *Journal of Forecasting* was created less than two years ago to serve the field. At present it has almost 2000 subscribers equally divided among the academic and business worlds. At a recent forecasting symposium attended by 1300 people 400 papers on all aspects of forecasting were delivered. As forecasting spreads its coverage and as its audience increases, the need for objectivity strengthens. Personal opinions and superstitious beliefs have to be replaced by hard scientific facts.

This book describes in detail a major empirical study dealing with forecasting accuracy which has provided a scientific basis for forecasting. It gives specific details on the 24 major time series methods that were covered by this competition. Additionally, it provides information about the methods used, the models employed, and the process by which the experts in each method forecast the real-life time series of the competition.

Forecasting accuracy is a major interest of anyone concerned with the future. Increasing forecasting accuracy could facilitate the saving of millions of dollars and becomes a major motivation for using formal forecasting methods. The topic of this book is, therefore, of uppermost importance to both practitioners and academics. The former has specific interest in actual and potential savings that can be achieved through more accurate forecasting. The latter is concerned in understanding why certain methods perform more accurately than others which will lead to ways of further improving forecasting accuracy. For both groups the empirical evidence presented in this book is indispensable. The competition has provided a unique opportunity for allowing a direct comparison of *all* major time series methods. Their advantages, limitations and disadvantages can be better understood. Decision and policy makers are presented with concrete information enabling more intelligent choices of the most appropriate method to use, the magnitude of forecasting errors to be expected, and the amount of uncertainty involved.

The conclusion of this book, that forecasting accuracy depends on the type of data and the forecasting situation considered, is creating a deep impact on the field of forecasting. As a consequence any monolithic approach to forecasting has been eliminated as a practical alternative. Knowing when and how to use the most appropriate forecasting method can provide considerable improvement in forecasting accuracy and be of direct usefulness to anyone involved with forecasting. Furthermore, the empirical evidence indicates that improved forecasting accuracy can often be achieved through simple methods. This increases the practical appeal of forecasting since simple methods are easy to use, require little data and training, and most importantly are cheap to utilize and can be applied in a routine automatic fashion.

This book is the combined effort of many researchers and experts in the field of forecasting. The results reported are the outcome of unaccountable hours of computer CPU time and several years of work. Several independent researchers have duplicated its findings and cross-checked the numerous tables and summary statistics. Its contribution to the topic of forecasting in general, and forecasting accuracy in particular is unique, by having utilized up to 1001 real-life time series and the best time series forecasters to obtain the results reported in this book.

Spyros Makridakis
Fontainebleau, October 1983

CHAPTER 1

Forecasting: State of the Art

Spyros Makridakis
INSEAD

The importance of forecasting has, without a doubt, grown tremendously during the last twenty years. The interest shown towards forecasting has come from both the academic world and from practitioners.

Academics have contributed to a proliferation of forecasting methods (even though there have been some notable exceptions from practitioners: e.g. Brown, 1963, 1977; McLaughlin, 1968; and Shiskin, 1961, among others). Practitioners, as users, have provided the *raison d'être* for forecasting which has found a captive market of eager buyers of forecasts and forecasting services.

Business executives and government policy-makers are constantly faced with uncertainty. The perception of such uncertainty has been increasing since the mid-1970s and has necessitated a more thorough and systematic consideration of the future. Predictions, provided by the various forecasting methods, are used as inputs for all types of planning, strategy formulation, policy-making, scheduling, purchasing, inventory control and a great majority of decision-making activities in general. There is no question that the role of forecasting is becoming central and its necessity indisputable. In the author's opinion, the biggest challenge currently facing the field is how to make forecasting as relevant and useful as possible.

Lately, there has been criticism and a great deal of discontent about large forecasting errors, or about the inability of forecasters to warn of forthcoming changes that caught almost everyone by surprise. At the same time, large errors, caused by a turbulent environment, unforeseen events, discontinuities and so forth have increased the demand for forecasting. If there is no uncertainty in the environment and everything turns out as expected, there is little need for forecasting. Ironically, in fast-changing times when the ability to predict is at its lowest, the perceived need for forecasting is at its highest. The demand for forecasting consultants increases during periods of recession or other crises. This is an experience observed by the author (see Makridakis and Wheelwright, 1979) and shared by almost all the forecasting consultants he knows.

A major reason for the criticism and discontent facing the field of forecasting has been *wrong* expectations from the users. Forecasting is not a substitute for

1

prophecy. Forecasters, unfortunately, do not possess crystal balls enabling them to look into the future. Forecasting errors are inevitable. What is extremely important, from a practical point of view, is to be able realistically to assess the advantages and limitations of forecasting methods (Hogarth and Makridakis, 1983) and take them into account while utilizing predictions for planning or other purposes.

Empirical studies are crucial in objectively assessing the pros and cons of various forecasting methods and the size of forecasting errors. They are the equivalent of the laboratory experimentation so popular among natural and physical scientists. From a practical point of view it is not satisfactory to believe in claims based on wishful thinking, personal interests, or selective information. Those who have developed, who sell, or are involved with certain forecasting methods, will undoubtedly advocate their method(s) to be the best. The purpose of empirical studies, on the other hand, is objectively to test the accuracy and other characteristics of forecasting methods in as scientific and objective a manner as possible. As such, empirical studies play an extremely important role in the field of forecasting. They aim at distinguishing myths from reality, and eliminating illusions and wishful thinking.

Sometimes the claim is advanced that empirical studies are not needed. It is argued that there is no substitute for good theory (e.g. see Priestley, 1979). According to such claims, if the results of empirical studies are at odds with theoretical predictions it is the latter and not the former that should be trusted. Obviously, such a claim is utterly false. Any theory is based on several assumptions. If one of these is not realistic the theory can predict events which are different from empirical observations. The best test of any good theory is its predictability (see Christ, 1951); otherwise, theories become academic constructs of no practical value, and have little use other than self-serving interests among those who advocate them.

In the field of forecasting there have been discrepancies between theoretical predictions and empirical results. The major reason for such discrepancies is that some of the theoretical assumptions do not hold. Any forecasting method is based on fitting a model to a set of data. Theoretically, the best method can be identified while fitting a model to available data. Systems of simultaneous equations will fit the data more accurately than single-equation models which in turn will be more accurate than time series (extrapolative) methods which do not include explanatory variables. Furthermore, within the domain of time series the more complex and statistically sophisticated methods (e.g. ARMA models) should be better than the simpler, and statistically naive approaches. There is no disagreement as far as the above is concerned.

Model fitting and forecasting, beyond the existing data used to develop the model, are not the same thing, however. Minimizing the model-fitting errors does not guarantee fewer errors in forecasting unless the assumption of constancy (see Makridakis, 1981) holds. This assumption is central to any statistical method and extremely relevant in forecasting. It simply states that structural changes in the data must not occur if the model-fitting results are to be accurately extrapolated beyond existing data. But there is no way to guarantee structural stability. The post-sample data can be structurally different from the data on which the forecasting model has been developed. If this is the case the best methods identified through theory do not have to be the best methods found empirically.

Unfortunately, in reality there are constant changes, structural shifts in the economy, changes in attitudes, political moves that alter established trends, new technological developments, and the like, which cause existing patterns to change and long-held relationships to shift. Forecasting must, therefore, accept that structural changes in the data are and will be taking place. Otherwise, it will not be a relevant and practical field. The major question, then, becomes how the various methods perform under a continuously changing environment. There is little interest in knowing which methods perform the best in fitting a model to a set of data. The most important and relevant aspect of forecasting is to know the methods which can minimize the post-sample forecasting errors.

This book reports the results of an empirical study organized by the author. The study was carried out as a forecasting competition in which many experts in the field of forecasting participated. The objective of the competition was to forecast up to eighteen time horizons ahead, using the specific method(s) of their expertise. The author analysed and compared the results. Chapter 4 provides a summary of the results published in the *Journal of Forecasting* as well as comments concerning these results. To avoid any accusations and to be as objective as it could humanly be, 'replicability' was made a ground rule. The results of the competition must be replicated by anyone so interested. To this extent a copy of the original data, and the forecasts by each method, can be obtained, at cost, by writing to Professor Robert Carbone. In addition, as many accuracy measures as possible were computed to avoid criticism that certain methods do better in some criteria than in others. Finally, it must be emphasized that several replications of the results have already been made, and many others are now in progress. The invitation to obtain a copy of the data and forecasts used in the competition has also appeared in the *Journal of Forecasting* and so far, more than thirty researchers have accepted the above invitation.

In the remainder of this chapter, an attempt will be made to summarize briefly the state of the art in forecasting. This will be done mainly by discussing empirical evidence now available and the author's interpretation of such evidence.

Forecasts can be obtained by (a) purely judgemental or intuitive approaches, (b) causal or explanatory methods such as regression or econometric models, (c) time series (extrapolative) methods, or (d) combinations of the above. Forecasting users consider several criteria in selecting among (a), (b), (c), or (d). Once this is done another choice is needed to select one of the many methods that exist within each of the above four categories.

Chapters 2 and 3 are two reprint articles that summarize a great deal of empirical evidence as far as forecasting accuracy is concerned. In addition, it should be emphasized that judgemental approaches do not necessarily produce more accurate forecasts than formal quantitative methods (Hogarth and Makridakis, 1981). Considerable evidence exists, that collaborates this statement beyond any reasonable doubt. A list of references is given in Chapter 3 which summarizes the evidence (mainly coming from judgemental psychology). Two additional review papers, written since 1979, further reinforce already existing evidence (Camerer, 1981; Dawes, 1979) in the domain of cognitive psychology. Practising forecasters cannot dismiss formal forecasting methods because they produce large errors. Unfortunately, the only other alternative, their own judgement, generally produces results which are, usually, even worse in repetitive forecasting situations.

Empirical comparisons have been made between causal or explanatory and time series (extrapolative) methods. Even though the evidence is not as unanimous as that between judgement and formal forecasting methods, the author of Chapter 2 concludes that it cannot be said that causal or explanatory methods produce forecasts which are more accurate than those of a time-series method. Lately, the argument has been advanced (McNees, 1982) that indeed this is true but holds only for the short term (less than one year). In the longer term, it is advocated by McNees that econometric models do better. This conclusion is an interesting one and deserves further attention through additional research studies; however, the comparisons should not be made between econometric models and the Box–Jenkins method only; they should be enlarged to include other time series methods which, as it is shown in Chapters 3 and 4, can outperform the Box–Jenkins approach. Another conclusion of Chapter 2 is that published empirical evidence does not support the statement that larger econometric models are more accurate than smaller ones.

The conclusions about large and small econometric models also hold with time series methods. Increasing complexity or statistical sophistication does not automatically mean an improvement in forecasting accuracy. This is an important finding of empirical studies which does not diminish, as some claim, but rather increases the potential usefulness of forecasting. If users are willing to pay considerable sums of money to obtain predictions from large and sophisticated models why should they object to obtaining equally accurate predictions much more cheaply and faster? Model simplicity is a positive characteristic as long as users believe that simple methods can produce accurate results. On the other hand, this does not deny that some sophisticated methods do better than simple methods. Lewandowski's FORSYS and Parzen's ARAMA models did extremely well in most of the accuracy measures used, time horizons, and types of data (see Tables 1, 2, and 3), as did Bayesian forecasting.

Other important criteria for selecting forecasting methods are the cost and complexity of the method (see Carbone and Armstrong, 1982). Obviously, simple methods are cheap to use and easy to understand; on the other hand, they are not always the most accurate. Thus higher cost and complexity need to be balanced against higher accuracy. Depending upon the specific forecasting situations in certain cases, simple methods might undoubtedly be the most appropriate, but in other cases this may not be true and sophisticated methods might be preferred.

How can the knowledge derived from the competition described in this book help in the selection of an appropriate forecasting method? The answer to this question can be derived in two different ways. First, the reader can look at the various tables shown in Chapter 4. This, it seems, is not an easy task since it requires long and careful study of the various accuracy measures, types of data, and forecasting horizons. The author believes that this should be the only reasonable alternative in practical situations when the choice of a given method, or the avoidance of another, might mean millions of dollars in benefits. For instance, if a user is interested in selecting the best method for monthly data which will minimize the MAPE for forecasts twelve months ahead, he or she will have to look at the appropriate table in Chapter 4. This is not too much work if a real desire for selecting the best method exists.

An alternative to a detailed study of the tables of the various results is to attempt to summarize these tables somehow. This can rarely be done, however, without problems and the need to make certain assumptions. A way of summarizing the results of the competition is by recording the best method, or the first and second best, and then presenting the results. This has been done in Tables 1 and 2, while Table 3 is a summary of Tables 1 and 2.

There are three conditions that affect forecasting accuracy. These are the time horizon of forecasting, the type of data used, and the accuracy measure computed. Various methods do consistently better or worse depending upon these three conditions. Looking at Table 1, for instance, it can be seen that Parzen's method is always the best for quarterly data, for each of the forecasting horizons, and MAPE; however, it does not do as well in the MSE measure where Lewandowski is the best three times and Holt's the fourth. Similarly, single exponential smoothing is the best method for forecasts one period ahead but does not do well for longer horizons.

Table 1 shows the best (B) method among those listed. Table 2 shows the second best (b). Even though this is one of the various possible ways of summarizing the many tables of results (see Chapter 4 for details) it nevertheless provides an overall summary since many users are interested in the best or second best method. On the other hand, a disadvantage of Tables 1 and 2 is that they give equal weight to all entries (e.g. being the best in the 'average of all forecasts' is more important than that of 'one-period ahead forecasts' which is one of the errors being averaged) in the 'average of all forecasts'. Furthermore, there is double counting in Tables 1 and 2, that is to say that a single series can be monthly, seasonal, and micro.

TABLE 1 Indication of the Best (B) Forecasting Method under Various Types of Series, Accuracy Measures, and Forecasting Horizons

Methods and forecasting horizons	All data				Yearly				Quarterly				Monthly				Seasonal				Non-seasonal				Micro				Macro				Total				Total of all four accuracy measures
	MAPE	MSE	AR	MD	MAPE	MSE	AR	MD	MAPE	MSE	AR	MD	MAPE	MSE	AR	MD	MAPE	MSE	AR	MD	MAPE	MSE	AR	MD	MAPE	MSE	AR	MD	MAPE	MSE	AR	MD	MAPE	MSE	AR	MD	
One period ahead forecast																																					
Deseas Sing EXP	B				B	B							B		B	B	B		B	B					B		B	B	B				4	0	3	3	10
Deseas Holt EXP			B		B	B	B	B		B					B	B			B	B					B						B		2	1	3	2	8
Holt-Winters							B	B			B	B	B	B										B									1	0	1	1	3
Autom AEP																			B	B													0	0	2	2	4
Bayesian																										B							0	0	0	0	—
Box–Jenkins		B				B								B				B												B			0	0	0	0	0
Lewandowski	B				B								B																				0	6	0	0	6
Parzen				B																	B												2	0	0	0	2
Combining A																																	1	0	0	1	—
Combining B																																	0	0	0	0	0
Others	B*				B	B	B	B	B*				B	B*			B	B*							B*							B*	4	3	1	1	9
Four period ahead forecast																																					
Deseas Sing EXP																																	0	0	0	0	0
Deseas Holt EXP			B				B	B			B	B															B				B		0	0	2	0	2
Holt-Winters																				B								B									—
Autom AEP																																	0	0	1	0	0
Bayesian		B				B									B	B										B						B	2	0	0	3	5
Box–Jenkins	B				B	B			B	B																							0	0	0	0	—
Lewandowski	B		B*		B	B	B*	B*													B	B	B	B									0	5	0	1	9
Parzen					B	B	B	B	B	B			B	B*	B	B			B	B	B	B	B	B			B					B	2	1	5	1	11
Combining A	B	B																B	B*	B*	B	B	B	B				B	B	B			3	1	4	3	4
Combining B					B*	B*	B*	B*					B*	B*			B	B*	B*	B*													1	0	3	0	2
Others	B*				B*	B*							B*	B*			B*	B*	B*						B*			B*					0	2	4	2	8

Long forecasting horizon (Yearly = 6, Quarterly = 8, Monthly = 18, Others = 12)

																							Total
Deseas Sing EXP																					0	0	0
Deseas Holt EXP	B	B																			0	1	2
Holt-Winters	B	B																		B	0	1	3
Autom AEP																					2	2	4
Bayesian							B	B											B		0	0	0
Box–Jenkins	B B B	B			B B	B B B	B B B B	B B	B B												2	5	19
Lewandowski							B														4	7	5
Parzen	B																				0	0	0
Combining A																					0	0	1
Combining B	B*	B*	B*	B*	B	B*	B*														1	1	8

Average of all forecasts

																			Total
Deseas Sing Exp							B											0	0	1
Deseas Holt EXP	B B	B B																1	1	3
Holt-Winters	B B	B B			B													0	2	4
Autom AEP	B		B		B													2	0	0
Bayesian																		0	4	6
Box–Jenkins	B B	B	B	B B	B B B B	B B B												2	2	9
Lewandowski			B	B														5	1	7
Parzen	B	B	B B	B	B B	B												0	0	6
Combining A																		0	4	2
Combining B	B*	B*	B*	B* B*	B* B*	B												1	2	5

Total of methods

																						Total
Deseas Sing EXP	1	0	0	0	0	0	2	0	0	0	0	0	0	0	0	0	0	2	0	3	3	3	11
Deseas Holt EXP	0	0	0	3	3	3	0	1	1	0	0	0	0	0	0	2	0	1	0	3	7	4	15
Holt-Winters	0	0	0	3	3	0	0	0	0	0	0	0	0	0	0	0	0	0	0	2	3	6	11
Autom AEP	0	0	0	0	0	1	0	0	0	1	0	0	0	0	0	0	1	0	1	6	2	2	4
Bayesian	0	1	0	0	0	0	0	0	0	0	3	0	3	3	0	0	0	0	0	0	0	3	16
Box–Jenkins	0	0	1	0	0	0	0	2	2	0	0	2	0	0	0	2	0	2	0	2	2	0	2
Lewandowski	0	3	1	0	3	4	2	0	0	2	3	3	2	3	3	1	0	1	6	6	17	10	43
Parzen	3	0	1	1	0	0	0	0	0	0	1	0	1	0	0	0	2	0	0	2	3	5	25
Combining A	0	0	0	0	0	0	2	0	2	2	0	2	0	0	0	0	0	4	0	2	0	3	12
Combining B	0	2	0	0	0	0	1	2	0	4	0	4	0	0	0	0	0	0	4	0	4	7	5
Others	0	1	1	0	4	4	0	4	4	2	0	2	1	1	0	0	0	2	2	6	12	5	30

B* means that the best forecasting method has been another method than the ten methods listed. In this particular case B denotes the best among the ten methods listed only.

TABLE 2 Indication of the Second Best (b) Forecasting Method under Various Types of Series, Accuracy Measures and Forecasting Horizons

Methods and forecasting horizons	All data				Yearly				Quarterly				Monthly				Seasonal				Non-Seasonal				Micro				Macro				Total				Total of all 4 accuracy measures
	MAPE	MSE	AR	MD	MAPE	MSE	AR	MD	MAPE	MSE	AR	MD	MAPE	MSE	AR	MD	MAPE	MSE	AR	MD	MAPE	MSE	AR	MD	MAPE	MSE	AR	MD	MAPE	MSE	SR	MD	MAPE	MSE	AR	MD	
One period ahead forecast																																					
Deseas Sing EXP				b																					b								1	0	0	0	2
Deseas Holt EXP	b		b																b				b				b						1	1	2	0	4
Holt-Winters		b																	b				b										0	0	1	1	2
Autom AEP								b													b											b	1	1	0	0	3
Bayesian																																	0	0	0	0	0
Box-Jenkins										b				b				b															0	2	0	0	2
Lewandowski									b																								0	0	0	2	2
Parzen												b				b				b				b									0	0	2	1	1
Combining A	b		b		b				b		b	b	b		b	b	b		b			b	b	b		b		b					3	0	2	3	8
Combining B			b	b			b				b	b			b				b	b			b				b						0	0	1	1	2
Others					b	b	b	b	b		b		b	b	b		b	b	b		b	b	b		b	b	b	b	b	b	b	b	4	4	4	3	15
Four period ahead forecast																																					
Deseas Sing EXP																			b				b									b	0	0	0	1	2
Deseas Holt EXP	b		b																b				b										0	0	2	1	3
Holt-Winters				b																	b		b										0	0	0	1	2
Autom AEP																																	0	0	0	0	0
Bayesian	b																											b					0	0	0	3	5
Box-Jenkins																																	0	0	0	0	0
Lewandowski																																	0	0	0	1	1
Parzen					b	b	b	b	b	b	b	b	b	b	b	b	b	b	b	b	b	b	b	b	b	b	b		b				4	4	3	1	9
Combining A	b	b	b		b		b		b		b		b		b						b		b		b	b	b			b			4	2	3	0	7
Combining B	b	b	b		b				b	b			b	b			b	b			b	b			b	b			b	b			1	2	2	0	3
Others					b					b			b	b			b	b	b		b				b	b	b	b	b	b		b	2	2	3	2	9

																									Total
Deseas Sing EXP																					1	0	1	0	2
Deseas Holt EXP																					1	1	1	0	3
Holt–Winters																					0	3	3	0	7
Autom AEP																					0	0	0	0	0
Bayesian																					0	0	0	1	1
Box–Jenkins																					2	0	0	0	2
Lewandowski																					0	0	0	4	4
Parzen																					0	1	1	1	3
Combining A																					0	0	0	1	1
Combining B																					0	0	2	0	2
Others																					4	5	2	4	15

Average of all forecasts

																									Total
Deseas Sing EXP																					2	0	0	0	2
Deseas Holt EXP																					0	0	2	1	3
Holt–Winters																					0	0	1	1	2
Autom AEP																					0	0	0	0	0
Bayesian																					2	0	0	0	2
Box–Jenkins																					1	1	0	0	2
Lewandowski																					0	2	2	1	5
Parzen																					1	0	0	2	3
Combining A																					0	1	2	1	2
Combining B																					2	0	0	2	4
Others																					2	2	2	2	8

Total of methods

																									Total
Deseas Sing EXP																					5	0	1	2	8
Deseas Holt EXP																					2	2	7	3	13
Holt–Winters																					1	5	6	1	13
Autom AEP																					0	0	2	1	3
Bayesian																					4	0	1	3	8
Box–Jenkins																					3	2	0	1	6
Lewandowski																					2	5	2	3	12
Parzen																					7	4	5	0	16
Combining A																					0	8	8	2	18
Combining B																					2	4	3	2	11
Others																					12	13	11	11	47

TABLE 3   Number of Times that Method Indicated Was Best or Second Best Performer (see Tables 1 and 2) for Each Type of Series and Accuracy Measure

Methods	Type of data																								Total			Accuracy measures												
	All			Yearly			Quarterly			Monthly			Seasonal			Non-seasonal			Micro			Macro						MAPE			MSE			AR			MD		
	Best	2nd Best	Total	Best	2nd Best	Total	Best	2nd Best	Total	Best	2nd Best	Total	Best	2nd Best	Total	Best	2nd Best	Total	Best	2nd Best	Total	Best	2nd Best	Total	Best	2nd Best	Total	Best	2nd Best	Total	Best	2nd Best	Total	Best	2nd Best	Total	Best	2nd Best	Total
Deseas Sing EXP	1	2	3	0	0	0	0	2	2	3	0	3	4	2	6	0	1	1	3	1	4	0	0	0	11	8	19	5	5	10	0	0	0	3	1	4	3	2	5
Deseas Holt EXP	2	2	4	8	1	9	0	1	1	0	0	0	0	1	1	0	3	3	0	1	1	4	4	8	15	13	28	3	1	4	1	2	3	7	7	14	4	3	7
Holt–Winters	0	5	5	8	1	9	0	0	0	0	3	3	3	0	3	0	3	3	0	0	0	0	0	0	11	13	24	2	1	3	1	0	1	3	5	8	5	7	12
Autom AEP	0	0	0	0	0	0	2	0	2	0	0	0	0	1	1	2	1	3	0	0	0	0	1	1	4	3	7	1	0	1	0	0	0	2	0	2	1	3	4
Bayesian	1	2	3	1	0	1	0	2	2	3	0	3	4	0	4	0	2	2	4	1	5	2	1	3	16	8	24	6	4	10	7	0	7	0	1	1	3	3	6
Box–Jenkins	1	1	2	0	0	0	4	0	4	0	1	1	0	3	3	2	0	2	0	0	0	0	1	1	2	6	8	0	3	3	0	2	2	0	0	0	2	1	3
Lewandowski	6	2	8	5	1	6	4	2	6	4	3	7	2	0	2	8	1	9	9	0	9	0	0	0	43	12	55	6	2	8	17	5	22	10	2	12	10	3	13
Parzen	4	2	6	3	4	7	0	4	4	2	4	6	0	3	3	5	3	8	0	1	1	5	0	5	25	16	41	14	2	16	3	2	5	5	4	9	3	8	11
Combining A	4	3	7	0	1	1	1	0	1	1	4	5	3	1	4	1	3	4	1	0	1	1	1	2	12	18	30	2	7	9	0	0	0	7	8	15	3	3	6
Combining B	0	1	1	0	0	0	1	2	3	2	3	5	2	0	2	0	2	2	0	2	2	0	2	2	5	11	16	0	2	2	4	4	8	0	3	3	1	2	3
Others	2	2	4	9	10	19	2	3	5	7	3	10	7	8	15	0	4	4	3	10	13	0	7	7	30	47	77	6	12	18	12	13	25	7	11	18	5	11	16

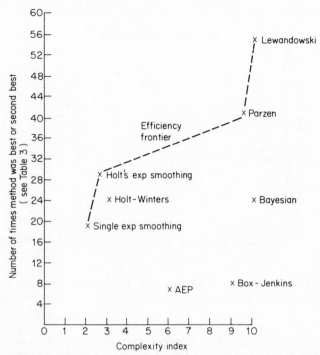

FIGURE 1　Efficiency frontier of forecasting methods (total of all forecasting horizons and types of series)

Figure 1 is an attempt to combine the 'TOTAL' results of Table 3 with the perceived complexity of the various methods being considered. (This is done as an alternative to Figure 2, Chapter 9). The complexity index has been supplied judgementally by the author for each method shown in Tables 1, 2, and 3:

(a)　Deseasonalized single exponential smoothing　2
(b)　Deseasonalized Holt's exponential smoothing　2·5
(c)　Holt–Winters exponential smoothing　3
(d)　AEP　6
(e)　Box–Jenkins　9
(f)　Parzen　9·5
(g)　Lewandowski　10
(h)　Bayesian forecasting　10

Figure 1 shows a kind of 'efficiency frontier' of forecasting methods. A forecasting user will have to make a trade-off between higher accuracy and more complexity. The choice will depend, ultimately, upon the specific situation being considered. In the final analysis, this will be influenced by the savings that could result from increased forecasting accuracy.

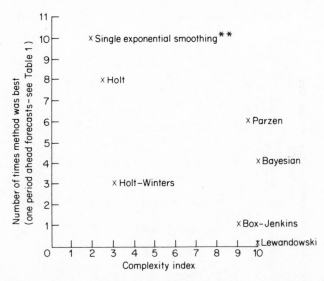

FIGURE 2   Efficient frontier of forecasting methods (one period ahead forecasts)
** Notice the efficient frontier is a single point, the single exponential smoothing

In addition to Figure 1 similar figures can be constructed on any of the three conditions presented in Tables 1, 2, or 3. For instance, Figure 2 shows the accuracy and complexity of the various methods for one-period ahead forecasts, while Figure 3 does so for four-periods ahead forecasting. Obviously, the efficiency frontier varies with each figure. Similar figures showing the relationship between accuracy and complexity can be constructed by individual forecasting users using either the summary results of Table 1 or 2 or the detailed results in Chapter 4. In the author's opinion this might be useful to help make the trade-offs between accuracy and complexity explicit. Clearly if anyone does not agree with the complexity index provided by this author he/she can modify it accordingly.

Tables 1, 2, and 3 provide overall summaries of the results. Looking at these tables as well as those of Chapter 4 can provide us with the following conclusions.

(1)   Lewandowski's method is by far the best for longer forecasting horizons (six periods ahead or longer). FORSYS does not extrapolate past trends directly, but rather it saturates them in a manner determined upon the randomness of the series (see Lewandowski, 1979). Lewandowski also does very well on the MSE criterion, outperforming all other methods. In addition, Lewandowski is the best with monthly and micro data.

(2)   Deseasonalizing the data through a simple classical decomposition procedure seems to work extremely well. As a matter of fact, a direct comparison of deseasonalizing the data, followed by the use of Holt's linear exponential smoothing, produces more accurate results than using

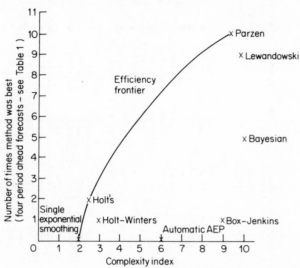

FIGURE 3   Efficient frontier of forecasting methods (one period ahead forecasts)

Winter's exponential smoothing which is equivalent to Holt's except that it takes seasonality directly into account.

(3) Deseasonalized single exponential smoothing is the best method overall when a one-period ahead forecast is needed. The next most appropriate method is Holt's.

(4) For four forecasting horizons ahead, Parzen's method is the best. This is also true with quarterly data.

(5) Differences in the various accuracy measures are very small as far as seasonal data is concerned. Most methods capture seasonality adequately and do well in nearly all accuracy measures concerned.

Another way of summarizing the results is by constructing Table 4. Table 4 lists the percentage by which each of the ten best methods in the competition is better than the other methods. The size of error is not important in such a comparison, there is no averaging, and pair-wise comparisons are made between only two methods at a time. The results of Table 4 show the clear superiority of Combining A which does well across all forecasting horizons. Holt's exponential smoothing does also very well with the exception of long horizons. Parzen does well on middle forecasting horizons.

Another interesting point in Table 4 is that not many of the differences are statistically significant. Lewandowski does clearly, statistically, worse than most methods for one-period ahead forecasting. Holt and Combining A do statistically better for one-period forecasts. Several significant differences also exist at the level of 'average of all forecasts' (because the number of observations is much larger) with Combining A showing a better statistically significant

TABLE 4 Percentage by which Method Listed Horizontally on Top of Table is Better (+) or Worse (−) than Method Listed on the Left-Hand Side of Table

Methods	Deseas Sing EXP smooth:	Deseas Holt EXP smooth:	Holt-Winter EXP smooth	Autom AEP	Bayesian forecast	Box–Jenkins ARIMA models	Lewandowski FORSYS	Parzen ARARMA models	Comb. A	Comb. B	Total No. of times method is better worse
One period ahead forecast											
Deseas Sing EXP	—	+4.1	−5.9	+2.3	−9.5**	−5.0	−12.2***	−6.8	+6.8	−6.3	6 3
Deseas Holt EXP	−4.1	—	−12.6***	−8.6*	−13.1**	−10.4**	−12.2***	−8.6*	−4.1	+0.5	8 1
Holt–Winters	+5.9	+12.6***	—	+2.3	−3.2	+0.5	−10.4***	+0.5	+8.6*	+7.7	2 7
Autom AEP	−2.3	+8.6*	−2.3	—	−5.0	−4.1	−10.4**	−5.9	+1.4	−1.4	7 2
Bayesian	+9.5**	+13.1***	+3.2	+5.0	—	+1.4	−5.0	+1.4	+13.1***	+9.5**	1 8
Box–Jenkins	+5.0	+10.4**	−0.5	+4.1	−1.4	—	−6.8	+0.5	+10.4**	+8.6*	3 6
Lewandowski	+12.2***	+12.2***	+10.4**	+10.4**	+5.0	+6.8	—	+6.8	+10.4**	+9.5**	0 9
Parzen	+6.8	+8.6*	−0.5	+5.9	−1.4	−0.5	−6.8	—	+12.2***	+11.3***	4 5
Combining A	−6.8	+4.1	−8.6*	−1.4	−13.1***	−10.4**	−10.4**	−12.2***	—	−2.3	8 1
Combining B	+6.3	−0.5	−7.7	+1.4	−9.5**	−8.6*	−9.5**	−11.3**	+2.3	—	6 3
Average	+3.61	+8.13	−2.72	+2.38	−5.69	−3.37	−9.3	−3.96	+6.79	+4.12	
Four period ahead forecast											
Deseas Sing EXP	—	+11.3**	+4.1	+2.3	+5.9	+6.8	+5.9	+7.7	+10.4**	+3.6	0 9
Deseas Holt EXP	−11.3**	—	−9.9*	−0.5	−0.5	−2.3	−2.3	+0.5	−3.2	−9.5**	8 1
Holt–Winters	−4.1	+9.9*	—	−0.5	+1.4	−5.0	+0.5	+2.3	+2.3	−4.1	4 5
Autom AEP	−2.3	+0.5	+0.5	—	+2.3	+0.5	−5.0	+4.1	+2.3	+1.4	2 7
Bayesian	−5.9	+0.5	−1.4	−2.3	—	+4.1	−3.2	+1.4	+0.5	−5.0	5 4
Box–Jenkins	−6.8	+2.3	+5.0	−0.5	−4.1	—	−5.0	+4.1	+1.4	−5.9	5 4
Lewandowski	−5.9	+2.3	−0.5	+5.0	+3.2	+5.0	—	+6.8	+0.5	−1.4	3 6
Parzen	−7.7	−0.5	−2.3	−4.1	−1.4	−4.1	−6.8	—	−5.9	−6.8	9 0
Combining A	−10.4**	+3.2	−2.3	−2.3	−0.5	−1.4	−6.8	+5.9	—	−5.0	7 2
Combining B	−3.6	+9.5**	+4.1	−1.4	+5.0	+5.9	+1.4	+6.8	+5.0	—	2 7
Average	−6.44	+4.33	−0.3	−0.48	+1.26	+1.06	−1.67	+4.4	+1.48	−3.63	

Twelve period ahead forecast

Deseas Sing EXP	—	−5.9	−1.5	−7.4	−1.5	+1.5	+2.9	+4.4	+0.7	5	4
Deseas Holt EXP	+5.9	−0.7	+0.7	0.0	+1.5	+5.9	−2.9	+7.4	+4.4	2	6
Holt–Winters	+1.5	0.0	0.0	0.0	−1.5	+4.4	0.0	+5.9	−1.5	4	3
Autom AEP	+7.4	−1.5	+2.9	−1.5	−4.4	+2.9	+4.4	+11.8**	+5.9	1	6
Bayesian	+4.4	+1.5	+1.5	+4.4	−2.9	+5.9	+10.3**	+5.9	+1.5	4	5
Box–Jenkins	+1.5	+1.5	+4.4	+2.9	−4.4	+4.4	+7.4	+4.4	+1.5	0	9
Lewandowski	−1.5	−5.9	−4.4	−5.9	−7.4	−2.9	+2.9	0.0	+1.5	6	2
Parzen	−2.9	+2.9	0.0	−4.4	−4.4	−5.9	−8.8*	+8.8*	−8.8*	6	2
Combining A	−4.4	−7.4	−5.9	−11.8**	−1.5	0.0	−8.8*	+14.7***	−14.7***	8	0
Combining B	−0.7	−4.4	+1.5	−5.9	−1.5	−1.5	+8.8*	+14.7***	—	5	4
Average	+1.24	−2.38	−0.58	−3.28	−2.12	+2.89	+2.78	+7.03	−1.71		

Average of all forecasts

Deseas Sing EXP	−3.01**	+3.01**	+0.85	−0.79	−0.65	+2.29*	+2.09	+5.69***	+1.90	2	7
Deseas Holt EXP	−0.85	+1.37	−1.37	−3.47***	−3.80***	−1.67	−1.31	+3.73***	−1.70	7	2
Holt–Winters	+1.37	+3.47***	—	−1.77	−3.36***	−3.08***	−0.90	+3.14**	−0.52	6	3
Autom AEP	+3.47***	+1.77	+1.77	+0.46	−0.46	+2.03	+4.97***	+4.97***	+2.09	1	8
Bayesian	+3.80***	+2.36*	+3.08**	−2.03	−0.92	+0.92	+3.40***	+2.75**	+0.72	0	9
Box–Jenkins	+1.67	+3.08**	−2.16*	−3.01**	−4.32***	+1.57	+1.77	−1.24	+0.72	3	6
Lewandowski	−2.88**	−2.16*	+0.98	−4.97***	−3.40***	−1.57	−1.18	+2.88**	−1.77	9	0
Parzen	−2.0	+1.31	−3.14**	−4.97***	−4.97***	−1.77	—	+2.88**	−1.11	5	4
Combining A	−5.69***	−3.73***	+0.52	−2.09	−2.75***	−2.75***	−2.88**	—	−5.10***	8	1
Combining B	−1.90	+1.70	—	−0.72	−0.72	−0.72	+1.11	+5.10***	—	4	5
Average	−2.11	+1.08	+0.32	−2.52	−2.40	−0.70	+0.78	+3.55	−0.53		

* denotes significant differences at a 10 per cent level.
** denotes significant differences at a 5 per cent level.
*** denotes significant differences at a 1 per cent level.

peformance than all other methods except one – Lewandowski – with which the difference is not significant.

The author is currently conducting research in order to be able to quantify the reasons giving rise to differences in forecasting accuracy and to find out whether or not such differences are statistically significant. Because of the inter-dependence of the various factors it is not very easy to isolate their separate impact and make statements that can be supported statistically. However, Tables 1, 2, 3, and 4 are an attempt to summarize the results and determine under what conditions various methods perform better than the rest. This the author believes to be a very important area which deserves further research effort, as it is very central to forecasting.

In the next three chapters of this book three articles will be reprinted. Chapter 2 is an article by Armstrong, which summarizes the empirical research that compares econometric models and time series methods on the one hand, and econometric models of various sizes and complexity on the other. Chapter 3 is an article by the author and Michèle Hibon which describes a previous study comparing time series methods. It also provides a summary of literature sources which compare judgemental and quantitative predictions. Another more recent and updated survey reference is Hogarth and Makridakis (1982). Finally, Chapter 4 is an article that originally appeared in the *Journal of Forecasting*. It provides a brief description and a summary, agreed by all participants in the forecasting competition described in this book, of the results of the competition.

The succeeding chapters are written individually by the participants in the forecasting competition. It is their own interpretation of the results and/or a more detailed description of the methods they used.

REFERENCES

Brown, R. G. (1963). *Smoothing, Forecasting and Prediction*, Englewood Cliffs, N. J., Prentice-Hall.

Brown, R. G. (1977). *Materials Management Systems: A Modular Library*, New York, Wiley.

Camerer, C. (1981). 'General conditions for the success of bootstrapping models', *Organisational Behavior and Human Performance*, **27**, 411–422.

Carbone, R. and Armstrong, J. S. (1982). 'Evaluation of extrapolative forecasting methods', *Journal of Forecasting*, **1**, No. 2, 215–218.

Christ, C. F. (1951). 'A test of an econometric model for the United States, 1921–1947', *Conference on Business Cycles*, New York, Natural Bureau of Economic Research, pp. 35–107.

Dawes, R. M. (1979). 'The robust beauty of improper linear models in decision making', *American Psychologist*, **34**, No. 7, 571–582.

Hogarth, R. and Makridakis, S. (1981). 'Forecasting and planning: an evaluation', *Management Science*, **27**, No. 2, 115–138.

Hogarth, R. and Makridakis, S. (1983). 'Limits to predictability: from superstition to science', INSEAD Working Paper, February.

Lewandowski, R. (1979). *La Prevision A Court Terme*, Paris, Dunod.

McLaughlin, R. L. (1968). *Short-Term Forecasting*, American Marketing Association Booklet.

McNees, S. K. (1982). 'The role of macroeconometric models in forecasting and policy analysis in the United States', *Journal of Forecasting*, 1, No. 1, 37–48.

Makridakis, S. (1981). 'Forecasting accuracy and the assumption of constancy', *Omega*, 9, No. 3, 307–311.

Makridakis, S. and Wheelwright, S. (1979). 'Forecasting the future and the future of forecasting', *TIMS Studies in Management Science*, Amsterdam: North-Holland, Vol. 12, pp. 329–352.

Priestley, M. B. (1979). 'Discussion of the Paper by Makridakis and Hibon', *Journal of the Royal Statistical Society*, Series A, 142, Part 2, 127–129.

Shiskin, J. (1961). 'Tests and Revisions of the Bureau of the Census Methods of Seasonal Adjustments', Bureau of the Census, Technical Paper No. 5.

CHAPTER 2

Forecasting with Econometric Methods: Folklore versus Fact

J. Scott Armstrong*
University of Pennsylvania

SUMMARY

Evidence from social psychology suggests that econometricians will avoid evidence that disconfirms their beliefs. Two beliefs of econometricians were examined: (1) Econometric methods provide more accurate short-term forecasts than do other methods; and (2) more complex econometric methods yield more accurate forecasts. A survey of 21 experts in econometrics found that 95 % agreed with the first statement and 72 % agreed with the second. A review of the published empirical evidence yielded little support for either of the two statements in the 41 studies. The method of multiple hypotheses was suggested as a research strategy that will lead to more effective use of disconfirming evidence. Although this strategy was suggested in 1890, it has only recently been used by econometricians.

INTRODUCTION

This paper is concerned with the use of econometric methods for forecasting in the social sciences. Although this is not the only use of econometric methods, it is one of the ways they are used; it is also the use that can most easily be validated. The paper examines only the predictive validity of econometric models. The importance of predictive validity has long been recognized by econometricians. Christ (1951) stated, "The ultimate test of an econometric model . . . comes with checking its predictions."

"Econometric methods" are defined in this paper as quantitative approaches that attempt to use causal relationships in forecasting. In particular, they refer to models based on regression analysis. This definition conforms to common usage of the term "econometric methods". "Folklore" is used here to reflect what

* The Stockholm School of Economics provided time and money for this project. Permission was granted by John Wiley & Sons to include sections from *Long-Range Forecasting: From Crystal Ball to Computer* (New York: Wiley-Interscience, 1978).

econometricians believe, as judged by what they do. "Fact" is based upon published empirical studies.

The first part of this paper draws upon evidence from social psychology to explain why folklore persists. Most of the evidence is based upon the behavior of people in general. However, there is evidence to suggest that scientists act as other people do when testing their favored hypotheses.

Two examples of the discrepancy between folklore and fact are provided in the second part of the paper. These are only two of a number of possible examples, but they deal with two important questions. First, do econometric methods provide the most accurate way to obtain short-range forecasts? Second, do complex econometric methods provide more accurate forecasts than simple econometric methods? The third part of the paper describes the method of multiple hypotheses. This method should help to overcome folklore.

THE PERSISTENCE OF FOLKLORE

Folklore persists because people who hold viewpoints on an issue tend to perceive the world so as to reinforce what they already believe; they look for "confirming" evidence and avoid "disconfirming" evidence. There is much literature on this phenomenon, commonly known as "selective perception".

The tendency for intelligent adults to avoid disconfirming evidence was demonstrated by Wason (1960, 1968). He provided three numbers (2, 4, 6) to subjects, and they were asked to determine what rule had been used to generate the three numbers. In order to gain additional information, the subjects were encouraged to generate other series of three numbers. The experimenter provided feedback on whether or not each new series was in agreement with the rule. What happened? The typical subject would think of a rule and then generate series that were consistent with that rule. It was unusual for a subject to try a series that was inconsistent with his own rule. Subjects who were told that their rules were incorrect were allowed to generate additional series. The majority of these subjects maintained the same rule that they had previously but stated it in different terms. (It is like magic; it will work if one can pronounce it correctly!)

In cases where disconfirming evidence is thrust upon people, they tend to remember incorrectly. Fischhoff and Beyth (1975), for example, found that subjects tended to remember their predictions differently if the outcome was in conflict with their prediction.

Wason's studies dealt with situations in which the person had no stake and no prior emotional attachment. When one has invested effort in supporting a particular viewpoint, the tendency to avoid disconfirming evidence would be expected to be stronger. The reward system in science encourages researchers to devote their energies to one viewpoint. The scientist gains recognition by being an advocate of a particular approach or theory. In such a case, the scientist can be expected to avoid disconfirming evidence.

Studies of scientists indicate that they are biased in favor of their own hypothesis. They interpret evidence so that it conforms to their beliefs. For example, in Rosenthal and Fode (1963) experimenters were provided with two equivalent samples of rats, but they were told that one sample was gifted and the other was disadvantaged. In the subsequent "scientific tests," the gifted rats learned tasks more quickly than did the disadvantaged rats.

The studies dealt with individuals rather than with groups. What happens when group pressures are involved – for example, when someone submits an article to be evaluated by his peers in the "marketplace of ideas"? What happens when learned societies, such as the Econometric Society, are formed to promote the advancement of the science? As the group pressures become stronger, one would expect stronger efforts to avoid evidence that disconfirms the group's opinions. Substantial literature shows how group judgment distorts reality. The study by Asch (1965) showed that most subjects would agree with the group that a given line B was longer than another line A, even though the reverse was obviously true.

In fact, the peer review process was studied in an experiment by Mahoney (1977). A paper was sent to 75 reviewers. Some reviewers received the paper along with results that were supportive of the commonly accepted hypothesis in this group. Other reviewers received a copy of the identical study except that the results were reversed so that they disconfirmed the prevailing hypothesis. Reviewers with confirming results thought the study was relevant and methodologically sound. Reviewers with disconfirming results thought the study was not relevant and that the methodology was poor. The confirming paper was recommended for publication much more frequently.

The studies cited above provide only a portion of the evidence. Other relevant studies include Pruitt (1961), Geller and Pitz (1968), Chapman and Chapman (1969), Rosenthal and Rosnow (1969), and Greenwald (1975). This evidence implies that scientists avoid disconfirming evidence. This tendency is stronger when the position is adopted by a group.

It is not surprising then, that great innovations in science have often met with resistance. (Barber [1961] describes some important historical examples). There is little reason to expect that "modern science" is different. For illustration, one might examine the treatment of Immanuel Velikovsky, a case that is being followed closely by sociologists (de Grazia, 1966). This treatment was not the result of a lack of interest or a lack of time; rather it was an active attempt to suppress Velikovsky's theories and to discredit him.

Social scientists are expected to be more prone to group opinion than are physical scientists. Thus, they would experience serious difficulties in adopting new findings. Are econometricians also resistant to innovations? In a critique of what is being done by econometricians, Bassie (1972) implies that they are. He claims that econometricians display much conformity to their preconceptions.

Two examples from econometrics are examined below. These examples were selected because they represent an important part in the life of an econometrician – and also because there seem to be discrepancies between folklore and fact. (Additional examples can be found in Armstrong [1978].)

SHORT-RANGE FORECASTING

Most textbooks on econometrics discuss short-range forecasting. Although seldom stated, the implication is that econometric methods provide more accurate short-range forecasts than other methods. Brown (1970, p. 441) asserted that econometric models were originally designed for short-range forecasting. Kosobud (1970, pp. 260–61), in a paper on short-range forecasting, referred to "...the growing body of evidence on the predictive value of econometric models". In a review of a book on short-range economy-wide forecasting, Worswick (1974, p. 118) said that "the value of econometric models in short-term forecasting is now fairly generally recognized". Various econometric services sell short-range forecasts, and one of their claims is improved accuracy. The press publishes short-range forecasts from well-known econometric models with the implication that these models will provide accurate forecasts.

Survey of econometricians

In order to go beyond the indirect evidence cited in the preceding paragraph, a questionnaire was mailed to experts in econometrics in late 1975. The survey was based on a convenience sample. Of 56 questionnaires that were sent out, 21 were completed. An additional eight were returned incomplete by respondents who said they lacked the necessary expertise. Thus, replies were received from over 40 % of the experts. The respondents were from some of the leading schools in econometrics–for example, M.I.T., Harvard, Wharton, Michigan State–and from well-known organizations that sell econometric forecasts. Many of the respondents are recognized as leading econometricians. (A listing of the sample was provided to the editors of the *Journal of Business*).

The questionnaire asked, "Do econometric methods generally provide more accurate or less accurate forecasts than can be obtained from competitive methods for short-term forecasting in the social sciences? Or is there no difference in accuracy?" A set of definitions was also provided.*

---

* These definitions were as follows: "(a) 'Econometric methods' include all methods which forecast by explicitly measuring relationships between the dependent variable and causal variables. (b) 'Competitive methods' would include such things as judgment by one or more 'experts' or extrapolation of the variable of interest (e.g., by relating the variable to 'time' such as in autoregressive schemes). (c) By 'do,' we mean that comparisons should be made between methods which appear to follow the best practices which are available at the current time. In other words, the methods should each be applied in a competent manner. (d) 'Short-term' refers to time periods during which changes are relatively small. Thus, for forecasts of the economy, changes from year to year are rather small, almost always less than 10 %. For some situations, however, one-year changes may be substantial. (e) 'Forecasts' refer to unconditional or 'ex ante' forecasts only. That is, none of the methods shall use any data drawn from the situation which is being forecast. Thus, for time series, only data prior to time t could be used in making the forecasts. (f) The 'social sciences' would include economics, psychology, sociology, management, etc. In short, any area where the behavior of people is involved."

TABLE 1   Survey of Experts on Accuracy of Short-Range
Econometric Predictions ($N = 21$)

Econometric predictions rated	Percentage
Significantly more accurate	33
Somewhat more accurate	62
No difference (or undecided)	0
Somewhat less accurate	5
Significantly less accurate	0

The results of the survey, presented in Table 1, were that 95 % of the experts agreed that predictions from econometric models are more accurate.

Respondents were asked now much confidence they had in their opinion on accuracy. Confidence was rated on a scale from 1 ("no confidence") to 5 ("extremely confident"). (If the question was not clear to respondents, they were instructed to report a low level of confidence). The average response was about 4. No one rated confidence lower than 3.0. Those who responded with "significantly more accurate" had the highest confidence level.

Another question asked how the respondent would rate himself "... as an expert on applied econometrics." Eight respondents rated themselves as "very much of an expert," six as "fairly expert," four as "somewhat of an expert," and two felt that they were "not much of an expert" (there was one nonresponse on this question). Those who rated themselves as more expert felt that econometric methods were more accurate: Five of the eight who rated themselves as "very much of an expert" felt that econometric methods were significantly more accurate, a rating that was significantly higher than the ratings by the other respondents ($P < 0.05$ using the Fisher Exact Test).

In general, the survey supported the anecdotal evidence. Experts are confident that short-range econometric predictions are more accurate than predictions from other methods.

Empirical evidence

Turning to "fact," an examination was made of all published empirical studies that I could find in the social sciences. This survey was conducted primarily by examining references from key articles and by searching through journals. Respondents to the expert survey were asked to cite evidence, but this yielded few replies.* Finally, early drafts of this paper were presented at conferences and were circulated for comments over a period of 4 years; this approach did lead to additional studies. A summary of all the studies is provided in this section.

* Two of the respondents who rated themselves highly as experts and who had the highest confidence in their ratings stated that they were not aware of any empirical evidence on this issue.

Christ (1951, 1956) provided disconfirming evidence on the accuracy of econometric predictions. In the 1951 study, econometric forecasts were better than "no change" forecasts on six occasions and worse on four. These were conditional or ex post forecasts; nevertheless, the results were not encouraging. The reaction to these findings was similar to previously mentioned occasions when disconfirming evidence was thrust upon scientists. Two of the discussants for Christ's paper were Lawrence Klein and Milton Friedman. Klein, whose model had been examined by Christ (1951, p. 121), stated that "... a competent forecaster would have used an econometric model ... far differently and more efficiently than Christ used his model". Friedman, however, was receptive. He said (Christ [1951], p. 112) that additional evidence would tend to strengthen Christ's conclusion and that "... the construction of additional models along the same general lines [as Klein's model] will, in due time, be judged failures".

Additional evidence on the predictive validity of econometric methods since Christ's papers is described here. Most of these studies are recent. Some are only of a suggestive nature because they compare ex post predictions of econometric models with ex ante predictions from alternative methods. Comparisons between extrapolations and ex post econometric forecasts were made by Kosobud (1970), Cooper (1972), Nelson (1972), Elliott (1973), Granger and Newbold (1974), Narasimham, Castellino, and Singpurwalla (1974), Levenbach, Cleary, and Fryk (1974), and Ibrahim and Otsuki (1976).* Extrapolations provided better forecasts than the econometric methods in all studies except Kosobud's and Levenbach's. In Levenbach's, there was a tie for the 1-year forecast, and the econometric model was better for the 2-year forecast. None of these eight studies claimed to find a statistically significant difference. A comparison between ex post econometric forecasts and judgmental forecasts was carried out by Kosobud (1970), Fair (1971), Haitovsky, Treyz, and Su (1974), and Rippe and Wilkinson (1974). Although the econometric forecasts were superior in all but Rippe and Wilkinson, none of these studies reported on statistical significance. However, sufficient data were provided in the Rippe and Wilkinson study to allow for such a test; my analysis of their results indicated that the econometric forecasts were significantly poorer than the judgmental forecasts. Thus, the analyses of 13 ex post studies with 14 comparisons did not provide evidence that econometric methods were superior.

To obtain direct evidence on the short-range predictive validity of econometric methods, a review was made of studies involving ex ante or unconditional forecasts. To qualify for inclusion, a study must have compared econometric and alternative methods where each was carried out in a competent manner. The

---

* One of these papers (Cooper 1972) had econometricians as discussants. The discussion was emotional and much effort was given to showing how the econometric forecasts might have been revised to yield a more favorable comparison. No attempt was made to show how the extrapolations might have been improved.

question of when a method was competently applied created some difficulty. The major effect of this restriction was to rule out studies where the alternative model was a "no-change" extrapolation. Some studies were retained (e.g., Ash and Smyth 1973), although the alternative models could have been improved.

In all, 12 studies involving 16 comparisons were found. These studies are summarized in Table 2. The criteria were taken from each study. In other words, they were the most appropriate criteria in the opinion of the researchers who did each study. Efforts were made to test for statistical significance where this had not been done in the published study. In general, serious difficulties were encountered; most of these studies did not provide sufficient data (e.g., Naylor, Seaks, and Wichern 1972), others failed to use comparable time periods, and still others suffered from small sample sizes.The most striking result was that *not one study was found where the econometric method was significantly more accurate.* Nor did the econometric method show any general superiority: Six comparisons showed the econometric method to be superior, three suggested no difference, and seven found that it was inferior.

To guard against biases that may have been held by the author and to ensure that this study could be replicated, two research assistants coded a sample of three studies. The coding was done independently (i.e., the coders did not meet each other) and it was done blindly (i.e., the coders were not aware of the hypotheses in this study). In each of the four comparisons from these studies (Vandome 1963; Markland 1970; Naylor *et al.* 1972) there was perfect agreement among the author and the two raters. In addition, five of the ex post prediction studies were coded (Kosobud 1970; Fair 1971; Cooper 1972; Elliott 1973; Granger and Newbold 1974). The only exception to perfect agreement occurred when one of the coders classified the econometric models as superior to extrapolation in Granger and Newbold. The agreement between the two raters and me on eight out of nine comparisons provides evidence that the ratings were reliable. (A copy of the instructions to the coders can be obtained from the author).

The 16 comparisons of predictive validity were in agreement with the 14 ex post comparisons. Econometric forecasts were not found to be more accurate.*

---

* These results do not imply that econometric methods are of no value in short-range forecasting. A number of studies (e.g., Granger and Newbold [1974] and Cooper and Nelson [1975]) suggest that econometric forecasts can be combined with other types of forecasts to yield forecasts that are superior to any one of the components. Econometric methods are also valuable because they provide greater accuracy in long-range economic forecasting. Three studies on long-range forecasting met the criteria stated for Table 2 (O'Herlihy *et al.* 1967; Armstrong 1968; Armstrong and Grohman 1972). Econometric methods were superior to other methods in each study. Furthermore, the relative superiority of the econometric method increased as the forecast horizon increased in two studies (Armstrong and Grohman 1972; Christ 1975). This finding conflicts with the viewpoints of many econometricians, however. For example, Wold (quoted as a discussant in NATO [1967], p. 48) implied that econometric methods are more appropriate for short-range than long-range forecasting because "the longer the forecast span, the more the actual course of events will be affected by minor influencing factors that are too numerous to be taken into account in a causal model."

TABLE 2   Accuracy of Econometric Methods for Short-Term Forecasting

Relative accuracy of econometric methods	Source of evidence	Forecast situation	Alternative forecasting method	Criteria for accuracy (RMSE = root mean square error; MAPE = mean absolute percentage error)	Test of statistical significance
Significantly more accurate ($P < .05$)	...	...	...	...	...
More accurate	Sims (1967)	Dutch economic indicators	Extrapolation	RMSE	None
	Ash and Smyth (1973)	U.K. economic indicators	Extrapolation	Theil's U	None
	McNees (1974)	U.S. economic indicators	Extrapolation	RMSE	None
	McNees (1974)	U.S. economic indicators	Judgmental	RMSE	None
	Haitovsky et al. (1974, table 7.3)	U.S. economic indicators	Judgmental	Average absolute error	None
	Christ (1975)	U.S. economic indicators	Extrapolation	RMSE	None
No difference	Sims (1967)	Norwegian economic indicators	Extrapolation	RMSE	None
	Ridker (1963)	Norwegian economic indicators	Extrapolation	(Five criteria used)	None
	Christ (1975)	U.S. economic indicators	Judgmental	RMSE	None
Less accurate	Vandome (1963)	U.K. economic indicators	Judgmental	Percentage changes	None
	Vandome (1963)	U.K. economic indicators	Extrapolation	MAPE	Armstrong*
	Naylor et al. (1972)	U.S. economic indicators	Extrapolation	Average absolute error	None
	McNees (1975)	U.S. economic indicators	Judgmental	Theil's U	None
	Cooper and Nelson (1975)	U.S. economic indicators	Extrapolation	RMSE/Theil's U	Armstrong*
	Liebling, Bidwell, and Hall (1976)	Nonresidential investment	Judgmental	MAPE	None
Significantly less accurate ($P < .05$)	Markland (1970)	Inventory control	Extrapolation	Coefficient of variation	Armstrong*

* Details on these tests can be obtained from Scott Armstrong, Wharton School, University of Pennsylvania.

SIMPLE VERSUS COMPLEX ECONOMETRIC METHODS

"Progress" in econometric methods appears to be reflected by an increase in complexity in the methods used to analyze data. Leser (1968) noted long-term tendencies toward the use of more variables, more equations, more complex functional forms, and more complex interactions among the variables in econometric models. This increase in complexity can be observed by examining various issues of *Econometrica* since 1933 or by examining textbooks. The inference is that, because more complex procedures provide more realistic ways to represent the real world, they should yield more accurate forecasts.

Some researchers imply that complexity will lead to greater accuracy. For example, Suits (1962, p. 105) states "... clearly the fewer the equations the greater must be the level of aggregation and the less accurate and useful the result." Of course, not all econometricians believe this. Bassie (1958, p. 81) proposed a general rule, "the more a function is complicated by additional variables or by nonlinear relationships, the surer it is to make a good fit with past data and the surer it is to go wrong sometime in the future".

Survey of econometricians

To gain further information on whether experts believe that increased complexity in econometric models leads to more accurate forecasts, my previously mentioned mail survey asked: "Do complex methods generally provide more accurate or less accurate forecasts than can be obtained from less complex econometric methods for forecasting in the social sciences? – or is there no difference in accuracy?"* As shown in Table 3, there was substantial agreement on the value of complexity; 72 % of the experts agreed and only 9 % disagreed.

TABLE 3   Survey of Experts on Complexity and Accuracy
($N = 21$)

Complex methods rated	Percentage
Significantly more accurate	5
Somewhat more accurate	67
No difference (or undecided)	19
Somewhat less accurate	9
Significantly less accurate	0

* The definitions were the same as provided in the first footnote. Complexity was defined as follows: "'Complexity' is to be thought of as an index reflecting the methods used to develop the forecasting model: (1) the use of coefficients other than 0 or 1, (2) the number of variables (more variables being more complex), (3) the functional relationship (additive being less complex than multiplicative; nonlinear more complex than linear, (4) the number of equations, (5) whether the equations involve simultaneity".

TABLE 4 Accuracy of Simple vs. Complex Methods

Relative accuracy of complex methods	Source of evidence	Forecast situation	Criterion for accuracy	Nature of comparison	Test of statistical significance
Significantly more accurate ($P < .05$)	...	...	...	...	...
More accurate	Stuckert (1958)	Academic performance	Percent correct	Units weights vs. regression	None
	McNees (1974)	GNP	Theil coefficient; RMSE, mean absolute error	Small vs. large models	None
	Grant and Bray (1970)	Personnel	Correlation coefficient	Unit weight vs. regression	Armstrong*
	Johnston and McNeal (1964)	Medicine	Correlation coefficient	Unit weight vs. regression	Authors
No difference	...	...	...	...	...
Less accurate	Dawes and Corrigan (1974)	Academic performance simulated data psychiatric ratings	Correlation coefficient	Unit weights vs. regression	Armstrong*
	Lawshe and Schucker (1959)	Academic performance	Percent correct	Unit weights vs. regression	None
	Reiss (1951)	Criminology	Percent correct	Few vs. many causal variables	None
	Wesman and Bennett (1959)	Academic performance	Correlation coefficient	Unit weight vs. regression	None
	Scott and Johnson (1967)	Personnel selection	Percent correct, correlation coefficient	Unit weight vs. regression	None
Significantly less accurate ($P < .05$)	Claudy (1972)	Simulated data (typical of psychological data)	Correlation coefficient	Unit weight vs. regression	Armstrong*
	Summers and Stewart (1968)	Political judgments	Correlation coefficients	Linear vs. nonlinear models	Armstrong*

* Details on these tests can be obtained from Scott Armstrong, Wharton School, University of Pennsylvania.

The experts were confident in their ratings on the value of complexity. The average confidence level was 4.0 (where 5 = "extremely confident").

Many factors could affect the relationship between complexity and accuracy. For example, Schmidt (1971), working with psychological data, found simple unit weights to be superior to regression weights for small sample sizes where there were many predictors. Furthermore, the relationship may not be a linear one; that is, complexity up to a modest level might be desirable, and beyond that it could be undesirable.

A specific question asked the experts to make any qualifications they felt important in assessing the relationship. Most respondents did qualify their answers, but it was difficult to find factors that were mentioned by more than one person.

Empirical evidence

To assess the value of complexity in econometric methods, an examination was made of all published empirical evidence that I could find in the social sciences. Some studies provided indirect evidence on the value of complexity. McLaughlin (1973) examined the accuracy of forecasts from 12 econometric services in the United States. These forecasts were made by models that differed substantially in complexity (although all were complex). There were no reliable differences in accuracy among these models: The rankings of accuracy for the models in 1971 were negatively correlated (0.3 Spearman rank correlation) with those for 1972. If there are no reliable differences, then no differences would be found between accuracy and complexity. I reanalyzed data from the study by Jorgenson, Hunter, and Nadiri (1970) and found a perfect negative correlation between complexity of the four models (ranked by the number of variables in the model) and the stability of the regression coefficients from one period to the next (ranked by Jorgenson *et al*); this lack of stability for more complex methods would suggest a loss in predictive validity. Friend and Taubman (1964) asserted that their simple model was superior to more complex models (unfortunately they did not include the data from their study and, furthermore, the study only examined ex post predictive validity). Fair (1971) found little difference between his simple model and the more complex Wharton model in a test of ex post predictive validity.

Direct evidence on the value of complexity was sought by using only studies with ex ante forecasts. Each of the models, whether simple or complex, was done in a competent manner. The results of this literature survey are summarized in Table 4.

To determine whether the coding of the studies in Table 4 was reliable, eight of the 11 studies (all but Johnston and McNeal [1964], Grant and Bray [1970], and McNees [1974]) were independently coded by two research assistants. The coding was blind in that the assistants were unaware of the hypotheses.

Discrepancies were noted on only two of these studies; one assistant coded Dawes and Corrigan (1974) to show that more complex methods were superior, and the other assistant reported complexity to be superior in Wesman and Bennett (1959). The studies in Table 4 suggest that complexity and accuracy are not closely related. No study reported a significant positive relationship between complexity and accuracy. Overall, seven comparisons favored less complexity and four favored more complexity.

The 11 studies that assessed predictive validity directly were in agreement with the five studies that provided indirect evidence: Added complexity did not yield improvements in accuracy. The empirical evidence does not support the folklore in this area.

MULTIPLE HYPOTHESES: AN ALTERNATIVE RESEARCH STRATEGY

The first part of this paper suggested that econometricians often act as advocates; they attempt to find evidence to support their viewpoint. Furthermore, group opinion is often used to judge truth. Under such conditions, it is likely that beliefs will persist if unsupported by empirical evidence.

An alternative to the use of advocacy is to adopt the method of multiple hypotheses (Chamberlin [1890] 1965; Platt 1964). Here, each scientist examines two or more reasonable hypotheses (or methods) at the same time. The role of the scientist is to determine which of the methods is most useful in the given situation. When two or more reasonable hypotheses are studied, it is less likely that the scientist will feel a bias in favor of "his" hypothesis—they are all "his" hypotheses. The orientation of the scientist is changed from one where he seeks to confirm a hypothesis to one where he seeks to disconfirm one or more hypotheses. Because the various hypotheses are tested within each study, there is less need to rely upon the opinions of other experts. The method of multiple hypotheses should help researchers to make more effective use of disconfirming evidence.

Although the method of multiple hypotheses would appear to be less prone to selective perception, and thus superior to the use of advocacy, surprisingly little evidence is available on this issue. This evidence, summarized in Armstrong (1979), provides modest support for multiple hypotheses over advocacy, although the advocacy strategy appears to be most common among social scientists.

CONCLUSIONS

Certain hypotheses about econometric methods have been accepted for years despite the lack of evidence. Ninety-five percent of the experts agreed that econometric methods are superior for short-range forecasting. An examination of the empirical literature did not support this belief: Econometric forecasts were not shown to be significantly better in any of the 14 ex post and 16 ex ante tests.

Furthermore, there was no tendency toward greater accuracy over these 30 tests. Similarly, 72% of the experts felt that complexity contributed to accuracy, but the examination of the literature did not support such a belief: Complex models were not significantly better in any of the five indirect and 11 direct tests.

Thrusting disconfirming evidence upon others provides an ineffective way of changing attitudes. Econometricians are more likely to be convinced by their own studies. The use of the method of multiple hypotheses provides a rational way for econometricians to test their beliefs.

In one sense the situation is encouraging. Twenty-three studies using the method of multiple hypotheses were found (see Tables 2 and 4). These studies are becoming more common; the oldest study was published in 1951 and almost half were published since 1970. This trend in research strategy should be useful in distinguishing folklore from fact.

REFERENCES

Armstrong, J. Scott. 1968. Long-range forecasting for international markets: the use of causal models. Pp. 222–27 in Robert L. King (ed.), *Marketing and the New Science of Planning*. Chicago: American Market Association.

Armstrong, J. Scott. 1978a. *Long-Range Forecasting: From Crystal Ball to Computer*. New York: Wiley-Interscience.

Armstrong, J. Scott. 1979. Advocacy and objectivity in science. *Management Science*, **25**: 423–28.

Armstrong, J. Scott, and Grohman, Michael C. 1972. A comparative study of methods for long-range market forecasting. *Management Science* **19**:211–21.

Asch, S. E. 1965. Effects of group pressure upon the modification and distortion of judgments. In Harold Proshansky and Bernard Seidenberg (eds.), *Basic Studies in Psychology*. New York: Holt, Rinehart & Winston.

Ash, J. C. K., and Smyth, D. J. 1973. *Forecasting the United Kingdom Economy*. Farnborough: Saxon House.

Barber, Bernard. 1961. Resistance by scientists to scientific discovery. *Science* **134**:596–602.

Bassie, V. Lewis. 1958. *Economic Forecasting*. New York: McGraw-Hill.

Bassie, V. Lewis. 1972. A note on scientific method in forecasting. Pp. 1211–18 in B. G. Hickman (ed.), *Econometric Models of Cyclical Behavior*. Vol. 2. Studies in Income and Wealth no. 36. New York: Columbia University Press.

Brown, T. Merritt. 1970. *Specification and Uses of Econometric Models*. New York: St. Martin's.

Chamberlin, T. C. (1890) 1965. The method of multiple working hypotheses. *Science* **148**:754–59.

Chapman, Loren J., and Chapman, J. R. 1969. Illusory correlation as an obstacle to the use of valid psychodiagnostic signs. *Journal of Abnormal Psychology* **74**:271–80.

Christ, Carl F. 1951. A test of an econometric model for the United States, 1921–47. Pp. 35–107 in *Conference on Business Cycles*. New York: National Bureau for Economic Research. (Discussion on pp. 107–29).

Christ, Carl F. 1956. Aggregate econometric models. *American Economic Review* **46**:385–408.

Christ, Carl F. 1975. Judging the performance of econometric models of the U.S. economy. *International Economic Review* **16** (1975):54–74.

Claudy, John G. 1972. A comparison of five variable weighting procedures. *Educational and Psychological Measurement* **32**:311–22.

Cooper, J. Phillip, and Nelson, Charles R. 1975. The ex ante prediction performance of the St. Louis and FRB-MIT-PENN econometric models and some results on composite predictors. *Journal of Money, Credit and Banking* **7**:1–32.

Cooper, Ronald L. 1972. The predictive performance of quarterly econometric models of the United States, pp. 813–926 in Bert G. Hickman (ed.), *Econometric Models of Cyclical Behavior*. Vol. 2. Studies in Income and Wealth no. 36. New York: Columbia University Press. (Discussion on pp. 926–47).

Dawes, Robyn M., and Corrigan, Bernard. 1974. Linear models in decision making. *Psychological Bulletin* **81**:95–106.

De Grazia, Alfred, Juergens, Ralph E., and Steccini, Livio C. 1966. *The Velikovsky Affair*. Hyde Park, N.Y.: University Books.

Elliott, J. W. 1973. A direct comparison of short-run GNP forecasting models. *Journal of Business* **46**:33–60.

Fair, Ray C. 1971. *A Short-Run Forecasting Model of the United States*. Lexington, Mass.: Heath.

Fischhoff, Baruch, and Beyth, Ruth. 1975. I knew it would happen: remembered probabilities of once-future things. *Organizational Behavior and Human Performance* **13**:1–16.

Friend, Irwin, and Taubman, P. 1964. A short-term forecasting model. *Review of Economics and Statistics* **46**:229–36.

Geller, E. Scott, and Pitz, Gordon F. 1968. Confidence and decision speed in the revision of opinion. *Organizational Behavior and Human Performance* **3**:190–201.

Granger, C. W. J., and Newbold, P. 1974. Economic forecasting: the atheist's viewpoint. In G. A. Renton (ed.), *Modelling the Economy*. London: Heinemann.

Grant, Donald L., and Bray, Douglas W. 1970. Validation of employment test for telephone company installation and repair occupations. *Journal of Applied Psychology* **54**:7–15.

Greenwald, Anthony G. 1975. Consequences of prejudice against the null hypothesis. *Psychological Bulletin* **82**:1–20.

Haitovsky, Yoel, Treyz, G., and Su, V. 1974. *Forecasts with Quarterly Macroeconomic Models*. New York: Columbia University Press.

Ibrahim, I. B., and Otsuki, T. 1976. Forecasting GNP components using the method of Box and Jenkins. *Southern Economic Journal* **42**:461–70.

Johnston, Roy, and McNeal, B. F. 1964. Combined MMPI and demographic data in predicting length of neuropsychiatric hospital stay. *Journal of Consulting Psychology* **28**:64–70.

Jorgenson, Dale W., Hunter, J., and Nadiri, M. I. 1970. The predictive performance of econometric models of quarterly investment behavior. *Econometrica* **38** 213–24.

Kosobud, Richard F. 1970. Forecasting accuracy and uses of an econometric model. *Applied Economics* **2**:253–63.

Lawshe, C. H., and Schucker, R. E. 1959. The relative efficiency of four test weighting methods in multiple prediction. *Educational and Psychological Measurement* **19**:103–14.

Leser, C. E. V. 1968. A survey of econometrics. *Journal of the Royal Statistical Society* **131**, series A: 530–66.

Levenbach, Hans, Cleary, J. P., and Fryk, D. A. 1974. A comparison of ARIMA and econometric models for telephone demand. *Proceedings of the American Statistical Association: Business and Economics Statistics Section*, pp. 448–50.

Liebling, Herman, I., Bidwell, Peter T., and Hall, Karen E. 1976. The recent performance of anticipations surveys and econometric model projections of investment spending in the United States. *Journal of Business* 49:451–77.

McLaughlin, Robert L. 1973. The forecasters' batting averages. *Business Economics* 3:58–59.

McNees, Stephen K. 1974. How accurate are economic forecasts? *New England Economic Review* (November-December), pp. 2–19.

McNees, Stephen K. 1975. An evaluation of economic forecasts. *New England Economic Review* (November-December), pp. 3–39.

Mahoney, Michael J. 1977. Publication prejudices: an experimental study of confirmatory bias in the peer review system. *Cognitive Therapy and Research* 1:161–75.

Markland, Robert E. 1970. A comparative study of demand forecasting techniques for military helicopter spare parts. *Naval Research Logistics Quarterly* 17:103–19.

Narasimham, Gorti, Castellino, V. F., and Singpurwalla, Nozer D. 1974. On the predictive performance of the BEA quarterly econometric model and a Box–Jenkins type ARIMA model. *Proceedings of the American Statistical Association: Business and Economics Section*, pp. 501–4.

NATO. 1967. *Forecasting on a Scientific Basis*. Lisbon: NATO.

Naylor, Thomas H., Seaks, T. G., and Wichern, D. W. 1972. Box–Jenkins methods: an alternative to econometric models. *International Statistical Review* 40:123–37.

Nelson, Charles R. 1972. The prediction performance of the FRB-MIT-PENN model of the U.S. economy. *American Economic Review* 5:902–17.

O'Herlihy, C., Fane, G., Gwilliam, K. M., and Ray, G. F. 1967. Long-term forecasts of demand for cars, selected consumer durables and energy. *National Institute Economic Review* 40:34–61.

Platt, John R. 1964. Strong inference. *Science* 146:347–53.

Pruitt, D. G. 1961. Informational requirements in making decisions. *American Journal of Psychology* 74:433–39.

Reiss, Albert J., Jr. 1951. The accuracy, efficiency and validity of a prediction instrument. *American Journal of Sociology* 56:552–61.

Ridker, Ronald G. 1963. An evaluation of the forecasting ability of the Norwegian national budgeting system. *Review of Economics and Statistics* 45:23–33.

Rippe, R. D., and Wilkinson, M. 1974. Forecasting accuracy of the McGraw-Hill anticipations data. *Journal of the American Statistical Association* 69:849–58.

Rosenthal, Robert, and Fode, K. L. 1963. The effect of experimenter bias on the performance of the albino rat. *Behavioral Science* 8:183–89.

Rosenthal, Robert, and Rosnow, Ralph L. 1969. *Artifact in Behavioral Research*. New York: Academic Press.

Schmidt, Frank L. 1971. The relative efficiency of regression and simple unit predictor weights in applied differential psychology. *Educational and Psychological Measurement* 31:699–714.

Scott, Richard D., and Johnson, R. W. 1967. Use of the weighted application blank in selecting unskilled employees. *Journal of Applied Psychology* 51:393–95.

Sims, Christopher A. 1967. Evaluating short-term macroeconomic forecasts: the Dutch performance. *Review of Economics and Statistics* 49:225–36.

Stuckert, R. P. 1958. A configurational approach to prediction. *Sociometry* 21:225–37.

Suits, Daniel B. 1962. Forecasting and analysis with an econometric model. *American Economic Review* 52:104–32.

Summers, David A., and Stewart, Thomas R. 1968. Regression models of foreign policy beliefs. *Proceedings of the American Psychological Association*, pp. 195–96.

Vandome, Peter. 1963. Econometric forecasting for the United Kingdom. *Bulletin of the Oxford University Institute of Economics and Statistics* 25:239–81.

Wason, P. C. 1960. On the failure to eliminate hypotheses in a conceptual task. *Quarterly Journal of Experimental Psychology* **12**:129–40.

Wason, P. C. 1968. On the failure to eliminate hypotheses – a second look. Pp. 165–74 in P. C. Wason and P. N. Johnson-Laird (eds.), *Thinking and Reasoning*. Baltimore: Penguin.

Wesman, A. G., and Bennett, G. K. 1959. Multiple regression vs. simple addition of scores in prediction of college grades. *Educational and Psychological Measurement* **19**:243–46.

Worswick, G. D. N. 1974. Preston, R. S., *The Wharton Annual and Industry Forecasting Model. Journal of Economic Literature* **12**:117–18.

Note: This paper was followed by commentary, in the same issue (*Journal of Business*, Vol. 4, 1978, pp. 565–593), by Gregory C. Chow, Richard F. Kosobud, Stephen K. McNees, Preston J. Miller, William E. Wecker, and Arnold Zellner. A reply by Armstrong, 'Econometric Methods and the Science Court', then followed (pages 595–600).

CHAPTER 3

Accuracy of Forecasting: An Empirical Investigation (with Discussion)

Spyros Makridakis and Michèle Hibon
INSEAD

ABSTRACT

In this study, the authors used 111 time series to examine the accuracy of various forecasting methods, particularly time-series methods. The study shows, at least for time series, why some methods achieve greater accuracy than others for different types of data. The authors offer some explanation of the seemingly conflicting conclusions of past empirical research on the accuracy of forecasting. One novel contribution of the paper is the development of regression equations expressing accuracy as a function of factors such as randomness, seasonality, trend-cycle and the number of data points describing the series. Surprisingly, the study shows that for these 111 series simpler methods perform well in comparison to the more complex and statistically sophisticated ARMA models.

INTRODUCTION

The ultimate test of any forecast is whether or not it is capable of predicting future events accurately. Planners and decision makers have a wide choice of ways to forecast, ranging from purely intuitive or judgemental approaches to highly structured and complex quantitative methods. In between, there are innumerable possibilities that differ in their underlying philosophies, their cost, their complexity and their accuracy. Unfortunately, since information about these differences is not usually available, objective selection among forecasting methods is extremely difficult. The major purpose of this paper is to deal with one important aspect of choosing a forecasting methodology: accuracy. Section 1 will survey past research on accuracy and will look into the reasons why the reported accuracies of different studies vary, often significantly. Section 2 will

Reproduced by kind permission of the Royal Statistical Society from Makridakis, S. and Hibon, M. (1979). Accuracy of forecasting: an empirical investigation. *J. R. Statist. Soc.*, Series A, **142**, Part 2, pp. 97–145.

report our own empirical findings on the accuracy of 111 time series and show how the results could be consistent with the conflicting conclusions of three major previous studies on the subject of accuracy. Section 3 will develop regression equations that express accuracy as a function of a number of factors related to the internal characteristics of the time series used. Section 4 will examine the reasons for variations in the relative accuracy of various methods and suggest needs for future research and Section 5 attempts a brief conclusion. The appendices describe the methods and technical details used to make the comparisons.

1.   RESEARCH FINDINGS ON ACCURACY PERFORMANCE

The first choice to be made in predicting the future is whether to use a formal forecasting method or to rely on judgemental processes. Evidence from the psychological literature asserts that in repetitive situations, quantitative methods outperform clinical judgement (Goldberg, 1970; Hogarth, 1975; Sarbin, 1943; Sawyer, 1966; Slovic, 1972). In his review of the literature, Meehl (1954, 1965), for example, concluded that he found only one case in which clinical judgement was superior to a statistical model. This case and a subsequent one (Libby, 1976) have been disputed by Goldberg (1976) who reversed the exceptional findings by simply transforming the data. In a recent article, Dawes (1977) says that he knows of no other findings that have been reported showing the superiority of clinical judgement.

The idea that decision makers' models can perform better than decision makers themselves is difficult to accept; acting on this idea arouses emotional reactions and suggests dehumanizing overtones. Slovic (1972), Kahneman and Tversky (1973), Tversky (1974), Tversky and Kahneman (1974), and Dawes (1977) have investigated the causes behind the inability of clinical judgement to outperform formal models. The main difficulties with judgemental predictions are stated to be lack of application of valid principles, anchoring effects, regression biases, the assumption that specific cases can be generalized (representation biases), lack of reliability and the basing of predictions on irrelevant information.

Outside the psychological literature, judgemental forecasts of earnings per share have been analysed in some detail. A number of researchers have compiled histories of the forecasts of earnings per share made by analysts (representing judgemental predictions) and compared them with the results obtained from quantitative methods (Green and Segall, 1967; Cragg and Malkiel, 1968; Elton and Gruber, 1972; Niederhoffer and Regan, 1972). In these four studies, the researchers conclude that analysts do not perform as well in forecasting earnings per share as do quantitative techniques. However, these conclusions are not unanimous.

Johnson and Schmitt (1974) claim that analysts can do better than quantitative methods provided that they have accurate economic and industrial information. But they make no attempt to find out if a model can also do better when this extra information is incorporated into it. Furthermore, several studies of professionally managed funds indicate that these funds do worse or the same as the market as a whole (Bauman, 1965; Fama, 1965; Jensen, 1968; O'Brien, 1970; Slovic, 1972).

Mabert (1975) reports a direct comparison between judgemental and quantitative methods of forecasting. He found that forecasts based on opinions of the sales force and corporate executives gave less accurate results over the five-year period covered by the study than did the quantitative forecasts of three methods (exponential smoothing, harmonic smoothing and Box–Jenkins). He also found that the quantitative techniques cost less and took less time than the subjective estimates. Adam and Ebert (1976), in a more complete study involving many subjects in a controlled setting, found that Winters' method produced forecasts that were statistically more accurate than those of human forecasters. The conclusion held true under different kinds of experimental variations.

When the forecasting accuracy of quantitative methods is compared, there is less agreement as to which method does best. Quantitative forecasting can be classified into two types: econometric (explanatory) and time series (mechanistic). A major reason for the popularity of ARMA models has been that several studies have found them to be at least as accurate as the complex econometric models of the USA economy. Cooper (1972, p. 920), for example, concludes that 'mechanical (ARMA) forecasting models can be constructed which predict economic variables about as well as econometric models'. Naylor *et al.* (1972, p. 153) report that 'the Box–Jenkins results were significantly better in all cases and, except for GNP, they provide better forecasts by a factor of almost two to one'. Similarly, Nelson (1972, p. 915) concludes that 'if the mean square error were an appropriate measure of loss, an unweighted assessment clearly indicates that a decision maker would have been best off relying simply on ARMA predictions in the post-sample period'.

Other studies by Christ (1951), Steckler (1968), Cleary and Fryk (1974), Narasimham *et al.* (1974, 1975), Cooper and Nelson (1975) and McWhorter (1975) reach similar conclusions even though the findings do not always suggest a clear superiority of ARMA models. The suggested reasons for the better performance of ARMA models are the inability of econometric models to accommodate structural changes in the economy (Christ, 1951; Steckler, 1968; Cooper, 1972; Nelson, 1972). Other authors suggest that this lack of accuracy of econometric models could be due to spurious correlations (Granger, 1969; Granger and Newbold, 1974); some even question the fundamental process of obtaining functional relationships in econometrics (Pierce, 1977). Econometricians disagree with these conclusions. They find fault with the methodology and the technicalities involved (Goldfield, 1972; McCarthy, 1972; Howrey *et al.*, 1974) and insist that the comparisons made are unfair to econometric models, whose main purpose is not predictive but explanatory (see Leser, 1968). In at least one study (Christ, 1975) found that ARMA models are the 'poorest' of those compared.

Comparisons between the accuracy of econometric models and anticipatory surveys of investment spending in the USA have also been reported. Liebling and Russell (1969) and Jorgenson *et al.* (1970) conclude that anticipation surveys were at least as successful (and frequently more so) in forecasting investment in the 1950s and 1960s. A recent study by Liebling *et al.* (1976, p. 451) reaches similar conclusions.

Within econometric models there is a fair amount of agreement as to the relative accuracy of individual techniques. Leser (1966), Jorgenson *et al.* (1970), Cooper (1972), Fromm and Klein (1973), Christ (1975) and McNees (1975) accept that no single econometric model is overwhelmingly superior to all others. McNees (1975, pp. 30–31) refers to comparisons of econometric models as follows: 'Differences ... typically did not fall into systematic patterns. Valid generalizations are extremely difficult to make.' Another finding seems to be that bigger econometric models with more equations do not necessarily produce more accurate forecasts than simpler models, involving only a few structural equations (von Hohenbalken and Tintner, 1962; Leser, 1966; Theil, 1966).

Comparing the relative accuracy of time series methods is more difficult because there are many more techniques to be compared and different researchers use different sets of methods when they make the comparisons. Time series methods can be placed into four groupings: smoothing, decomposition, filtering and ARMA (Makridakis, 1976). Some early studies compare the accuracy of smoothing methods, and some later ones compare smoothing and ARMA (Box–Jenkins) models. Unfortunately, nothing has been reported comparing all four groups of methods or, even better, including econometric models as well as judgemental predictions in such a comparison.

In a study reported by Kirby (1966), three different time series methods were compared: moving averages, exponential smoothing and regression (trend fitting). Kirby found that in terms of month-to-month forecasting accuracy, the exponential smoothing methods did best, with moving averages and exponential smoothing giving similar results when the forecasting horizon was increased to six months. The regression model (trend fitting) included in that study was the best method for longer-term forecasts of one year or more.

In a study reported by Levine (1967), the same three forecasting methods examined by Kirby were compared. Levine concluded that although the moving average method had the advantage of simplicity, exponential smoothing offered the best potential accuracy for short-term forecasting. Other studies reported by Gross and Ray (1965), Raine (1971) and Krampf (1972) have arrived at similar conclusions.

There is considerable disagreement concerning comparisons between ARMA and smoothing methods. Newbold and Granger (1974, p. 143) report: 'The

Box–Jenkins forecasts do seem to be better than those derived from two fully automatic procedures – the Holt–Winters method and stepwise autoregression – for a sizeable majority of the time series in our sample.'

A similar conclusion was reached by Reid (1969, p. 266), who reported: 'Thus the Box–Jenkins method is clearly better than Brown's general exponential smoothing, even when the latter is modified for serially correlated errors, for practically all series.'

A study by Groff (1973, p. 30), however, arrived at a rather different conclusion: 'The forecasting errors of the best of the Box–Jenkins models that were tested are either approximately equal to or greater than the errors of the corresponding exponentially smoothed models for most series.'

Similarly, Geurts and Ibrahim (1975, p. 187), although they examined a single series only, found that the 'exponentially smoothed models patterned on Brown's model and the Box–Jenkins approach seem to perform equally well'.

Other studies dealing with accuracy (McNees, 1976; Makridakis and Wheelwright, 1978) report results showing that the most accurate method varies from one set of data to another and from one time period to the next. These seemingly conflicting findings regarding the accuracy record of forecasting methods are a disturbing factor for both academics and practitioners when they have to choose between alternative forecasting methods.

2. FORECASTING ACCURACY FOR 111 TIME SERIES

The 111 time series used in this study were collected from a variety of sources, including several countries, industries and companies. These time series also represent different periods of time and time intervals (monthly, quarterly or yearly). Some of the series were seasonal and others were not. In addition, the length of both the series and seasonality (for seasonal series only) varies. This selection was made in order to minimize any bias that could arise from the use of a single source of data. However, it cannot be considered a random or even representative choice because the majority of the series come from French sources, have monthly time horizons and involve observations taken during the 1970s.

Taking n_j as the number of data points in the jth series, we used $n_j - 12$ points to develop a forecasting model, and subsequently 12 forecasts were obtained. The error, e_{tj}, is defined as

$$e_{tj} = X_t - \hat{X}_{tj}, \qquad t = n_j - 11, n_j - 10, \ldots, n_j, \tag{1}$$

where X_t is the actual value at period t, $\hat{X}_{tj}$ is the value forecast by the jth method, and e_{tj} is the forecast error. Accuracy, undoubtedly, is related negatively to e_t, except that it is up to the decision maker to assign different loss functions involving e_t. The most common measures of accuracy are the mean square error

(MSE), Theil's *U*-coefficient (Theil, 1966) and the mean absolute percentage error (MAPE).

The MSE involves a quadratic loss function and is preferred when more weight is to be given to big errors. Its disadvantage is that it does not allow for comparisons across methods, since it is an absolute measure related to a specific series. The *U*-coefficient is a relative measure, it assumes a quadratic loss function and allows comparisons with the naive ($\hat{X}_t = X_{t-1}$) or random walk model. In addition, it has several other properties that make its use attractive (Theil, 1966, pp. 21–36). Its disadvantage is that its interpretation is more difficult than the MAPE. Moreover, the *U*-statistic has no upper bound, so a few very large values can easily distort the comparisons. We will present the forecasting errors for the 111 series when both Theil's *U*-statistic and the MAPE were used. In addition, a ranking of the percentage of time that Naive 1, Naive 2 and ARMA are better than all other methods (without respect to the magnitude of the error) will be given.

A forecasting model using $n_j - 12$ data points was estimated for each of twelve time series methods and one naive method (Naive 1, $\hat{X}_t = X_{t-1}$). In addition, the data ($n_j - 12$ values) were deseasonalized and a forecasting model involving the eight nonseasonal methods was estimated. Finally, Naive 2 was defined as the most recent seasonally adjusted observation ($\hat{X}_t = X'_{t-1}$ where X'_{t-1} is seasonally adjusted). In all, forecasting models were developed for twenty-two methods (see Appendix A for a description of each method and the technical details of developing a forecasting model and obtaining predictions). No attempt was made to choose an appropriate method for each time series. (That is not necessary for several methods – Harrison's, Winters', adaptive filtering and ARMA – that can deal with all types of data). The major difficulty arises with nonseasonal methods when the data involved are seasonal. In these latter cases, however, the data were seasonally adjusted (nonseasonal data obviously did not need adjustment). After this adjustment there was no need to choose an appropriate method for each time series and introduce biases into the results through such a choice.

Once a forecasting model was developed, several MAPE measures were computed. First, the MAPE of fitting a model to the data was found by the following equation:

$$\text{MAPE}_{oj} = \frac{1}{n_j - 12} \sum_{t=1}^{n_j - 12} \frac{|X_t - \hat{X}_{t,j}|}{X_t}(100), \qquad j = 1, 2, 3, \ldots, 22, \qquad (2)$$

where $\hat{X}_{t,j}$ is the one-period-ahead forecast of period t by the jth method. The MAPE_{oj} tells us how well method j does in fitting a model to *existing data* and is of questionable value since some model can always be found (as long as n_j is finite) to make equation (2) equal to zero.

Mean absolute percentage errors up to the period k ($k = 1, 2, 3, 4, 5, 6, 9, 12$) were also calculated for each of the twenty-two methods and the 111 series. These errors, denoted as $MAPE_{ij}$, are in effect average absolute cumulative errors from period $n - 11$ to $n - 11 + k$ – that is, the average cumulative forecasting error of up to k periods ahead. In our opinion, this way of measuring accuracy is of higher practical relevance than multiple lead-time forecasting errors (for example, in budgeting, production planning, inventory management and so on). Furthermore, no research has been carried out, establishing that multiple lead-time forecasts produce smaller errors than average cumulative ones.

The mean absolute percentage errors of forecasting for each of 111 series were found by:

$$MAPE_{ij} = \frac{1}{i} \sum_{k=1}^{i} \frac{|X_{n-12+k} - \hat{X}_{n-12+k,j}|}{X_{n-12+k}} (100) \qquad (3)$$

where $i = 1, 2, 3, 4, 5, 6, 9, 12; j = 1, 2, 3, \ldots, 22$.

Table 1 shows the average of the mean absolute percentage errors for the fitted models and MAPEs for 1, 2, 3, 4, 5, 6, 8 and 12 periods ahead. The average MAPEs have been calculated over all 111 series as:

$$\text{average } MAPE_{ij} = \frac{1}{111} \sum_{s=1}^{111} MAPE_{sij}, \qquad (4)$$

where $i = 1, 2, 3, 4, 5, 6, 9, 12; j = 1, 2, 3, \ldots, 22$.

Table 1 indicates that if a single user had to forecast for all 111 series, he would have achieved the best results by using exponential smoothing methods after adjusting the data for seasonality (the seasonal indices being calculated by a decomposition method – see Appendix B). Thus a combination of decomposition (which cannot provide forecasts on its own) and exponential smoothing would have produced the best results (assuming a linear loss error function). The table also shows that the MAPEs of simpler methods are surprisingly close to those of the more statistically sophisticated ones. A final observation is that if the trend fitting and harmonic smoothing methods are excluded, the MAPEs of the remaining methodologies within each category (nonseasonal, seasonal and seasonally adjusted) are rather similar.

In order to get a wider perspective on these results, we should ask if the MAPEs of Table 1 are in agreement with those reported elsewhere. Unfortunately, not many direct comparisons are possible. To our knowledge, no one has used seasonally adjusted data with the exception of Geurts and Ibrahim (1975) who tested only a single series. Their conclusion agrees with the results of Table 1. Groff (1973) used more variations of exponential smoothing models than we did; however, both the magnitude of percentage errors he found and his conclusion that exponential smoothing models are at least as good as the Box–Jenkins methodology agree with our conclusion about the comparison between Winters' exponential smoothing and the ARMA models (the only two similar methods employed by both studies). There is also a certain amount of agreement between our conclusion and those of Reid (1969) regarding Harrison's

TABLE 1 The Average of the Mean Absolute Percentage Errors (MAPE) of all Series (111)

	Forecasting Method	Model Fitting	Forecasting Horizons							
			1	2	3	4	5	6	9	12
Original Data: Nonseasonal Methods	1 Naive 1	21.9	15.5	18.4	20.4	27.9	28.8	28.6	32.2	34.1
	2 Single moving average	19.5	13.8	16.4	18.7	27.2	28.2	27.8	30.7	32.3
	3 Single exponential smoothing	19.5	14.4	16.6	19.0	27.3	28.2	27.9	31.3	33.3
	4 Adaptive response rate exponential smoothing	21.4	13.5	15.4	18.0	25.8	26.4	26.0	28.6	30.5
	5 Linear moving average	22.2	17.1	20.3	23.6	34.2	36.5	37.1	44.1	49.6
	6 Brown's linear exponential smoothing	20.2	13.2	15.8	18.4	26.5	27.7	27.3	31.2	34.7
	7 Holt's (2 parameters) linear exp. smoothing	20.5	13.3	15.6	18.1	26.2	27.7	27.5	30.5	32.5
	8 Brown's quadratic exponential smoothing	20.8	13.6	15.9	18.1	26.2	28.4	29.0	36.4	43.3
	9 Linear trend (regression fit)	22.5	19.0	19.8	22.3	30.8	31.3	30.6	34.8	38.0
Seasonal and Nonseasonal Methods	10 Harrison's harmonic smoothing	11.0	26.4	26.3	27.6	27.4	28.0	29.3	32.2	34.2
	11 Winters' linear and seasonal exp. smoothing	10.9	13.8	14.8	15.4	16.2	17.1	18.4	21.3	23.6
	12 Adaptive filtering	11.7	15.6	16.7	16.8	18.9	18.7	19.5	22.9	24.5
	13 Autoregressive moving average (Box-Jenkins)	10.6	14.7	15.0	15.7	16.6	17.1	18.1	21.6	24.3
Seasonally Adjusted Data: Nonseasonal Methods	14 Naive 2	10.0	14.5	15.0	15.1	15.3	15.6	16.6	19.0	21.0
	15 Single moving average	8.4	12.9	13.6	13.7	13.8	14.3	15.3	17.7	19.8
	16 Single exponential smoothing	8.5	12.8	13.4	13.8	14.0	14.3	15.6	18.1	20.2
	17 Adaptive response rate exponential smoothing	8.5	13.0	14.0	14.5	14.7	15.2	16.2	18.5	20.4
	18 Linear moving average	9.1	15.0	15.6	16.3	16.6	17.4	18.6	22.6	26.4
	19 Brown's linear exponential smoothing	8.5	12.9	14.3	14.6	14.9	15.9	17.1	20.3	23.5
	20 Holt's (2 parameters) linear exp. smoothing	9.0	12.0	12.8	13.2	13.7	14.8	16.0	19.7	23.0
	21 Brown's quadratic exponential smoothing	8.7	12.5	14.0	14.7	15.6	17.0	18.6	23.6	28.9
	22 Linear trend (regression fit)	11.4	19.6	20.4	21.1	21.1	21.9	22.8	25.3	27.4

harmonic smoothing,* in that both feel that its performance is disappointingly bad. There is further agreement in the findings of Gross and Ray (1965), Kirby (1966) and Levine (1967), and our own findings concerning the fact that exponential smoothing methods are in general more accurate than moving average ones in forecasts up to five or six periods ahead, after which they become the same or worse. Rather striking differences appear between our conclusion and those of Newbold and Granger (1974), and Reid (1969), who have found that Box–Jenkins models do better in most cases than exponential smoothing ones.

We tried classifying the data into categories to see if that would produce significantly different results. The classification involving the extent of randomness – that is, fluctuations that cannot be classified as seasonal or caused by changes in the level of economic activity – produced the most significant differences. Table 2, for example, shows the averages of the MAPE for 64 series whose randomness is less than 10 percent. Comparing Tables 1 and 2, we see that ARMA does relatively better in Table 2 than in Table 1. If we go one step further and compute the average MAPEs of the 31 series with less than 5 per cent randomness (see Table 3), we see that ARMA models do considerably better than Naive 1 and better than Winters' method. This result is consistent with the findings of Newbold and Granger (1974) that ARMA models do better for shorter forecasting horizons, even though they did not do as well in percentages in our studies (see Tables 4, 5 and 6) as Newbold and Granger (1974) found or as Reid (1977) reported.

Tables 1, 2 and 3 leave little doubt that the extent of randomness in these 111 data series does influence the relative performance of forecasting methods. Furthermore, it could explain discrepancies in research findings. Groff (1973) used sales series whose randomness is high (the extent of randomness could be even higher in our series, many of which include data from the 1974–75 major recession), and his conclusions are in agreement with those of Table 1. Reid (1969, 1977) used macro-series whose randomness is even smaller than those in Table 3. His results are not in perfect agreement with those of Table 3, but they are in the same direction. Newbold and Granger (1974) do not give MAPE measures nor do they provide a breakdown of the macro- and micro-series used, or indications of the magnitude of percentage errors involved; however, we believe that their series were mostly of a macro type, having little randomness, but more than that in the series used by Reid. It is interesting to note that the percentage of times the Box–Jenkins method did better than exponential smoothing is much higher in Reid's study than in Newbold and Granger's. The latter results should be more in agreement with those of Table 3.

The MAPE is one of several alternative measures of expressing accuracy. Tables 4, 5 and 6 use another approach involving relative ranking of the methods used in relation to Naive 1, Naive 2 and the ARMA models respectively. They

---

* It should be noted that it was *not* the improved version of Harrison's smoothing (i.e. SEATREND) that was used in this study (see Appendix A).

TABLE 2 The Average of the Mean Absolute Percentage Errors (MAPE) for Series Whose Randomness is Less Than 10 Percent (64 Series)

Forecasting Method	Model Fitting	Forecasting Horizons							
		1	2	3	4	5	6	9	12
Original Data: Nonseasonal Methods									
1 Naive 1	10.4	9.3	11.0	11.4	12.6	13.6	13.8	17.3	18.6
2 Single moving average	9.6	8.4	10.6	11.4	12.5	13.5	13.7	17.0	18.1
3 Single exponential smoothing	9.5	8.5	10.4	11.2	12.3	13.3	13.5	16.9	18.2
4 Adaptive response rate exponential smoothing	10.3	8.9	10.4	11.1	12.4	13.4	14.0	17.2	18.2
5 Linear moving average	10.5	9.1	12.7	14.1	15.8	17.4	18.2	23.6	26.9
6 Brown's linear exponential smoothing	9.6	7.8	10.4	11.6	12.8	14.2	14.4	18.4	20.8
7 Holt's (2 parameters) linear exp. smoothing	9.8	7.7	10.4	11.2	12.4	13.5	13.7	16.7	17.9
8 Brown's quadratic exponential smoothing	9.9	8.8	11.2	12.7	14.3	16.4	17.8	24.4	30.4
9 Linear trend (regression fit)	10.7	15.3	15.5	16.1	17.2	18.3	18.6	22.4	23.4
Seasonal and Nonseasonal Methods									
10 Harrison's harmonic smoothing	6.2	16.6	17.2	17.7	18.1	18.7	19.5	21.8	23.1
11 Winters' linear and seasonal exp. smoothing	5.9	7.2	8.0	8.4	9.2	10.0	11.0	13.6	15.4
12 Adaptive filtering	5.9	6.7	8.1	8.8	9.5	10.1	10.7	13.2	14.4
13 Autoregressive moving average (Box-Jenkins)	5.3	6.5	7.3	8.1	8.7	9.5	10.3	14.0	16.8
Seasonally Adjusted Data: Nonseasonal Methods									
14 Naive 2	5.2	6.5	6.5	6.9	7.5	8.2	8.9	11.5	13.6
15 Single moving average	4.6	6.4	6.9	7.3	7.8	8.6	9.3	11.8	13.8
16 Single exponential smoothing	4.6	6.3	6.7	7.1	7.5	8.3	9.0	11.5	13.6
17 Adaptive response rate exponential smoothing	5.2	6.6	7.6	8.0	8.7	9.6	10.6	13.1	14.8
18 Linear moving average	4.7	8.2	9.0	9.5	10.0	11.0	11.8	14.8	18.1
19 Brown's linear exponential smoothing	4.4	6.4	7.5	8.0	8.6	9.6	10.3	14.0	15.7
20 Holt's (2 parameters) linear exp. smoothing	4.6	5.8	7.1	7.7	8.2	9.2	10.0	12.9	15.7
21 Brown's quadratic exponential smoothing	4.5	6.2	7.7	8.4	9.3	10.7	11.8	16.2	20.9
22 Linear trend (regression fit)	6.4	14.0	13.7	14.0	14.4	15.3	16.0	18.3	19.7

TABLE 3   The Average of the Mean Absolute Percentage Errors (MAPE) for Series Whose Randomness is Less Than 5 Percent (31 Series)

	Forecasting Method	Model Fitting	Forecasting Horizons								
			1	2	3	4	5	6	9	12
Original Data: Nonseasonal Methods	1 Naive 1	5.0	4.9	5.9	6.8	7.8	8.8	10.2	11.9	12.6
	2 Single moving average	4.8	4.7	5.7	6.6	7.6	8.7	10.2	11.8	12.5
	3 Single exponential smoothing	4.9	4.7	5.7	6.6	7.7	8.8	10.2	11.8	12.6
	4 Adaptive response rate exponential smoothing	5.9	6.0	7.6	7.9	8.8	9.6	10.9	12.2	13.0
	5 Linear moving average	2.5	3.7	4.0	4.9	5.5	6.3	7.3	9.1	10.0
	6 Brown's linear exponential smoothing	4.7	4.6	6.0	6.8	8.0	9.2	10.3	13.1	15.3
	7 Holt's (2 parameters) linear exp. smoothing	4.7	4.8	5.9	6.6	7.6	8.6	9.9	11.0	11.4
	8 Brown's quadratic exponential smoothing	5.1	4.9	6.9	8.9	11.4	14.0	16.9	24.2	32.1
	9 Linear trend (regression fit)	6.7	10.7	10.7	10.8	11.0	11.7	12.8	13.7	14.4
Seasonal and Non-seasonal Methods	10 Harrison's harmonic smoothing	4.3	9.4	9.6	10.2	10.5	10.9	11.5	12.3	12.9
	11 Winters' Linear and seasonal exp. smoothing	3.5	3.3	3.7	4.8	5.4	6.0	6.8	8.0	8.6
	12 Adaptive filtering	3.0	3.6	4.3	5.0	5.5	5.9	6.6	7.3	7.7
	13 Autoregressive moving average (Box-Jenkins)	2.5	3.1	3.3	4.3	4.8	5.5	6.4	8.3	10.5
Seasonally Adjusted Data: Nonseasonal Methods	14 Naive 2	2.9	2.7	3.2	4.3	5.1	6.0	6.9	8.7	10.1
	15 Single moving average	2.7	3.3	3.5	4.6	5.4	6.2	7.1	8.9	10.2
	16 Single exponential smoothing	2.8	2.9	3.4	4.4	5.2	6.0	7.0	8.8	10.1
	17 Adaptive response rate exponential smoothing	3.3	3.2	4.4	5.3	6.3	7.0	8.1	9.7	11.0
	18 Linear moving average	2.5	3.7	4.0	4.9	5.5	6.3	7.3	9.1	10.0
	19 Brown's linear exponential smoothing	2.3	3.1	3.4	4.1	4.7	5.3	6.0	7.2	7.7
	20 Holt's (2 parameters) linear exp. smoothing	2.4	2.7	3.2	3.9	4.5	5.1	5.9	7.2	8.1
	21 Brown's quadratic exponential smoothing	2.4	3.1	3.5	4.2	5.0	5.5	6.9	9.4	11.7
	22 Linear trend (regression fit)	4.5	9.5	9.4	9.9	10.2	10.7	11.3	12.2	13.1

TABLE 4 Percentage of Time that Naive 1 is Better than Other Methods Listed (111 Series)

Forecasting Method	Model Fitting	Forecasting Horizons							
		1	2	3	4	5	6	9	12
Original Data: Nonseasonal Methods									
1 Naive 1									
2 Single moving average	6.4	34.5	30.9	34.5	37.3	37.3	37.3	36.4	30.9
3 Single exponential smoothing	8.2	35.5	30.9	35.5	37.5	38.2	38.2	36.4	35.5
4 Adaptive response rate exponential smoothing	39.1	41.8	47.3	50.0	54.5	53.6	55.5	50.9	50.0
5 Linear moving average	53.6	53.6	55.5	58.2	60.9	62.7	61.8	62.7	65.5
6 Brown's linear exponential smoothing	26.4	43.6	47.3	48.2	52.7	52.7	54.5	51.8	56.4
7 Holt's (2 parameters) linear exp. smoothing	30.0	43.6	48.2	48.2	49.1	50.0	51.8	50.0	44.5
8 Brown's quadratic exponential smoothing	34.5	43.6	44.5	46.4	46.4	50.0	51.8	54.5	56.4
9 Linear trend (regression fit)	44.5	57.3	56.4	55.5	55.5	55.5	58.2	50.9	52.7
Seasonal and Non-seasonal Methods									
10 Harrison's harmonic smoothing	18.2	55.5	57.3	60.9	54.5	57.3	56.4	50.9	50.0
11 Winters' linear and seasonal exp. smoothing	12.7	42.7	45.5	48.2	40.0	40.9	41.8	32.7	30.9
12 Adaptive filtering	9.1	41.8	43.6	44.5	40.0	36.4	41.8	38.2	36.4
13 Autoregressive moving average (Box-Jenkins)	1.8	40.9	36.4	42.7	36.4	35.5	40.9	36.4	32.7
Seasonally Adjusted Data: Nonseasonal Methods									
14 Naive 2	1.8	34.5	35.5	28.2	23.6	25.5	27.3	19.1	17.3
15 Single moving average	1.8	34.5	37.3	32.7	30.0	30.9	31.8	23.6	20.9
16 Single exponential smoothing	1.8	32.7	35.5	31.8	30.9	32.7	33.6	26.4	21.8
17 Adaptive response rate exponential smoothing	13.6	38.2	45.5	43.6	38.2	36.4	37.3	30.9	29.1
18 Linear moving average	7.3	47.3	47.3	47.3	40.0	39.1	40.0	33.6	33.6
19 Brown's linear exponential smoothing	1.8	37.3	41.8	39.1	31.8	35.0	39.1	32.7	30.5
20 Holt's (2 parameters) linear exp. smoothing	2.7	34.5	38.2	34.5	30.9	30.0	36.4	27.3	25.5
21 Brown's quadratic exponential smoothing	3.6	35.5	41.8	38.2	35.5	33.6	37.3	35.5	37.3
22 Linear trend (regression fit)	21.8	53.6	54.5	51.8	45.5	45.5	47.3	39.1	36.4

TABLE 5 Percentage of Time that Naive 2 is Better than Other Methods Listed (111 Series)

	Forecasting Method	Model Fitting	Forecasting Horizons							
			1	2	3	4	5	6	9	12
Original Data: Nonseasonal Methods	1 Naive 1	82.7	50.0	49.1	56.4	60.9	59.1	57.3	65.5	67.3
	2 Single moving average	78.2	51.8	53.6	53.6	60.9	62.7	60.9	65.5	67.3
	3 Single exponential smoothing	79.1	50.9	51.8	56.4	62.7	61.8	62.7	66.4	70.9
	4 Adaptive response rate exponential smoothing	94.5	52.7	56.4	58.2	67.3	69.1	69.1	73.6	74.5
	5 Linear moving average	85.5	60.9	68.2	65.5	70.0	75.5	71.8	72.7	71.8
	6 Brown's linear exponential smoothing	80.9	47.3	54.5	58.2	63.6	67.3	63.6	73.6	78.2
	7 Holt's (2 parameters) linear exp. smoothing	83.6	45.5	53.6	54.5	62.7	64.5	60.9	68.2	70.0
	8 Brown's quadratic exponential smoothing	81.8	52.7	55.5	56.4	60.0	65.5	61.8	67.3	70.0
	9 Linear trend (regression fit)	87.3	65.5	61.8	64.5	68.2	72.7	70.0	72.7	75.5
Seasonal and Non-Seasonal Methods	10 Harrison's harmonic smoothing	54.5	63.6	70.9	71.8	69.1	70.0	70.9	71.8	70.0
	11 Winters' Linear and Seasonal Exp. Smoothing	67.3	49.1	57.3	54.5	51.8	53.6	55.5	57.3	57.3
	12 Adaptive Filtering	71.8	47.3	60.0	60.0	61.8	62.7	60.9	61.8	60.0
	13 Autoregressive moving average (Box-Jenkins)	59.1	46.4	53.6	55.5	57.3	51.8	58.2	61.8	60.9
Seasonally Adjusted Data: Nonseasonal Methods	14 Naive 2									
	15 Single moving average	5.5	33.6	39.1	40.0	39.1	43.6	42.7	42.7	37.3
	16 Single exponential smoothing	4.5	40.9	44.5	50.0	47.3	47.3	48.2	48.2	48.2
	17 Adaptive response rate exponential smoothing	40.0	42.7	59.1	63.6	63.6	64.5	67.3	62.7	60.0
	18 Linear moving average	32.7	53.6	58.2	60.0	61.8	61.8	63.6	62.7	66.4
	19 Brown's linear exponential smoothing	17.3	45.5	52.7	55.5	55.5	56.4	56.4	58.2	59.1
	20 Holt's (2 parameters) linear exp. smoothing	18.2	47.3	47.3	49.1	50.0	51.8	54.5	51.8	50.9
	21 Brown's quadratic exponential smoothing	20.0	46.4	54.5	55.5	56.4	55.5	53.6	61.8	61.8
	22 Linear trend (regression fit)	49.1	61.8	66.4	65.5	61.8	63.6	61.8	60.9	59.1

TABLE 6 Percentage of Time that the ARMA Method is Better than Other Methods Listed (111 Series)

	Forecasting Method	Model Fitting	Forecasting Horizons							
			1	2	3	4	5	6	9	12
Original Data: Nonseasonal Methods	1 Naive 1	98.2	59.1	63.6	57.3	63.6	64.5	59.1	63.6	67.3
	2 Single moving average	92.7	49.1	56.4	53.6	60.9	64.5	61.8	65.5	66.4
	3 Single exponential smoothing	90.9	53.6	57.3	55.5	60.0	62.7	62.7	65.5	66.4
	4 Adaptive response rate exponential smoothing	95.5	49.1	56.4	56.4	60.0	62.7	64.5	62.7	67.3
	5 Linear moving average	97.3	63.6	63.6	63.6	69.1	70.0	70.0	70.0	70.9
	6 Brown's linear exponential smoothing	92.7	42.7	51.8	50.0	53.6	59.1	59.1	67.3	70.9
	7 Holt's (2 parameters) linear exp. smoothing	90.9	50.0	55.5	50.0	54.5	62.7	61.8	62.7	66.4
	8 Brown's quadratic exponential smoothing	91.8	48.2	51.8	50.9	57.3	61.8	63.6	65.5	73.6
	9 Linear trend (regression fit)	90.9	60.9	61.8	63.6	65.5	69.1	68.2	71.8	72.7
Seasonal and Non-seasonal Methods	10 Harrison's harmonic smoothing	49.1	62.7	68.2	72.7	70.0	70.0	73.6	71.8	65.5
	11 Winters' linear and seasonal exp. smoothing	69.1	47.3	49.1	51.8	50.0	50.0	50.9	46.4	45.5
	12 Adaptive filtering	70.9	49.1	54.5	52.7	55.5	56.4	56.4	58.2	51.8
	13 Autoregressive moving average (Box-Jenkins)									
Seasonally Adjusted Data: Nonseasonal Methods	14 Naive 2	40.9	53.6	46.4	44.5	42.7	48.2	41.8	38.2	39.1
	15 Single moving average	19.1	46.4	42.7	43.6	39.1	42.7	39.1	37.3	38.2
	16 Single exponential smoothing	19.1	50.0	42.7	42.7	38.2	43.6	39.1	41.8	40.0
	17 Adaptive response rate exponential smoothing	30.9	44.5	46.4	42.7	40.9	47.3	48.2	44.5	41.8
	18 Linear moving average	26.4	55.5	50.9	50.9	49.1	47.3	49.1	51.8	51.8
	19 Brown's linear exponential smoothing	19.1	42.7	48.2	44.5	49.1	49.1	47.3	49.1	46.4
	20 Holt's (2 parameters) linear exp. smoothing	20.9	41.8	43.6	42.7	40.0	46.4	44.5	46.4	45.5
	21 Brown's quadratic exponential smoothing	20.0	41.8	44.5	44.5	45.5	47.3	47.3	50.9	59.1
	22 Linear trend (regression fit)	41.8	63.6	59.1	62.7	60.0	59.1	57.3	53.6	50.0

TABLE 7 Theil's *U*-Coefficient

		Forecasting Method	Model fitting[*]	Forecasting Horizons								
				1	2	3	4	5	6	9	12	Average
Original Data: Nonseasonal Methods	1	Naive 1	A	1.00	1.00	1.00	1.00	1.00	1.00	1.00	1.00	1.00
			10	1.00	1.00	1.00	1.00	1.00	1.00	1.00	1.00	1.00
			5	1.00	1.00	1.00	1.00	1.00	1.00	1.00	1.00	1.00
	2	Single moving average	A	0.88	1.46	1.37	1.46	1.51	1.84	1.37	1.20	2.14	1.54
			10	0.91	1.47	1.46	1.47	1.48	1.46	1.48	1.08	2.73	1.58
			5	0.95	1.77	1.04	0.97	1.00	1.19	1.02	0.97	1.01	1.12
	3	Single exponential smoothing	A	0.88	1.47	1.27	1.35	1.43	1.60	1.15	1.07	1.15	1.31
			10	0.91	1.26	1.32	1.31	1.27	1.27	1.14	1.00	1.13	1.21
			5	0.97	1.41	1.01	0.99	1.02	1.06	1.00	1.00	1.00	1.06
	4	Adaptive response rate exponential smoothing	A	0.97	1.32	1.49	1.54	1.75	1.62	1.60	1.43	2.41	1.64
			10	1.02	1.13	1.61	1.53	1.97	1.60	1.94	1.17	2.68	1.70
			5	1.13	1.14	1.87	1.06	1.38	1.27	1.16	1.25	1.10	1.28
	5	Linear moving average	A	0.96	2.56	2.59	2.85	2.96	4.51	2.52	5.13	10.78	4.24
			10	0.96	1.76	3.02	2.83	3.08	3.16	2.10	4.27	11.49	3.96
			5	0.95	2.07	3.55	2.56	2.40	3.11	2.20	6.10	4.84	3.35
	6	Brown's linear exponential smoothing	A	0.86	1.49	1.79	1.85	1.79	2.04	1.66	2.39	4.93	2.24
			10	0.87	1.15	2.17	2.12	2.05	2.13	2.04	2.78	6.59	2.63
			5	0.86	1.50	2.51	1.68	1.93	2.35	1.81	4.38	3.41	2.45
	7	Holt's (2 parameters) linear exp. smoothing	A	0.87	1.16	1.58	1.68	1.71	2.80	1.73	1.65	5.23	2.19
			10	0.87	0.98	1.79	1.83	1.95	2.01	1.89	1.43	7.07	2.37
			5	0.84	1.15	1.89	1.28	1.59	2.22	1.65	1.51	1.66	1.62
	8	Brown's quadratic exponential smoothing	A	0.90	1.46	1.89	2.15	2.34	2.90	1.93	4.88	9.51	3.38
			10	0.92	1.49	2.33	2.53	3.02	2.69	2.17	6.26	12.15	4.08
			5	0.93	1.94	2.73	2.40	4.03	3.73	2.29	10.90	11.77	4.98
	9	Linear trend (regression fit)	A	1.03	2.84	2.57	2.44	2.45	3.04	2.12	1.86	3.67	2.62
			10	1.16	3.01	3.00	2.87	2.87	2.75	2.48	1.60	4.32	2.86
			5	1.45	3.69	4.18	2.36	1.91	2.28	1.67	1.95	1.57	2.45
Seasonal and Non seasonal Methods	10	Harrison's harmonic smoothing	A	0.75	6.61	3.60	3.55	3.43	5.04	2.82	2.49	5.92	4.18
			10	0.88	7.48	3.75	4.39	3.68	3.18	3.08	2.05	8.12	4.47
			5	1.16	11.97	4.08	3.96	2.57	2.20	2.54	2.27	1.98	3.95
	11	Winters' linear and seasonal exp. smoothing	A	0.66	1.47	1.76	1.55	2.21	2.66	1.65	1.76	3.53	2.07
			10	0.69	1.36	1.75	1.52	2.27	2.12	1.79	1.60	4.82	2.15
			5	0.78	1.58	1.88	1.35	1.84	2.54	1.57	1.88	1.19	1.73
	12	Adaptive filtering	A	0.69	1.93	1.97	1.46	2.23	2.26	1.67	1.74	3.37	2.08
			10	0.70	1.36	1.62	1.47	2.22	2.41	1.69	1.63	4.36	2.12
			5	0.75	1.41	2.14	1.13	1.63	2.66	1.37	1.50	1.45	1.66
	13	Autoregressive moving average (Box-Jenkins)	A	0.61	1.51	1.68	1.73	1.91	2.22	1.77	1.88	3.04	1.97
			10	0.58	1.45	1.45	1.50	1.82	1.87	1.69	1.75	4.17	1.96
			5	0.57	1.91	1.48	1.36	1.63	2.37	1.44	1.78	1.21	1.65
Seasonally Adjusted Data: Nonseasonal Methods	14	Naive 2	A	0.66	1.62	1.09	1.14	1.30	1.68	1.20	1.17	1.00	1.28
			10	0.64	1.03	0.92	1.14	1.40	1.31	1.32	1.14	1.00	1.16
			5	0.70	0.99	0.72	1.10	1.27	1.49	1.04	1.28	1.00	1.11
	15	Single moving average	A	0.57	1.61	1.35	1.25	1.47	1.93	1.30	1.35	2.87	1.64
			10	0.57	1.42	1.15	1.24	1.45	1.43	1.33	1.17	3.91	1.64
			5	0.66	1.87	0.80	1.14	1.28	1.60	1.05	1.31	1.00	1.26
	16	Single exp. smoothing	A	0.57	1.66	1.27	1.13	1.32	1.68	1.23	1.25	1.05	1.32
			10	0.58	1.22	1.13	1.03	1.36	1.37	1.31	1.14	1.05	1.20
			5	0.68	1.51	0.77	1.11	1.27	1.49	1.04	1.28	1.00	1.18
	17	Adaptive response rate exponential smoothing	A	0.63	1.59	1.61	1.39	1.83	2.17	1.58	1.65	3.11	1.87
			10	0.66	1.09	1.62	1.31	2.02	1.82	1.74	1.34	3.92	1.86
			5	0.79	1.01	1.61	1.23	1.50	1.70	1.26	1.51	1.12	1.37
	18	Linear moving average	A	0.58	1.77	2.30	1.80	1.96	2.29	2.01	2.41	7.19	2.72
			10	0.56	1.43	2.67	1.79	2.39	2.32	2.38	1.92	10.35	3.16
			5	0.58	1.51	3.00	1.16	1.77	2.44	1.53	1.92	2.36	1.96
	19	Brown's linear exponential smoothing	A	0.53	1.49	1.95	1.51	1.88	2.41	1.55	1.64	5.42	2.23
			10	0.51	1.14	2.09	1.52	2.02	2.19	1.67	1.49	7.94	2.51
			5	0.53	1.41	2.11	1.02	1.62	2.15	1.31	1.57	1.21	1.55
	20	Holt's (2 parameters) linear exp. smoothing	A	0.55	1.41	1.57	1.37	1.79	2.59	1.52	1.61	4.95	2.10
			10	0.52	0.94	1.65	1.46	1.96	2.06	1.69	1.21	6.47	2.18
			5	0.53	1.06	1.51	1.03	1.64	2.04	1.47	1.15	1.34	1.40
	21	Brown's quadratic exponential smoothing	A	0.55	1.67	1.96	1.59	2.11	3.03	1.81	1.67	6.13	2.50
			10	0.53	1.22	2.14	1.65	2.34	2.68	2.06	1.62	8.71	2.80
			5	0.56	1.60	2.18	1.15	1.67	2.59	1.17	1.33	2.25	1.74
	22	Linear trend (regression fit)	A	0.76	2.96	2.85	2.60	2.48	3.68	2.13	2.23	5.13	3.01
			10	0.88	2.86	3.09	3.00	2.81	3.06	2.26	1.97	6.94	3.25
			5	1.19	3.30	4.05	2.68	1.88	2.52	1.70	2.46	1.25	2.48

[*] A = all series (111).
10 = series whose randomness is less than 10% (64).
5 = series whose randomness is less than 5% (31).

TABLE 8 Theil's U-Coefficient Adjusted

		Forecasting Method	Model fitting	1	2	3	4	5	6	9	12	Average
							Forecasting Horizons					
Original Data: Nonseasonal Methods	1	Naive 1	A	1.00	1.00	1.00	1.00	1.00	1.00	1.00	1.00	1.00
			10	1.00	1.00	1.00	1.00	1.00	1.00	1.00	1.00	1.00
			5	1.00	1.00	1.00	1.00	1.00	1.00	1.00	1.00	1.00
	2	Single moving average	A	0.86	1.01	1.05	1.06	1.07	1.11	1.01	1.00	1.06	1.04
			10	0.91	1.00	1.09	1.07	1.05	1.06	1.05	0.96	1.02	1.04
			5	0.95	0.99	1.01	0.97	1.00	1.03	1.02	0.97	1.01	1.00
	3	Single exponential smoothing	A	0.86	1.00	1.01	1.09	1.07	1.06	1.03	1.01	1.05	1.04
			10	0.91	0.97	1.05	1.07	1.08	1.06	1.03	0.99	1.00	1.03
			5	0.97	0.94	1.01	0.99	1.02	1.03	1.00	1.00	1.00	1.00
	4	Adaptive response rate exponential smoothing	A	0.97	1.00	1.02	1.09	1.07	1.09	1.03	0.99	1.10	1.05
			10	1.02	1.02	1.12	1.05	1.20	1.11	1.13	0.98	1.06	1.08
			5	1.13	1.05	1.30	1.00	1.19	1.09	1.06	1.03	1.07	1.10
	5	Linear moving average	A	0.96	1.16	1.18	1.29	1.28	1.33	1.25	1.34	1.43	1.28
			10	0.96	1.11	1.23	1.27	1.28	1.29	1.24	1.26	1.28	1.24
			5	0.95	1.03	1.15	1.26	1.24	1.28	1.28	1.23	1.23	1.21
	6	Brown's linear exponential smoothing	A	0.86	0.92	1.04	1.11	1.09	1.11	1.04	1.11	1.28	1.09
			10	0.87	0.89	1.09	1.13	1.14	1.17	1.12	1.06	1.20	1.10
			5	0.86	0.96	1.07	1.04	1.09	1.10	1.11	1.09	1.07	1.07
	7	Holt's (2 parameters) linear exp. smoothing	A	0.87	0.95	1.02	1.09	1.05	1.15	1.03	1.04	1.19	1.07
			10	0.87	0.89	1.09	1.10	1.10	1.07	0.99	0.96	1.11	1.04
			5	0.84	0.99	1.08	1.01	0.99	0.99	0.98	0.91	1.03	1.00
	8	Brown's quadratic exponential smoothing	A	0.90	0.96	1.01	1.11	1.12	1.25	1.16	1.22	1.43	1.16
			10	0.92	0.94	1.08	1.15	1.20	1.28	1.22	1.23	1.35	1.18
			5	0.93	0.96	1.07	1.19	1.22	1.31	1.28	1.23	1.29	1.19
	9	Linear trend (regression fit)	A	1.03	1.22	1.14	1.14	1.10	1.15	1.07	1.07	1.20	1.14
			10	1.16	1.35	1.26	1.17	1.19	1.23	1.11	1.00	1.14	1.18
			5	1.45	1.56	1.31	1.13	1.06	1.12	0.96	0.95	1.13	1.15
Seasonal and Nonseasonal Methods	10	Harrison's harmonic smoothing	A	0.75	1.18	1.22	1.23	1.14	1.23	1.21	1.09	1.27	1.20
			10	0.88	1.15	1.26	1.23	1.26	1.24	1.20	1.08	1.18	1.20
			5	1.16	1.11	1.09	1.11	1.10	1.16	0.97	0.98	1.04	1.07
	11	Winters' linear and seasonal exp. sm.	A	0.68	0.94	0.97	0.99	0.97	1.03	1.08	0.99	1.09	1.01
			10	0.69	0.88	0.96	0.98	1.03	1.01	1.08	0.93	1.04	0.99
			5	0.78	0.89	0.86	0.92	0.83	0.86	0.82	0.86	0.93	0.87
	12	Adaptive filtering	A	0.69	1.04	1.03	1.06	1.02	1.10	1.08	1.04	1.19	1.08
			10	0.70	1.01	0.98	1.08	1.05	1.03	1.04	0.95	1.10	1.03
			5	0.75	1.00	0.94	0.88	0.88	0.83	0.80	0.71	0.82	0.86
	13	Autoregressive moving average (Box-Jenkins)	A	0.61	0.97	0.94	1.01	0.97	0.96	1.03	1.03	1.10	1.00
			10	0.58	0.87	0.88	1.05	1.01	0.93	1.08	0.97	1.06	0.98
			5	0.57	0.86	0.81	1.04	0.82	0.86	0.86	0.90	0.85	0.88
Seasonally Adjusted Data: Nonseasonal Methods	14	Naive 2	A	0.66	0.94	0.87	0.92	0.89	0.94	0.98	0.94	1.00	0.93
			10	0.64	0.88	0.77	0.93	0.93	0.90	1.01	0.95	1.00	0.92
			5	0.70	0.82	0.72	0.94	0.91	0.89	0.88	0.92	1.00	0.89
	15	Single moving average	A	0.57	0.93	0.88	0.90	0.87	0.94	0.98	0.95	1.02	0.93
			10	0.57	0.86	0.83	0.95	0.94	0.93	1.02	0.95	0.97	0.93
			5	0.66	0.90	0.76	0.98	0.92	0.90	0.86	0.92	1.00	0.91
	16	Single exp. smoothing	A	0.57	0.91	0.87	0.90	0.87	0.89	0.98	0.93	1.02	0.92
			10	0.58	0.85	0.82	0.93	0.94	0.89	1.02	0.95	1.02	0.93
			5	0.68	0.87	0.76	0.95	0.91	0.89	0.88	0.92	1.00	0.90
	17	Adaptive response rate exponential smoothing	A	0.63	0.90	0.98	1.00	0.93	0.97	1.03	0.93	1.08	0.98
			10	0.66	0.83	1.01	1.01	1.08	0.95	1.12	0.95	1.03	1.00
			5	0.79	0.87	1.09	1.00	1.14	0.91	1.00	0.94	1.08	1.00
	18	Linear moving average	A	0.58	1.01	0.96	1.02	0.92	1.09	1.08	1.12	1.29	1.06
			10	0.56	0.96	0.93	0.99	1.00	1.07	1.09	1.04	1.17	1.03
			5	0.58	0.89	0.77	0.89	0.90	0.95	0.96	0.97	0.97	0.91
	19	Brown's linear exponential smoothing	A	0.53	0.89	0.90	0.95	0.93	1.02	1.04	1.01	1.16	0.99
			10	0.51	0.80	0.86	0.98	1.03	1.03	1.05	0.94	1.09	0.97
			5	0.53	0.77	0.70	0.84	0.80	0.82	0.79	0.86	0.88	0.81
	20	Holt's (2 parameters) linear exp. smoothing	A	0.55	0.85	0.84	0.90	0.85	1.02	0.96	1.02	1.16	0.95
			10	0.52	0.76	0.86	0.89	0.95	0.96	0.98	0.92	1.10	0.93
			5	0.53	0.73	0.71	0.79	0.77	0.79	0.74	0.82	0.96	0.79
	21	Brown's quadratic exponential smoothing	A	0.55	0.87	0.90	0.95	0.92	1.05	1.06	1.11	1.26	1.02
			10	0.53	0.79	0.88	0.96	1.01	1.04	1.08	1.09	1.19	1.01
			5	0.56	0.77	0.67	0.88	0.82	0.85	0.83	1.04	1.14	0.87
	22	Linear trend (regression fit)	A	0.76	1.16	1.11	1.15	1.05	1.09	1.09	1.02	1.12	1.10
			10	0.88	1.21	1.22	1.21	1.21	1.16	1.14	0.98	1.04	1.15
			5	1.19	1.39	1.20	1.19	1.13	1.05	0.88	0.89	0.96	1.09

* A = all series (111).
  10 = series whose randomness is less than 10% (64).
  5 = series whose randomness is less than 5% (31 series).

show the percentage of time that Naive 1, Naive 2 or ARMA is better than the remaining methods. Such a comparison takes no account of the magnitude of the errors; it only indicates when the error of one method is smaller than that of the other. (The comparisons shown in Tables 4, 5 and 6 are given because previous research findings have been expressed in a similar form).

The final comparison uses Theil's U-statistic, which is based on a quadratic loss function, and provides a comparison over a 'no-change' model – that is, the Naive 1 or random walk method. Tables 7 and 8 show the average U-coefficients for all series, 64 series whose randomness is less than 10 percent and 31 series with less than 5 percent randomness. A value of 1 in these coefficients means that Naive 1 is as good as the method it is compared with. A U value of greater than 1 indicates that Naive 1 does better than the forecasting method it is compared with. Finally, when U is smaller than 1, the other method does better than Naive 1. Table 8 differs from Table 7 in that the U-coefficients have been adjusted (any value greater than 2 has been set to 2). This adjustment allows for more meaningful comparisons, since a few high U-values tend to distort the averages shown in Table 7.

The next section will formally test the hypothesis that randomness is indeed an important factor influencing accuracy and the relative performance of different forecasting methods. Furthermore, it will look for additional factors that affect accuracy and relative performance, and will attempt to specify and measure such relationships through regression equations.

3. EXPLAINING VARIATIONS IN THE ACCURACY OF FORECASTING

Section 2 has shown that at least one factor (randomness) affects accuracy and may explain differences in the relative performance of various methods. To go one step further, we assumed that the accuracy of a forecasting method depends upon several factors and that these factors could be isolated and quantified, and their influence measured.

In this and an earlier study (Makridakis and Vandenburgh, 1975), several sets of factors were explored for their ability to influence accuracy. Two sets produced the best regression equations (that is, statistically significant coefficients, highest R^2). Both involve the MAPEs as the dependent variable and two combinations of independent variables. The first set relates the MAPE of the fitted model, $n_j - 12$ values, to combinations of the mean absolute percentage change of the trend-cycle, randomness, seasonality (as calculated by the decomposition method shown in Appendix B) and the number of data points. The second relates the MAPEs of the forecasts (1, 2, 3, 4, 5, 6, 9 and 12 periods ahead) to the mean absolute percentage change in the trend-cycle, randomness, seasonality, the absolute percentage change of the trend-cycle at the last period of data ($n_j - 12$), and the number of periods ahead of forecasting. The coefficients of both sets of

TABLE 9 Regression Coefficients

(Dependent variable is the mean absolute percentage error (m.a.p.e.) of model fitting)

		Forecasting Method	R^2	Constant	No. of Data	Mean Absolute % Change			Standard Error of Estimate	F-Test
						Trend Cycle	Random-ness	Season-ality		
Original Data: Nonseasonal Methods	1	Naive 1	.98	0			.95 (38.29)*	1.02 (51.51)		2051
	2	Single moving average	.95	1.40 (3.24)			.69 (18.08)	.92 (33.84)	2.48	974
	3	Single exponential smoothing	.96	1.35 (3.06)			.71 (17.99)	.90 (36.80)	2.58	1081
	4	Adaptive response rate exponential smoothing	.93	2.02 (3.37)			.75 (13.94)	.95 (28.43)	3.49	646
	5	Linear moving average	.94	1.55 (2.74)			.80 (15.88)	1.04 (28.96)	3.26	724
	6	Brown's linear exponential smoothing	.95	1.04 (2.15)			.76 (17.65)	.94 (34.77)	2.83	984
	7	Holt's (2 parameters) linear exp. smoothing	.97	.79 (2.03)			.80 (23.20)	.97 (39.33)	2.26	1393
	8	Brown's quadratic exponential smoothing	.95	1.24 (2.36)			.77 (16.44)	.97 (33.43)	3.05	895
	9	Linear trend (regression fit)	.76	0		2.72 (4.83)	.64 (5.4)	.90 (14.03)	7.01	104
Seasonal and non seasonal Methods	10	Harrison's harmonic smoothing	.76	1.06 (1.75)		1.65 (5.66)	.58 (10.54)	.10 (4.04)	3.27	108
	11	Winters' linear and seasonal exp. smoothing	.90	3.61 (5.55)	−.02 (−3.72)		.87 (27.71)	.06 (3.72)	2.00	310
	12	Adaptive filtering	.86	2.09 (2.36)	−.01 (−1.82)		.99 (22.49)	.10 (4.60)	2.70	212
	13	Autoregressive moving average (Box-Jenkins)	.84	2.30 (2.44)	−.02 (−1.86)		.94 (21.25)	.07 (3.13)	2.89	178
Seasonally Adjusted Data: Nonseasonal Methods	14	Naive 2	.96	.61 (2.94)		.33 (3.49)	.90 (46.22)		1.15	1384
	15	Single moving average	.92	.59 (2.29)		.71 (5.82)	.68 (28.22)		1.47	619
	16	Single exp. smoothing	.93	.71 (3.00)		.62 (5.49)	.69 (31.15)		1.36	730
	17	Adaptive response rate exponential smoothing	.92	1.04 (3.86)		.73 (5.76)	.72 (28.45)		1.53	627
	18	Linear moving average	.91	.58 (1.83)		.30 (2.03)	.82 (27.66)		1.80	518
	19	Brown's linear exponential smoothing	.93	.48 (1.88)		.28 (2.31)	.77 (32.04)		1.47	695
	20	Holt's (2 parameters) linear exp. smoothing	.95	.82 (3.74)			.83 (42.29)		1.34	1788
	21	Brown's quadratic exponential smoothing	.93	.55 (2.16)		.31 (2.55)	.78 (32.57)		1.46	722
	22	Linear trend (regression fit)	.72	1.72 (2.87)		1.44 (4.72)	.62 (12.29)		3.33	126

* Numbers in parentheses are the t-tests of regression coefficients.

regressions and the related statistics can be seen in Tables 9 and 10 respectively. Almost all coefficients are statistically significant at a 99 percent level; only a few exceptions involve a 95 percent confidence level, and these usually refer to the regression constant.

In total, there are 44 regression equations. Two for each of the 22 methods – one when the MAPEs of the fitted model are the dependent variable and one when the MAPEs of the forecasts are the dependent variable.

For example, the regression equation for the MAPE of the fitted model for ARMA (see Table 9) is

$$\text{MAPE}_{o(\text{ARMA})} = 2.30 - 0.02X_1 + 0.94X_3 + 0.97X_4, \tag{5}$$

where X_1 is the number of data points in the jth series, X_3 is the mean absolute percentage change in randomness of the jth series, and X_4 is the mean absolute percentage change in seasonality of the jth series; the $R^2 = 0.84$, $\hat{\sigma}_u = 2.8$, and the F-statistic $= 178$.

Similarly, the corresponding equation of Winters' model is:

$$\text{MAPE}_{o(\text{Winters})} = 3.61 - 0.02X_1 + 0.89X_3 + 0.06X_4, \tag{6}$$

with an $R^2 = 0.90$, $\hat{\sigma}_u = 2.0$, and the F-statistic $= 310$.

In equation (5), when randomness increases by 1 percent (all other things being equal), the MAPE of the model will increase by 0.94 percent. In equation (6), this increase is only 0.89 percent. However, the constant in equation (6) is considerably bigger than in equation (5), which also explains why ARMA models are relatively more accurate than Winters', when data with a low level of randomness are involved.

Another interesting observation is the magnitude of $\hat{\sigma}_u$. In equation (5), this value is 2.8, and in equation (6) it is 2.0. This means that the chances of going above or below the average of the regression equation are higher in equation (5) than in equation (6). In other words, a conservative forecaster who would like to take risk into account would be more inclined towards equation (6) than equation (5), all other things being equal. On the other hand, there is more room for improvement through equation (5) (the ARMA method) than through equation (6), if we assume that the user has some control over the accuracy of the time series to be forecast.

Table 10 lists the regression coefficients and related statistics of the regression equations involving the MAPEs of the forecasts as the dependent variable. The form of the equation (with a few exceptions) is:

$$\text{MAPE}_{ij} = a + b_2X_2 + b_3X_3 + b_4X_4b_5X_5 + b_6X_6, \tag{7}$$

where a is the constant, b_2, b_3, b_4, b_5 and b_6 are the regression coefficients, X_2 is the mean absolute percentage change in the trend-cycle of the jth data series, X_3 is the mean absolute percentage change in the randomness of the jth data series, X_4 is the mean absolute percentage change in the seasonality of the jth data

TABLE 10 Regression Coefficients
(Dependent variable is the mean absolute percentage error (MAPE) of forecasts)

		Forecasting Method	R^2	Constant	Mean Absolute % change			Absolute % Change of Trend-Cycle at Period $n = 12$	Number of Periods Ahead Forecasting	Standard Error of Estimate	F-Test
					Trend-Cycle	Random-ness	Seaso-nality				
Original Data: Nonseasonal Methods	1	Naive 1	.47	-5.91 (-4.68)*	2.22 (4.81)	1.11 (11.69)	.12 (2.48)	.66 (9.74)	1.44 (9.52)	15.06	145
	2	Single moving average	.42	-4.02 (-3.47)	1.71 (4.03)	.73 (8.68)	.30 (6.96)	.45 (7.23)	1.51 (10.90)	13.80	121
	3	Single exponential smoothing	.48	-5.79 (-5.12)	2.51 (6.06)	.81 (9.84)	.25 (5.82)	.54 (8.97)	1.54 (11.36)	13.45	149
	4	Adaptive response rate exponential smoothing	.38	-2.19 (-2.02)	2.27 (5.70)	.83 (10.53)	.20 (4.78)	.13 (2.26)	1.26 (9.68)	12.89	101
	5	Linear moving average	.37	-5.69 (-3.27)		1.07 (9.05)	.63 (10.25)	.22 (2.18)	2.51 (11.42)	22.00	123
	6	Brown's linear exponential smoothing	.41	-5.02 (-4.12)	.99 (2.21)	.79 (8.87)	.29 (6.57)	.42 (6.47)	1.86 (12.73)	14.62	117
	7	Holt's (2 parameters) linear exp. smoothing	.40	-3.91 (-3.19)	1.43 (3.19)	.85 (9.06)	.18 (4.05)	.43 (6.50)	1.60 (10.84)	14.74	109
	8	Brown's quadratic exponential smoothing	.34	-5.46 (-3.68)	1.66 (3.04)	.72 (6.65)	.18 (3.45)	.37 (4.71)	2.49 (13.97)	17.76	84
	9	Linear trend (regression fit)	.36	-4.10 (-2.71)	5.30 (9.53)	.39 (3.61)	.46 (8.15)	.40 (5.06)	1.57 (8.70)	18.01	92
Seasonal and Non seasonal Methods	10	Harrison harmonic smoothing	.38	0	5.50 (9.88)	.62 (5.48)	.37 (7.02)	.63 (7.39)	.75 (5.09)	18.80	103
	11	Winters' linear and seasonal exp. smoothing	.43	-2.45 (-2.69)	1.84 (5.52)	.65 (9.77)	.09 (3.02)	.50 (10.16)	.91 (8.42)	10.77	122
	12	Adaptive filtering	.48	-3.06 (-3.33)	1.52 (4.54)	.82 (11.85)	.17 (5.30)	.49 (9.63)	.91 (8.33)	10.95	155
	13	Autoregressive moving average (Box-Jenkins)	.43	-3.85 (-4.09)	2.08 (5.87)	.69 (10.00)	.19 (5.83)	.37 (6.97)	.98 (8.68)	11.14	123
Seasonally Adjusted Data: Nonseasonal Methods	14	Naive 2	.47	-2.33 (-2.93)	2.17 (7.83)	.78 (14.19)		.43 (10.33)	.62 (6.47)	9.48	182
	15	Single moving average	.43	-1.31 (-1.81)	1.98 (7.46)	.54 (10.27)	.15 (6.06)	.22 (5.68)	.72 (8.25)	8.65	126
	16	Single exp. smoothing	.46	-1.76 (-2.44)	2.18 (8.26)	.64 (12.07)	.12 (4.85)	.22 (5.71)	.65 (7.67)	8.48	141
	17	Adaptive response rate exponential smoothing	.41	0	2.15 (8.18)	.53 (9.44)	.14 (5.74)	.17 (4.24)	.66 (9.36)	8.89	114
	18	Linear moving average	.39	-2.57 (-2.52)	1.62 (4.32)	.68 (9.24)	.08 (2.38)	.56 (9.92)	1.07 (8.83)	12.00	105
	19	Brown's linear exponential smoothing	.46	-2.37 (-2.86)	.95 (3.13)	.77 (12.80)	.09 (3.17)	.44 (9.88)	.92 (9.28)	9.84	142
	20	Holt's (2 parameters) linear exp. smoothing	.40	-2.75 (-3.21)	1.57 (5.02)	.65 (9.97)	.08 (2.86)	.26 (5.60)	1.07 (10.46)	10.18	111
	21	Brown's quadratic exponential smoothing	.41	-3.95 (-3.88)	1.61 (4.46)	.62 (9.04)		.65 (12.38)	1.40 (11.46)	12.12	144
	22	Linear trend (regression fit)	.32	0	4.33 (9.61)	.31 (3.41)	.25 (5.99)	.46 (6.77)	.82 (6.86)	15.16	77.30

*/Numbers in parentheses are the t-tests of regression coefficients.

series, X_5 is the absolute percentage change in the trend-cycle of the jth data series at period $n - 12$, X_6 is the number of periods ahead of forecasts (that is, 1, 2, 3, 4, 5, 6, 9 or 12). Obviously, intrapolation for the values 7, 8, 10 and 11 is possible, and maybe extrapolation beyond 12 might be justified.

The equation relating the forecasting performance of the ARMA method is

$$\text{MAPE}_{i(\text{ARMA})} = -3.85 + 2.08X_2 + 0.69X_3 + 0.19X_4 + 0.37X_5 + 98X_6 \quad (8)$$

with the $R^2 = 0.43$, $\hat{\sigma}_u = 11.14$, and the F-statistic $= 123$.

The corresponding value of Winters' smoothing is

$$\text{MAPE}_{i(\text{Winters})} = -2.45 + 1.84X_2 + 0.65X_3 + 0.09X_4 + 0.50X_5 + 91X_6 \quad (9)$$

with the $R^2 = 0.43$, $\hat{\sigma}_u = 10.77$, and the F-statistic $= 122$.

The parameters of equations (5) and (6), or (8) and (9), as well as those corresponding to other methodologies, could provide some revealing information about the relative forecasting performance of the various methodologies. Unlike the regression coefficients of Table 9, those of Table 10 have relatively higher standard errors, which raises questions as to the statistical significance of the differences in the coefficients. The approach of specifying and measuring the relationship, however, is valid even though its usefulness can be enlarged by the following improvements, which would require further work:

1. Some random selection of the series used must be made.
2. The sample size (number of series used) must be increased considerably.
3. More independent variables are probably required so that a higher percentage of total variations in m.a.p.e.s could be explained by the forecasting equations.
4. In addition to MAPE, quadratic accuracy measures such as Theil's U-coefficients should be introduced as the dependent variable. However, there will be difficulty in specifying appropriate quadratic measures to describe the different series.

If we ignore for the moment the fact that some of the differences may not be significant, the examination of the coefficients of equations (8) and (9) can reveal that the constant corresponding to the ARMA method has a smaller value than that of Winters' model. Similarly, the coefficient corresponding to the trend-cycle change of the last period is smaller in ARMA than in the regression equation of Winters' model whose coefficients are smaller for the remaining variables. This means that the ARMA method will be relatively more accurate when the numerical values of the variables are small and when there are stronger fluctuations in the level of the econometric activity (trend-cycle *in the last* period of data that is, ARMA models will predict trend-cycle changes more accurately than Winters' model). Winters' method, on the other hand, will do better, in terms of smaller MAPE, when there are high mean absolute percentage changes in trend-cycle, randomness and seasonality.

4.   THE CAUSES OF VARIATIONS IN ACCURACY

After noting these variations in accuracy, we went on to examine possible causes for these differences. The MAPE of model fitting is determined by a different process from that which influences the m.a.p.e. of forecasts. The variations in the former can be precisely explained, in the sense that the R^2 values of the regression equations are usually larger than 0.90 (see Table 9), by two or three factors that account for most of the changes in the level of MAPE. The extent of randomness in the series (see Appendix B) seems to be the overwhelming factor influencing the accuracy of all methods used. That is not surprising, of course, except that we would expect the regression coefficients of randomness of the more sophisticated methods to be smaller than those of the simpler methods. But the opposite is true. Nonseasonal methods, as expected, do include significant coefficients for the variable 'seasonality'; seasonal methods and seasonally adjusted data do not, which indicates that both groups of methods sufficiently eliminate seasonality so that it does not, statistically, influence accuracy in the model-fitting process. Apart from linear trend fitting and Harrison's harmonic smoothing, all methods that do not use seasonally adjusted data seem to predict the trend-cyle fairly well. Even Naive 1 is quite effective, which suggests that period-to-period changes in the trend-cycle are not important enough, in the fitting phase, to produce statistically significant coefficients. The equations for the seasonally adjusted data include a significant coefficient for the variable 'trend-cycle'. That is somewhat surprising and might be caused by the decomposition process which seems to result in nonrandom residuals (Burman, 1965; Cleveland and Tiao, 1976). Finally, only three of the more sophisticated methods have a significant coefficient for the number of data points in the series. The values of these coefficients, however, are small and the t-statistics are very high, which indicates that the number of data points is not as important a factor as might have been thought.

Given the magnitude of R^2 of the regression equations in Table 9, it can be asserted that these equations can be used to predict the level and explain the variations in the MAPE of the model fitting fairly well. In addition, the standard errors of the estimate are not large. Thus it can be said that these equations can even be used to forecast the MAPE of the fitting phase of various series/methods. Obviously, such information can be extremely helpful in selecting forecasting methodologies, if it is assumed that the decision makers are concerned with linear errors in the model-fitting phase.

The R^2 of the regression equations for the MAPE of forecasts do not explain as much of the total variation in accuracy. (The average of all R^2 of Table 9 is 0.91 while that of Table 10 is only 0.41.) Furthermore, the standard errors of estimate are rather large, which makes the use of these equations unreliable for predicting the MAPE of forecasts, since the confidence limits will be large. Finally, it is surprising to see how little the factors included in the equations and the magnitude of their coefficients vary in the great majority of methods. No general patterns can be inferred, and differences between groups of methods are hard to explain from a statistical point of view. There are two additional factors of importance in the equations in Table 10: the first is the absolute percentage change in the trend-cycle in the last period of data $(n - 12)$. This factor does make sense and is consistent with experience and findings in the literature (McNees, 1976; Makridakis and Majani, 1977); it is included as a sign of impending change in the level of econometric activity. The second factor is the number of periods ahead of forecasts $(k = 1, 2, 3, 4, 5, 6, 9, 12)$. The interpretation of this factor is obvious in that as the time horizon of forecasting increases, accuracy decreases. The variable that is missing from Table 10 is 'number of data points'. The influence of this variable was either nonsignificant statistically, or when it was significant, its sign was positive, which made no sense. It was therefore excluded.

Many other factors (autocorrelation coefficients, standard error, coefficient of variation, mean absolute percentage change in the original series, seasonal indices, intercept and slope of regression, equation (trend fitting), the absolute change in trend-cycle at periods $n - 13$, $n - 14$, $n - 15$, and so forth) were tested, but (i) they were statistically nonsignificant, (ii) they resulted in R^2 that were inferior to those shown in Tables 9 and 10, or (iii) they were multicollinear with other variables. None of this additional information available up to period $n - 12$ could be used to understand or explain differences in variations in accuracy, so it is not presented here.

Naive 1 and, to a lesser extent, the exponential smoothing methods seem to do well because they 'hedge' their forecasts towards the middle. In this respect the chances of large errors are smaller when the pattern changes. The more sophisticated ARMA methods, on the other hand, attempt to follow the pattern as closely as possible. When there is no change from the previous pattern they are very accurate, which can be clearly seen by examining well-behaved series. When they miss, however, the errors are large. These large errors are shown in several ways. The coefficients of variation of the MAPE of, for example, Naive 1, single exponential smoothing. Holt's exponential smoothing, Winters' exponential smoothing, adaptive filtering and ARMA are 1.15, 1.29, 1.2, 1.27, 1.65 and 1.55 respectively. These coefficients indicate that the MAPE of the more sophisticated methods of adaptive filtering and ARMA are more dispersed around the mean values than the smoothing methods. Furthermore, Naive 1 has the smallest coefficient. With similar reasoning it can be seen that Naive 1 does the best in almost all cases when the unconstrained quadratic loss function is used (see Table 7). Smoothing methods on the original data are also relatively more accurate than on the same data they have been seasonally adjusted, even when the series with high randomness are excluded. This higher accuracy means that seasonal adjustments, which attempt to predict the seasonal pattern, can be inaccurate, which may cause large errors. The relative accuracy comparison changes in Table 8, which adjusts the value of Theil's U-statistic (it sets it equal to

2 when it exceeds 2). The extent of large random values then becomes less important, and the seasonal methods (excluding Harrison's harmonic smoothing) do better than Naive 2 and about the same as the methods using seasonally adjusted data. When a linear loss function is used (see Tables 1, 2 and 3), the difference is more pronounced. For example, in Table 3 the ARMA method is at least as good as any of the remaining methods in terms of the average of the MAPEs and considerably better than Naive 1.

It may be that when high randomness is present in data series, more sophisticated methods such as ARMA may overfit a model to these data. This overfit could occur while the mean square error is being minimized when stationarity is being achieved through and/or seasonal differencing, or when an ARMA (p, q) model is selected. Lack of randomness in residuals, for instance, does not always mean better forecasting results. In our study, 28 of the first identified appropriate ARMA (p, q) models for the 111 series did not produce random residuals, which means that the model fitted was not adequate. Alternative models were later selected, and in the final analysis the residuals from all series were random. Surprisingly, however, the new average MAPEs for the forecasts were no better than when 28 series did not have random residuals. Smoothing methods, on the other hand, do not require random residuals; instead the optimal model is determined by minimizing the MSE. Decomposition methods are not even concerned with minimizing mean square errors, or any other kind of loss function, except possibly implicitly through the averaging process they employ.

Finally, the difference between *model fitting* and *forecasting* as well as the type of loss function should be mentioned. In Table 7 which is based on an unconstrained standardized quadratic loss function, the seasonal methods do as well as those using seasonally adjusted data and much better than the nonseasonal methods. In forecasting, however, that is not true. The best methods vary depending upon the loss function assumed and the amount of randomness present in the series. It is also interesting to note that the MAPE of the forecasts is often smaller than the MAPE of model fitting (see Tables 1, 2 and 3). That is not so with Tables 7 and 8 which utilize a quadratic accuracy measure.

5.   CONCLUSIONS

A decision maker using these 111 series who wanted to apply a single forecasting method would have obtained very different results depending upon what loss function he wanted to minimize and whether he wanted to minimize the errors in the model fitting or in the forecasting phase. Overall, however, he would have done as well by using simpler rather than more sophisticated methods.

Further research will be required to shed light on the 'mystery' of why, under certain circumstances, simpler methods do as well or better than sophisticated ones. Obviously, this result is contrary to expectations and the previous experience of the authors. We believe it probably happened because the series used include observations from the 1974–75 major recession. At the same time, this occurrence is not an isolated instance; it has been reported in several other studies (Chatfield and Prothero, 1973; Groff, 1973; Dawes and Corrigan, 1974; Geurts and Ibrahim, 1975; McWhorter, 1975; Narasimham *et al.*, 1975; McCoubrey and McKenzie, 1976).

In addition to more research, several other improvements should be considered:

(i)   Combining quantitative forecasts (Bates and Granger, 1969; Newbold and Granger, 1974) and subjective and quantitative predictions in a Bayesian framework (Pankoff and Roberts, 1968) may improve forecasting efficiency.

(ii) Developing alternative forecasting approaches, based on different methodologies such as anticipatory surveys, Delphi and so forth (Friend and Jones, 1964; Ripe *et al.*, 1976), can improve results.

(iii) Being able to deal with situations caused by unusual events (for example, the oil embargo, legislation; see Box and Tiao, 1975) or to anticipate changes in the structural pattern of the data (Box and Tiao, 1976) will undoubtedly result in lower errors.

(iv) Switching methods (for example, using an exponential smoothing model or Naive 1 if a recession is imminent) may also reduce forecasting errors or provide more conservative forecasts.

In the final analysis, however, the findings of this and other studies cannot be generalized until more is known about the reasons and causes affecting the accuracy of forecasting. We hope that determining and measuring the factors affecting accuracy, as described in Section 3, can be of great help in finding the most accurate forecasting methods for specific situations. What is needed is an understanding of *when* and *under what circumstances* one method is to be preferred over the others.

ACKNOWLEDGEMENTS

The authors would like to thank Professors Ahmet Aykac, Robert Fildes, Robin Hogarth, Jean Claude Larréché, Harry Roberts, Bernard Tate and many others, too numerous to mention, for their comments and useful suggestions.

APPENDIX A – THE METHODS

This appendix presents the methods used to make the analyses described in the body of the paper.

Two sets of errors were calculated for each method. The first was arrived at by fitting a model to the first $n - 12$ values of each of the 111 series and calculating the error e_t as follows:

$$e_t = X_t - \hat{X}_t, \qquad (A\text{-}1)$$

where X_t is the actual value, and $\hat{X}_t$ is the one-period-ahead forecasted value.

Three so-called errors of 'model fitting' were also calculated as in (a)–(c) below, where all summations go from 1 to $n - 12$.

(a) The mean average percentage error (MAPE) $= (n - 12)^{-1}\Sigma(|e_t|/X_t)(100)$

$$(A\text{-}2)$$

(b) The mean square error (MSE) $= (n - 12)^{-1}\Sigma e_t^2$. $\qquad (A\text{-}3)$

(c) Theil's U-statistic: $= \sqrt{\{\Sigma e_t^2/\Sigma(X_t - X_{t+1})^2\}}$. $\qquad (A\text{-}4)$

The percentage of time the error in method i was smaller than in method m was also recorded.

The second set of errors involves the last 12 values, which were utilized as post-sample measures to determine the magnitude of the errors. The three measurements shown in equations (A-2), (A-3), (A-4), as well as the percentage of time method i was better than method m, were also computed for up to k forecasting horizons, starting at period $n - 11$. For convenience, k takes the values of 1, 2, 3, 4, 5, 6, 9 and 12 only. In no instance have the last $n - 11$ values been used to develop a forecasting model or estimate its parameters.

The following methods were used in the comparison. A more complete description of them given by Makridakis and Wheelwright (1978) also provides additional references.

(1) *Naive 1*

$$\text{Model fitting:} \quad \hat{X}_{t+1} = X_t, \qquad (A\text{-}5)$$

where $t = 1, 2, 3, \ldots, n - 12$.

$$\text{Forecasts:} \quad \hat{X}_{n-12+k} = X_{n-12}, \qquad (A\text{-}6)$$

where $k = 1, 2, 3, 4, 5, 6, 9, 12$.

(2) *Simple moving average*

$$\text{Model fitting:} \quad \hat{X}_{t+1} = \frac{X_t + X_{t-1} + X_{t-2} + \cdots + X_{t-N+1}}{N}, \qquad (A\text{-}7)$$

where N is chosen so as to minimize Σe_t^2, again summing over t from 1 to $n - 12$.

$$\text{Forecasts:} \quad X_{n-12+k} = \frac{X_{n-12+k} + X_{n-12+k-2} + \cdots + X_{n-12+k-N}}{N}. \qquad (A\text{-}8)$$

When the subscript of X on the right-hand side of (A-8) is larger than $n - 12$, the corresponding forecasted value is substituted.

(3) *Single exponential smoothing*

$$\text{Model fitting:} \quad \hat{X}_{t+1} = \alpha X_t + (1 - \alpha)\hat{X}_t, \tag{A-9}$$

where α is chosen so as to minimize Σe_t, the mean square error where again summing is over t from 1 to $n - 12$.

$$\text{Forecasts:} \quad \hat{X}_{n-12+k} = \alpha X_{n-12} + (1 - \alpha)\hat{X}_{n-12+K-1}. \tag{A-10}$$

(4) *Adaptive response rate exponential smoothing*

The equations are exactly the same as in (A-9) and (A-10), except α varies with t. The value of α_t is found by

$$\alpha_t = |E_t/M_t|, \tag{A-11}$$

where $E_t = \beta e_t + (1 - \beta)E_{t-1}$ and $M_t = \beta|e_t| + (1 - \beta)M_{t-1}$. β is set at 0.2.

(5) *Linear moving average*

$$\text{Model fitting:} \quad S'_t = \frac{X_{t-1} + X_{t-2} + \cdots + X_{t-N}}{N},$$

$$S''_t = \frac{S'_{t-1} + S'_{t-2} + \cdots + S'_{t-N}}{N}, \qquad \hat{X}_{t+1} = a_t + b_t, \tag{A-12}$$

where $a_t = 2S'_t - S''_t$ and $b_t = 2(N - 1)^{-1}(S'_t - S''_t)$.

The value of N is chosen so as to minimize the mean error.

$$\text{Forecasts:} \quad \hat{X}_{n-12+k} = a_{n-12} + b_{n-12}(k). \tag{A-13}$$

(6) *Brown's one-parameter linear exponential smoothing*

$$\text{Model fitting:} \quad S'_t = \alpha X_t + (1 - \alpha)S'_{t-1}, \quad S''_t = \alpha S'_t + (1 - \alpha)S''_{t-1}, \quad \hat{X}_{t+1} = a_t + b_t, \tag{A-14}$$

where $a_t = 2S'_t - S''_t$ and $b_t = (1 - \alpha)^{-1}(S'_t - S''_t)$.

The value of α is chosen so as to minimize the mean square error.

$$\text{Forecasts:} \quad \hat{X}_{n-12+k} = a_{n-12} + b_{n-12}(k). \tag{A-15}$$

(7) *Holt's two-parameter linear exponential smoothing*

$$\text{Model fitting:} \quad S_t = \alpha X_t + (1 - \alpha)(S_{t-1} + T_{t-1}), \tag{A-16}$$

$$T_t = \beta(S_t - S_{t-1}) + (1 - \beta)T_{t-1}, \tag{A-17}$$

$$\hat{X}_{t+1} = S_t + T_t. \tag{A-18}$$

The values of α and β are chosen so as to minimize the mean square error. Marquardt's (1963) nonlinear optimization algorithm is employed for this purpose.

$$\text{Forecasts:} \quad \hat{X}_{n-12+k} = S_{n-12} + T_{n-12}(k). \tag{A-19}$$

(8) *Brown's one-parameter quadratic exponential smoothing*

$$\text{Model fitting:} \quad S'_t = \alpha X_t + (1 - \alpha)S'_{t-1}, \tag{A-20}$$

$$S''_t = \alpha S'_t + (1 - \alpha)S''_{t-1}, \tag{A-21}$$

$$S'''_t = \alpha S''_t + (1 - \alpha)S'''_{t-1}, \tag{A-22}$$

$$\hat{X}_{t+1} = a_t + b_t + 1/2c_t, \tag{A-23}$$

where

$$a_t = 3S'_t - 3S''_t + S'''_t,$$

$$b_t = \alpha\{2(1 - \alpha)^2\}^{-1}\{(6 - 5\alpha)S'_t - (10 - 8\alpha)S''_t + (4 - 3\alpha)S'''_t\}$$

and

$$c_t = \alpha(1 - \alpha)^{-2}(S'_t - 2S''_t + S'''_t).$$

The value of α is chosen so as to minimize the mean square error.

$$\text{Forecasts:} \quad \hat{X}_{n-12+k} = a_{n-12+k} + b_{n-12+k}(k) + 1/2c_{n-12+k}(k)^2. \tag{A-24}$$

(9) *Linear regression trend fitting*

$$\text{Model fitting:} \quad \hat{X}_t = a + bt, \tag{A-25}$$

where $t = 1, 2, 3, \ldots, n - 12$, and a and b are chosen so as to minimize the sum of the square errors by solving the normal equations:

$$a = \frac{\Sigma X}{n - 12} - b\frac{\Sigma t}{n - 12}, \qquad b = \frac{(n - 12)\Sigma tX - t\Sigma X}{(n - 12)\Sigma t^2},$$

where all summations go from $n - 12$.

$$\text{Forecasts:} \quad \hat{X}_{n-12+k} = a + b(n - 12 + k). \tag{A-26}$$

(10) *Harrison's harmonic smoothing*

(a) *Removal of trend-cycle.* The trend-cycle is removed by calculating an L-period moving average (where L is the length of seasonality) and dividing it into the time series X_t. Using the moving average eliminates seasonality and randomness. Thus,

$$M_t = T_t C_t, \tag{A-27}$$

where T_t denotes trend and C_t denotes cycle.

(b) *Calculation of crude seasonal estimates.* Equation (A-27) can be divided into X, yielding

$$R^*_t = \frac{X_t}{M_t} = \frac{S_t T_t C_t R_t}{T_t C_t} = S_t R_t, \tag{A-28}$$

where S_t denotes seasonality and R_t denotes randomness. The R^*_t's are estimates of seasonality with some randomness remaining.

In order to eliminate randomness, R_t^* is arranged in such a way that similar months can be averaged together. Letting r_{ij} denote the ith year and the jth period (month), one could average all the periods together as follows:

$$r_j = \Sigma r_{ij} m_j^{-1}, \tag{A-29}$$

summing over i from 1 to m_j, where m_j is the number of observations available for the jth period and the r_j are the crude seasonal estimates.

The standard deviation of the crude seasonal estimate is

$$s = \sqrt{\left\{ \sum_{i=1}^{m_j} \sum_{j=1}^{L} (r_{ij} - r_j)^2 \right\} \Big/ \left(\sum_{j=1}^{L} m_j - L \right)}. \tag{A-30}$$

(c) *Obtaining smoothed seasonal estimates through Fourier analysis.* The smoothed seasonal estimates, r_j, can be obtained by applying

$$\hat{r}_j = 1 + \sum_{k=1}^{L} a_k \cos(kd_j) + b_k \sin(kd_j), \qquad j = 1, 2, \ldots, L, \tag{A-31}$$

where

$$d_j = \frac{2(j-1)\pi}{L/2} - \pi, \qquad j = 1, 2, \ldots, L, \tag{A-32}$$

$$a_k = \frac{1}{L/2} \sum_{j=1}^{L} r_j \cos(KX_j), \tag{A-33}$$

$$b_k = \frac{1}{L/2} \sum_{j=1}^{L} r_j \sin(KX_j). \tag{A-34}$$

In equations (A-33) and (A-34), the crude seasonal estimates, r_j, and the actual values, X_j, are used to estimate the Fourier coefficients, which in turn are used in equation (A-31) to calculate the smoothed seasonal estimates. Only those $\hat{r}_j$ in equation (A-31) that are significantly different from zero are included in the final calculations; the remainder are discarded.

(d) *Replacement of the outliers.* If the difference between r_{ij}, the original estimates of seasonality, and $\hat{r}_j$, the smoothed estimates, is greater than 2.50 (see (A-30)), the corresponding X_t value is replaced by some more likely value (for example, $\hat{r}_j R_t^*$). When outliers are replaced, the effect of unusual events such as strikes, wars and total breakdowns is eliminated from the series, which allows a more realistic estimate.

After the outliers are replaced, steps (a), (b) and (c) above are recomputed with the replaced values. Thus, new smoothed seasonal estimates are calculated with equation (A-31).

(e) *Measurement of the adequacy of the seasonal fit.* The adequacy of the seasonal fit is tested by an *F*-test. The *F*-value is computed as the ratio of the mean square errors between the variations caused by seasonal effects, M_s, and that caused by unexplained residual factors, M_r.

$$F\text{-test} = M_s/M_r.$$

Forecasts: $\hat{X}_{n-12+k} = T_{n-12+k}\hat{r}_j,$ \hfill (A-35)

where T_{n-12+k} is an estimate of the trend-cycle at period $n-12$ and $\hat{r}_j$ is the corresponding seasonal index for period $n-12+k$.

T_{n-12+k} is calculated by different methods depending on whether or not there is a statistically significant trend in the data.

(11) *Winters' three-parameter linear and seasonal exponential smoothing*

$$\text{Model fitting:}\quad S_t = \alpha\frac{X_t}{I_{t-L}} + (1-\alpha)(S_{t-1} + T_{t-1}),\hfill (A\text{-}36)$$

$$T_t = \gamma(S_t - S_{t-1}) + (1-\gamma)T_{t-1}, \qquad I_t = \beta\frac{X_t}{S_t} + (1-\beta)I_{t-L},$$

$$\hat{X}_{t+1} = (S_t + T_t)I_{t-L+1},$$

where L is the length of seasonality.

The values of α, β and γ are chosen so as to minimize the mean square error by Marquardt's (1963) nonlinear optimization steepest descent algorithm.

$$\text{Forecasts:}\quad \hat{X}_{n-12+k} = (S_{n-12} + kT_{n-12})I_{n-12+k}.\hfill (A\text{-}37)$$

(12) *Adaptive filtering*

$$\text{Model fitting:}\quad X_t = \phi_{1t}X_{t-1} + \phi_{2t}X_{t-2} + \phi_{3t}X_{t-3} + \cdots + \phi_{pt}X_{t-p}.\hfill (A\text{-}38)$$

The values of the parameters ϕ_{it} are modified by equation (A-39) so as to minimize the mean square error.

$$\phi_{ti} = \phi_{(t-1)i} + 2Ke_t^*X_{t-i}^{**},\hfill (A\text{-}39)$$

where K is a learning constant set equal to $1/p$ and e_t^* and X_{t-1}^* are standardized values of e_t and X_{t-1} correspondingly.

The order of the autoregressive equation (A-38) is set automatically. Thus, $p = 3$ for nonseasonal data and $p = L$ for seasonal stationary data. If the data are nonstationary and seasonal, the first difference is found, then p is set equal to L.

Forecasts:

$$\hat{X}_{n-12+k} = \phi_{1(n-12)}X_{n-11+k} + \phi_{2(n-12)}X_{n-10+k} + \cdots + \phi_{p(n-12)}X_{n-p+k}.$$
$$\hfill (A\text{-}40)$$

If the first difference has been taken, equation (A-39) is adjusted accordingly. When the subscript of X on the right-hand side of equation (A-40) exceeds $n-12$, the forecasted value $\hat{X}$ is substituted in equation (A-40).

(13) *Autoregressive moving average (Box–Jenkins) methodology*
Model fitting:

$$\hat{X}_t = \phi_1 X_{t-1} + \phi_2 X_{t-2} + \cdots + \phi_p X_{t-p} - \theta_1 e_{t-1} - \theta_2 e_{t-2} - \cdots - \theta_q e_{t-q}.$$

(A-41)

The parameters of equation (A-41) are estimated, using Marquardt's (1963) nonlinear optimization procedure so as to minimize the mean square error. Initial estimates of the parameters are provided to start the algorithm by solving the Yule–Walker equations for the ϕ_i. An iterative procedure is used for initial estimates of ϕ_i. If there is no convergence in the iterative process (which is unusual), the values of ϕ_i are set at 0.2, 0.3, 0.4, ...

The orders of p and q are determined by examining the autocorrelations and partial autocorrelations of the first $n - 12$ data points of each series. The determination was made judgementally. If the data were seasonal, a multiplicative seasonal model was specified by incorporating seasonal parameters in equation (A-41).

Before applying equation (A-41), two further steps were taken. (i) Two transformations were applied to the data to assure stationarity in variance. Square root and logarithmic transformations were compared to no transformation, and the one that produced the most homogeneous variance was chosen. The method of range mean plot described by Box and Jenkins (1973) was employed to select the best transformation. It was found, however, that the difference in accuracy between the transformed and untransformed data was marginal, if any, and was restricted to the longer forecasting horizons of 9 and 12 periods ahead. (ii) Next, regular (short) and seasonal (long) differences were applied to the data, if necessary, so as to achieve stationarity of the mean.

Once the best transformation (or no transformation) had been made and the appropriate level of differences chosen, expression (A-41) or its seasonal extension was used to estimate the optimal ARMA parameters. Then the autocorrelation of residuals was examined. If the autocorrelations showed random residuals, the tentatively identified model was considered adequate; otherwise, another model was specified. It may be of interest that 75 per cent of the tentatively identified models produced random residuals the first time. About half of the remaining 25 percent were correctly identified the second time, and only one-fifth of the remainder the third time. It took 6 or 7 trials to achieve random residuals for all of the series. The biggest difficulty in correct identification arose through having specified the wrong level of differencing and not getting enough clues from the autocorrelations to perceive the error. The residuals were considered random when all of their autocorrelations, up to 24 time lags, were within the limits of $\pm 1.96\{\sqrt{(n-12)}\}^{-1}$ and the Box–Pierce statistic was smaller than the corresponding value of the χ^2 distribution for the lags 1–12 and 1–24. There were rarely autocorrelations outside the 95 percent confidence limits, but in no case was the Box–Pierce greater than the corresponding χ^2 value (for both time lags 1–12 and 1–24).

Only the ARMA method required a considerable amount of judgemental input; other methods were run automatically. Finally, no backforecasting was used to estimate initial values for the error terms e_{t-i}.

Forecasting: Forecasts were made after the estimates were retransformed and the differences taken into consideration. Forecasted values were used when subscripts of X on the right-hand side of equation (A-42) were larger than $n - 12$.

$$\hat{X}_{n-12+k} = \phi_1 X_{n-11+k} + \phi_2 X_{n-10+k} + \cdots + \phi_p X_{n-p+k}$$
$$- \theta_1 e_{n-11+k} - \theta_2 e_{n-10+k} - \theta_q e_{n-q+k}. \tag{A-42}$$

(14) *Seasonally adjusted for 22 methods*

Model fitting: The $n - 12$ *data* of each of the 111 series were first adjusted for seasonality, as

$$X'_t = X_t/S'_t, \tag{A-43}$$

where X'_t is the seasonally adjusted value at period t, and S'_j is the corresponding seasonal factor for period t.

Forecasts: The forecasts for $\hat{X}'_{n-11}$, $\hat{X}'_{n-10}, \ldots, \hat{X}'_n$ were reseasonalized as

$$\hat{X}_{n-12+k} = \hat{X}_{n-12+k}(S'_j). \tag{A-44}$$

Equations (A-43) and (A-44) were applied for all nonseasonal methods 1 to 9 above.

The indices S'_j were computed (see Appendix B) by applying the simple ratio-to-moving average decomposition procedure. It was found that the seasonal indices and the one-year-ahead forecasts of the seasonal indices computed by the Census II decomposition method of the USA Bureau of Labor Statistics did not produce better forecasting results than those of the simple ratio-to-moving average method.

APPENDIX B – DECOMPOSING A TIME SERIES

If a multiplicative relationship is assumed of the form:

$$X_t = S_t T_t C_t R_t, \tag{A-45}$$

where S_t is the seasonality at period t, T_t is the trend at period t, C_t is the cycle at period t and R_t is the randomness at period t, then by computing a centred moving average of equation (A-45), whose length is equal to the length of seasonality, we get

$$M_t = T_t C_t. \tag{A-46}$$

Dividing equation (A-46) into equation (A-45) yields

$$\frac{X_t}{M_t} = \frac{S_t T_t C_t R_t}{T_t C_t} = S_t R_t \tag{A-47}$$

Averaging equation (A-47) will eliminate randomness and yield seasonal estimates, S_j'. Dividing S_j' into equation (A-45) yields

$$X_t^* = T_t C_t R_t. \tag{A-48}$$

Computing a low order moving average (the one used was a double 3×3 moving average) on the X_t^* values of equation (A-48) results in the elimination of randomness and yields another set of trend-cycle values M_t' or

$$M_t' = T_t C_t. \tag{A-49}$$

Finally, dividing equation (A-49) into equation (A-48) will isolate randomness, R_t. Thus, randomness is defined as whatever fluctuation remains after seasonality and trend-cycle fluctuations have been removed. It is a residual term in this sense and is equal to

$$\frac{X_t^*}{M_t'} = \frac{T_t C_t R_t}{T_t C_t} = R_t. \tag{A-50}$$

The concept of randomness as used throughout this paper is defined by equation (A-50) above.

The seasonal indices, S_j', were used in equation (A-43) and (A-44) for adjusting and readjusting for seasonality. Furthermore, mean absolute percentage changes for each of the components of the time series were computed as

$$\text{MAPC } Y = 100 \left(\frac{1}{n-12} \sum_{t=2}^{n-12} \frac{|Y_t - Y_{t-1}|}{Y_{t-1}} \right), \tag{A-51}$$

where Y_t can be any of the components of the time series. The MAPC of the components are used as the independent variables of the regression equations shown in Tables 9 and 10.

REFERENCES

Adam, E. E. and Ebert, R. J. (1976). A comparison of human and statistical forecasting. *AIITE Trans.*, **8**, 120–127.

Bates, J. M. and Granger, C. W. J. (1969). Combination of forecasts. *Operat. Res. Q.*, **20**, 451–468.

Bauman, W. S. (1965). The less popular stocks versus the most popular stocks. *Financial Analysts J.*, **21**, 61–69.

Box, G. E. P. and Jenkins, G. M. (1973). Some comments on a paper by Chatfield and Prothero and on a review by Kendall. *J. R. Statist. Soc.*, **139**, 295–336.

Box, G. E. P. and Tiao, G. C. (1975). Intervention analysis with applications to economic and environmental problems. *J. Amer. Statist. Ass.*, **70**, 70–79.

Box, G. E. P. and Tiao, G. C. (1976). Comparison of forecast and actuality. *Appl. Statist.*, **25**, 195–200.

Burman, J. P. (1965). Moving seasonal adjustment of economic time series. *J. R. Statist. Soc.* A, **128**, 534–558.

Chatfield, C. and Prothero, D. L. (1973). Box–Jenkins seasonal forecasting: problems in a case study. *J. R. Statist. Soc.* A, **136**, 295–336.

Christ, C. F. (1951). A test of an econometric model of the United States, 1921–1974. In *Conference on Business Cycles*. New York: National Bureau of Economic Research.

Christ, C. F. (1975). Judging the performance of econometric models of the U.S. economy. *Int. Econ. Rev.*, **16**, (1), 57–81.

Clearly, J. P. and Fryk, D. A. (1974). A comparison of ARIMA and econometric models for telephone demand. *Proc. Amer. Statist. Ass.*, Business and Economics Section, pp. 448–450.

Cleveland, W. P. and Tiao, G. C. (1976). Decomposition of seasonal time series: a model for the census X-11 program. *J. Amer. Statist. Ass.*, **71**, 581–587.

Cooper, J. P. and Nelson, C. R. (1975). The ex ante performance of the St. Louis and FRB-MIT-PENN econometric models and some results on composite predictors. *J. of Money, Credit and Banking*, February issue, pp. 1–32.

Cooper, R. L. (1972). The predictive performance of quarterly econometric models of the United States. In *Econometric Models of Cyclical Behavior* (B. G. Hickman, ed.) New York: National Bureau of Economic Research.

Cragg, J. and Malkiel, B. (1968). The consensus and accuracy of some predictions of the growth in corporate earnings. *J. of Finance*, March issue, pp. 67–84.

Dawes, R. M. (1977). In *Shallow Psychology* (J. S. Carroll and J. W. Payne, eds). Hilldale, N. J.: Laurence Erlbaunn.

Dawes, R. M. and Corrigan, B. (1974). Linear models in decision making. *Psychol. Bull.*, **81**, 95–106.

Elton, E. J. and Gruber, M. J. (1972). Earnings estimates and the accuracy of expectational data. *Management Sci.* B, **18**, 409–424.

Fama, E. F. (1965). The behaviour of the stock market prices. *J. Business*, **38**, 34–105.

Friend, J. and Jones, R. C. (1964). Short-run forecasting models incorporating anticipatory data. In *Models of Income Determination* (NBER). Princeton: Princeton University Press.

Fromm, G. and Klein, L. R. (1973). A comparison of eleven econometric models of the United States. *Amer. Econ. Ass.*, May issue, 385–401.

Geurts, M. D. and Ibrahim, I. B. (1975). Comparing the Box–Jenkins approach with the exponentially smoothed forecasting model application to Hawaii tourists. *J. Marketing Res.*, **12**, 182–188.

Goldberg, L. R. (1970). Man versus model of man: a rationale, plus some evidence, for a method of improving clinical inferences. *Psychol. Bull.*, **73**, 422–432.

Goldberg, L. R. (1976). Man versus model of man: just how conflicting is that evidence? *Organizational Behavior and Human Performance*, **16**, 13–22.

Goldfeld, S. M. (1972). The predictive performance of quarterly econometric models of the United States: Comments. In *Econometric Models of Cyclical Behavior* (B. G. Hickman, ed.). (Studies in Income and Wealth, no. 36). New York: Columbia University Press for NBER.

Granger, C. W. J. (1969). Investigating causal relations by econometric models and cross spectral methods. *Econometrica*, **37**, 424–438.

Granger, C. W. J. (1974). *Spectral Analysis of Economic Time Series*. Princeton: Princeton University Press.

Granger, C. W. J. and Newbold, P. J. (1974). Spurious regressions in econometrics. *J. Econometrics*, **2**, 111–120.

Green, D. and Segall, J. (1967). The predictive power of first-quarter earnings reports. *J. Business*, **40**, January issue, 44–55.

Groff, G. K. (1973). Empirical comparison of models for short-range forecasting. *Management Sci.*, **20**(1), 22–31.

Gross, D. and Ray, J. L. (1965). A general purpose forecasting simulator. *Management Sci.* B, **11**(6), 119–135.

Hogarth, R. M. (1975). Cognitive processes and the assessment of subjective probability distributions. *J. Amer. Statist. Ass.*, **70**, 271–290.

von Hohenbalken, B. and Tintner, G. (1962). Econometric models of the O.E.E.C. member countries, the United States and Canada, and their application to economic policy. *Weltwirtschaftliches Archiv*, **89**, 29–86.

Howrey, E. P., Klein, L. R. and McCarthy, M. D. (1974). Notes on testing the predictive performance of econometric models. In *Econ. Rev.*, **15**, 2, 376–377.

Jensen, M. C. (1968). The performance of mutual funds in the period 1945–1956. *J. of Finance*, **23**, 389–416.

Johnson, T. E. and Schmitt, T. G. (1974). Effectiveness of earnings per share forecasts. *Financial Management*, summer issue, 64–72.

Jorgenson, D. W., Hunter, J. and Nadiri, M. I. (1970). A comparison of alternative econometric models of quarterly investment behavior. *Econometrica*, **38**, 187–213.

Kahneman, D. and Tversky, A. (1973). On the psychology of prediction. *Psychol. Rev.*, **80**, no. 4, 237–251.

Kirby, R. M. (1966). A comparison of short and medium-range statistical forecasting methods. *Management Sci.* B, **15**, no. 4, 202–210.

Krampf, R. F. (1972). The turning point problem in smoothing models. Ph.D. Thesis, University of Cincinnati.

Leser, C. E. V. (1966). The role of macro-models in short-term forecasting. *Econometrica*, **34**, 862–872.

Leser, C. E. V. (1968). A survey of econometrics. *J. R. Statist. Soc.* A, **131**, 530–566.

Levine, A. H. (1967). Forecasting techniques. *Management Accounting*, January issue, 86–95.

Libby, R. (1976). Man versus model of man: some conflicting evidence. *Organizational Behaviour and Human Performance*, **16**, 1–12.

Liebling, H. I. and Russell, J. M. (1969). Forecasting business investment by anticipation surveys and econometric models in 1968–69. *1969 Proceedings of Business and Econometric Statistics Section*, American Statistical Association, pp. 250–255.

Liebling, H. I., Bidwell, P. T. and Hall, K. E. (1976). The recent performance of anticipation surveys and econometric model projections of investment spending in the United States. *J. of Business*, **49**, no. 4, 451–478.

Mabert, V. A. (1975). Statistical versus sales force – executive opinion short-range forecasts: a time-series analysis case study. Krannert Graduate School, Purdue University (working paper).

McCarthy, M. D. (1972). The predictive performance of quarterly econometric models of the United States: comments. In *Econometric Models of Cyclical Behavior* (B. G. Hickman, ed.). (Studies in Income and Wealth, no. 36). New York: Columbia University Press for NBER.

McCoubrey, C. A. and McKenzie, W. Jr (1976). Forecasting collections using expert opinions in a deterministic model. ORSA/TIMS Conference, Miami, Fl.

McNees, S. K. (1975). An evaluation of economic forecasts. *New England Econ. Rev.*, November–December issue.

McNees, S. K. (1976). The forecasting performance in the early 1970s. *New England Econ. Rev.*, July–August 29–40.

McWhorter, A. Jr (1975). Time-series forecasting using the Kalman filter: an empirical study. *Proc. Amer. Statist. Ass.*, Business and Economics Section, pp. 436–446.

Makridakis, S. (1976). A survey of time series. *Int. Statist. Rev.*, **44**, 29–70.

Makridakis, S. and Majani, B. (1977). Can recession be predicted? *Long-range Planning J.*, April issue.

Makridakis, S. and Vandenburgh, H. M. (1975). The accuracy and cost of nonseasonal time-series forecasting methods. *INSEAD Research Papers Series*, no. 143.

Makridakis, S. and Wheelwright, S. (1978). *Interactive Forecasting: Univariate and Multivariate Methods*. San Francisco: Holden-Day.

Marquardt, D. W. (1963). An algorithm for least squares estimation of nonlinear parameters. *Soc. Indust. Appl. Math.*, **11**, 431–441.

Meehl, P. E. (1954). *Clinical versus Statistical Prediction: A Theoretical Analysis and Review of the Literature*. Minneapolis: University of Minnesota Press.

Meehl, P. E. (1965). Clinical versus statistical prediction. *J. Exper. Res. in Personality*, **1**, 27–32.

Narasimham, G. V. L. *et al.* (1974). On the predictive performance of the BEA quarterly econometric model and a Box–Jenkins type of ARIMA model. *Proc. Amer. Statist. Assoc.* Business and Economics Section, pp. 448–450.

Narasimham, G. V. L. *et al.* (1975). A comparison of predictive performance of alternative forecasting techniques: time series versus an econometric model. *Proc. Amer. Statist. Ass.*, Business and Economics Section, pp. 459–464.

Naylor, T. H. *et al.* (1972). Box–Jenkins methods: an alternative to econometric forecasting. *Int. Statist. Rev.*, **40**, 123–137.

Nelson, C. R. (1972). The prediction performance of the FRB–MIT–PENN model of the U.S. economy. *Amer. Econ. Rev.*, **62**, 902–917.

Newbold, P. and Granger, C. W. J. (1974). Experience with forecasting univariate time series and the combination of forecasts. *J. R. Statist. Soc.* A, **137**, 131–165.

Niederhoffer, V. and Regan, D. (1972). Summarized in *Barron's Magazine*, December issue.

O'Brien, J. W. (1970). How market theory can help investors set goals, select investment managers and appraise investment performance. *Financial Analysts J.*, **26**, 91–103.

Pankoff, L. D. and Roberts, H. V. (1968). Bayesian synthesis of clinical and statistical prediction. *Psychol. Bull.*, **70**, 762–773.

Pierce, D. A. (1977). Relationships—and the lack thereof—between economic time series, with special reference to money, reserves, and interest rates. *J. Amer. Statist. Ass.*, **72**, 11–27.

Raine, J. E. (1971). Self-adaptive forecasting considered. *Decision Sci.*, **2**, no. 2, 264–279.

Reid, D. J. (1969). A comparative study of time-series prediction techniques on economic data. Ph.D. Thesis, University of Nottingham.

Reid, D. J. (1977). A survey of statistical forecasting techniques with empirical comparisons. Paper presented at the *I.E.E. Symposium on Statistical Model Building for Precision and Control*.

Ripe, R., Wilkinson, M. and Morrison, D. (1976). Industrial market forecasting with anticipatory data. *Management Sci.*, **22**, 639–651.

Sarbin, T. R. (1943). Contribution to the study of actuarial and individual methods of prediction. *Amer. J. Sociol.*, **48**, 593–602.

Sawyer, J. (1966). Measurement and prediction, clinical and statistical. *Psychol. Bull.*, **66**, 178–200.

Slovic, P. (1972). Psychological study of human judgement: implications for investment decision making. *J. of Finance*, **37**, 779–799.

Slovic, P. (1974). Limitations of the mind of man: implications for decision-making in
the nuclear age. *Oregon Research Institute Research Bulletin*, **11**, no. 17.
Steckler, H. O. (1968). Forecasting with econometric models: an evaluation.
Econometrica, **36**, 437–463.
Theil, H. (1966). *Applied Economic Forecasting*, pp. 26–32. Amsterdam: North Holland.
Tversky, A. (1974). Assessing uncertainty (with Discussion). *J. R. Soc.* B, **36**, 148–159.
Tversky, A. and Kahneman, D. (1974). Judgement under uncertainty: heuristics and
biases. *Science*, **185**, 1124–1131.

DISCUSSION OF THE PAPER BY PROFESSOR MAKRIDAKIS AND DR HIBON

Dr W. G. GILCHRIST (Sheffield City Polytechnic): It gives me pleasure to propose the vote of thanks to this paper. Modern man is fascinated with the subject of forecasting. The authors' first sections list a large number of references of people's attempts to compare methods. One reference which is not given is that from the Operational Research Society Forecasting Study Group. About 10 or 11 years ago it gathered together various sets of data and sent them out to different firms, asking them to forecast these data using their standard methodology. The reference is not in the paper because, having received the results, the Study Group was so horrified by them that they were brushed under the carpet, the carpet patted down, and no more was said about them. It is all very well, as statisticians, to say that there is a proper methodology for doing forecasting, that it is all laid out nicely in Box–Jenkins and various other books and papers, but we have to accept that most firms have vast numbers of time series to forecast, they have standard methods which are in their computer packages and they use these more or less regardless. Therefore, although we might say this is not the way to do it, the facts of life are such that this is the way most people do it: they take their data in large quantities and push it through their own standard methods. Therefore, there is real value in seeing how these very commonly used methods work when large varieties of different sorts of time series are put through them – because this is what happens in practice.

In the Operational Research Society exercise some things went disastrously wrong because of series which clearly became highly unstable just at the end – and the answer should have been that there should not be a forecast, but we should think about the situation. Another major source of error arose because seasonal data were used with non-seasonal methods. The authors carefully broke up the data according to randomness, looking at the subset of data with low randomness. I thought that they should have done the same with the seasonal and non-seasonal data. Looking at the Tables, the non-seasonal methods were given very short shrift because of the fact that most of the data that went through them, I suspect, were essentially seasonal data.

It also emerged from the Operational Research Society exercise that there was a need to do a fair amount of forecasting before it was possible properly to assess its quality. I was slightly concerned about the number of forecast errors used by the authors as the basis for their arguments. It was argued this evening that the problem is one of reidentifying the model – but, if this is what people do in practice, perhaps this is what the authors should have done. What concerns me is that analysis is based on the 12 errors $e_1, e_2, \ldots, e_{12}$. When forecasting one time unit into the future, if I understood correctly, it was just the single error e_1 that was used as a basis for comparing approaches and methods. Using only that one error for one step ahead slightly concerns me. The authors adopted a methodology of averaging the absolute percentage errors for sets of different lead times. This has effects which I do not really understand. Clearly, it creates quantities which are highly correlated to each other, which makes their interpretation difficult.

If we plot, for some of the better methods, mean absolute percentage error (MAPE) against lead-time (forecast horizon), one obtains remarkably linear plots. I am interested to know how linear they really are. The averaging operation will tend to make it much more difficult to observe deviations from linearity if they are fairly small.

I should, therefore, like to ask the authors to tell us a little more about this methodology of using mean absolute percentage errors because I feel that it does not help in understanding the effect of lead-time.

I rather query some of the terminology. When the authors talk about 'fitting', in fact, they are using one-step ahead forecasting, but forecasts based on using all the data up to the last 12 to fit the model or to optimize the parameters. I should like to know what the model fitting results would have been if they had used the actual residuals between the fitted model and the observations at that point.

One other aspect of methodology I want to query is one which their predecessors who compared methods also adopted. It worries me that when comparisons are made the dice are very much loaded in favour of the autoregressive moving average (ARMA) models. A whole family of models is first defined, this is followed by choosing the one that fits best. In the authors' case, they also looked at various possible transformations. The chosen model is used for that particular set of data, and different models are used for different sets of data. When we turn to the polynomial models, a constant mean is used on all 111 models – linear trend on all 111, quadratic on all 111 and so on – and they are all then compared. It seems to me that we are not really comparing like with like. If one methodology is adopted for Box–Jenkins, perhaps we ought to consider for each time series which of the family of polynomials ought to be used, whether transformations should be used etc. The family of ARMA models would then be compared with the family of polynomial models.

Again, returning to the authors' general approach, it is fair to say that they are comparing what people use in practice. It is amazing to me, however, that after all this exercise in identifying models, transforming and so on, that the autoregressive moving averages come out so badly. I wonder whether it might be partly due to the authors not using the backwards forecasting approach to obtain the initial errors.

The approach and objective of the authors in trying to ask the questions: what is it that produces a good method, what is it that produces a bad one, and what characteristics of a set of data might guide us in choosing our methods, is worthwhile. Although it is possible to criticize the way in which it has been done in the paper, as a first attempt it is an exercise that is well worth doing. If some method could be devised for characterizing sets of data which could then be plugged into automatic forecasting methods this might improve those methods.

May I thank the authors for an interesting paper, for raising some interesting issues and for underlining once again that, in reality, methods which are simple and fairly robust, fairly responsive to the non-stationary unstable world in which we live, are well worth considering, even though they lack the statistical finesse of the methods we like to teach our students. I should like to propose the vote of thanks.

Professor M. B. PRIESTLEY (University of Manchester Institute of Science and Technology): The subject of forecasting has aroused considerable interest over the last fifteen years or so and it has now acquired a massive literature, most of which is concerned with its theoretical aspects. It is therefore timely and refreshing to have presented to us the results of a substantial empirical study of various forecasting techniques, and in this respect the authors have rendered the Society a valuable service.

As in all empirical studies, however, we must resist the temptation to read too much into the results of the analyses. Let us clarify some basic concepts. *There is no such thing as a forecasting 'method'; there is no such thing as an* ARMA *(or Box–Jenkins) forecasting 'method'.* There *is* something called a 'least-squares' forecasting method, and this, in fact, provides the basis for virtually all theoretical studies. What the authors refer to as the 'Box–Jenkins method' is simply a recursive algorithm for calculating *least-squares* forecasts, and, if the model for the series was known precisely, the Box–Jenkins 'method' would lead to exactly the same answer as the Wiener 'method', the Kolmogorov 'method' or even the Kalman filter 'method'. (All these 'methods' are simply different computational procedures for calculating the same quantity, namely, the least-squares forecast of a future value from linear combinations of the past data.)

It is also extremely important to distinguish very clearly between a *model* and a *forecasting procedure. A model (such as an* ARMA *scheme) does not automatically lead to a particular forecasting formula.* A model represents a stochastic description of a series, and given a particular model it is up to the user to decide how to use this stochastic description so as to obtain that form of forecast which is optimal according to a chosen loss function. Of course, if we choose the mean-square error as our loss function, and if we are dealing with an ARMA model in which residuals are independent (rather than merely uncorrelated), then the optimal forecast is very easily constructed from the model by simply setting to zero the values of those residuals, ε_t, which correspond to future time points. This procedure produces the conditional expectation of the variable to be forecast (given the past data), and whilst the conditional expectation is optimal for loss functions other than the mean square error (for example, it is optimal for any loss function which is an even power of the forecast error, e_t, provided the conditional distribution is symmetric), it is certainly not optimal for an arbitrary choice of loss function and arbitrary distribution. To see this we need only remind ourselves of the well-known result that the expression $E[|X - c|]$ is minimized by choosing c as the median (not the mean) of X. This would certainly affect the form of the forecasts corresponding to the loss function, $E[|e_t|]$, if the series is non-Gaussian.

Thus, before one can start to consider the optimal forecast derived from a particular model one must first specify the chosen form of loss function. The authors apparently follow the reverse procedure, i.e. they first calculate the forecast, and consider afterward how best to assess its accuracy. In fact, the forecast formula used for ARMA models (A-42) is the one which is appropriate for the *mean square error* loss function, and the same is true for most of the other forecasting 'methods' used, since their associated parameters (such as, for example, the parameter α in the 'single exponential smoothing' technique) are all estimated by minimizing the mean square error.

The question now arises as to whether it is valid to assess the relative accuracies of the different methods according to the MAPE criterion which the authors select as their main basis for comparison. (They dismiss the mean square error criterion by saying that it cannot be used *across* series, but it can certainly be used to compare different forecasting techniques applied to the same series.) In principle, the optimal MAPE forecast could be determined from a given ARMA model, although it would be extremely difficult to obtain an analytical expression. However, it is certainly safe to say that it is by no means obvious that the optimal MAPE forecast would be identical with the optimal least-squares forecast! This prompts one to ask the question: are the authors' general conclusions heavily dependent on their use of the MAPE criterion?

Suppose that instead of using the MAPE they had used the mean-square error. In this case the optimal (linear) forecast is uniquely determined by the second-order properties of the series, or equivalently, by its model, and there is no scope for debating whether or not method A is superior to method B. *If the series conforms to an* ARMA *model, and the model has been fitted correctly, then the forecast based on this* ARMA *model must, by definition, be optimal.* (Apart from the ARMA model, all the other forecasting methods considered are of an *ad hoc* nature. The ARMA method involves model fitting and its performance depends to a large extent on the ability of the user to identify correctly the underlying model.)

I am afraid that I really cannot believe that the regression models described by the authors in Section 3 reflect some fundamental law of nature. Suppose we know (or generate) a seasonal + trend + residual series from a seasonal **ARMA** model. If we again use the mean square error as a criterion then the forecast constructed from the model *must* be optimal – irrespective of the percentage variations due to the different components. The 'single exponential smoothing' technique (more commonly referred to as 'exponentially weighted moving averages') is optimal in the mean sense for a series whose first differences satisfy an MA(1) model. *If* the series is of this form, then single exponential smoothing will be superior to a forecast obtained from, for example, an (incorrectly fitted) ARMA(2, 3) model – but this tells us nothing about the relative merits of the two forecasting techniques. If the authors argue that these theoretical considerations do not apply to their study because in many cases the structures of their series were changing over time, then what they are really saying is that it is the *degree of non-stationary* (suitably measured) which mainly affects the forecasting performance. However, it may be noted that the optimal least-squares forecasts can be constructed for non-stationary series provided their structure is changing 'smoothly' over time – see Abdrabbo and Priestley (1967).

The authors tell us that the regression equations were obtained by starting with a large number of 'independent' variables, $(X_1, X_2, X_3, \ldots)$ and then dropping those which were deemed to be 'non-significant'. It would be interesting to know how variables were originally included, how these were selected and whether this selection was determined before or after the forecasting results were available. However, the fact that the authors started out with a large number of variables suggests that the 'significance' of those finally chosen could be somewhat dubious. (Ideally, the authors should have considered the F_{max} test rather than the standard F test.) I note also that in Section 4, third paragraph, the authors disclose the fact that the variable corresponding to the 'number of data points' was dropped in the regression model for 'forecasts' because, when significant, its coefficient had the 'wrong sign'. This also suggests that there may be some spurious features in the construction of these regression models.

I do not believe that it is very fruitful to attempt to classify series according to which forecasting techniques perform 'best'. The performance of any particular technique when applied to a particular series depends essentially on (a) the model which the series obeys; (b) our ability to identify and fit this model correctly and (c) the criterion chosen to measure forecasting accuracy.

Despite these reservations I am impressed by the magnitude of the authors' study and by the care which they have taken in scrutinizing the literature and comparing their qualitative results with those of previous authors. I was pleased to hear that the authors are willing to make available the full results of their analysis to others who may be interested in studying forecasting, in which case the substance of their study will provide a very valuable background for further work.

I have much pleasure in seconding the vote of thanks.

The vote of thanks was carried by acclamation.

Mr J. P. BURMAN (Bank of England): I should also like to thank the authors for a very interesting paper. The results are, indeed, provoking, but I find it difficult to draw conclusions because they are presented in summary form. It might almost be said that it is a 'black box'. For instance, we do not know how long the series are. In the case of Box–Jenkins models, it is my experience, and I am sure that of many others, that a considerable amount of effort is required for model identification before a successful result is obtained. This is why I think it is difficult in a study on a large scale of over 100 series to apply these sorts of models successfully – they are not really suitable for automatic handling.

Dr Gilchrist has already made the point about distinguishing between seasonal and non-seasonal series. As I understand it, the first block of data in all the Tables contains two entirely disparate things: the application of non-seasonal methods to non-seasonal series and the application of non-seasonal methods to seasonal series. In the third block is the application of the non-seasonal methods to seasonally adjusted series. Presumably, the series that had no seasonal pattern are simply repeated from the first group.

Secondly, about the naive forecasts and stationarity. I assume that, although the authors talk about ARMA models, in fact they also include a good many non-stationary series. Therefore, my concept of a naive forecast for a series which requires first differencing would be

$$\Delta X_t = \Delta X_{t-1}.$$

I wonder, therefore, whether the authors, when referring to the Naive 1 and Naive 2 methods, have applied the appropriate naive method depending on the degree of differencing.

Nevertheless, it is surely surprising and disappointing that for the less noisy series the Naive 2 method gives the best results for nearly all the forecasting horizons, in terms of the MAPE – although, as Professor Priestley said, this may not be the appropriate criterion, given the way in which the series were fitted. I am drawing this conclusion from Table 2.

Does this cast some doubt on the adequacy of the seasonal adjustment? The method of seasonal adjustment described in Appendix B seems to be a version of X-9 – that is, the method two before the present X-11 which most people use. It carries out smoothing by a 3 × 3 moving average. I wonder why they did not use the standard X-11 with 3 × 5 moving averages, and whether they included any modification for extremes.

There is a statement in Section 4, paragraph 5:

Smoothing methods on the original data are also relatively more accurate than on the same data after they have been seasonally adjusted, even when series with high randomness are excluded.

However, this is not true of the MAPE comparisons in Tables 2 and 3, nor of those based on the adjusted Theil coefficient in Table 8.

Dr C. CHATFIELD (Bath University): Today's paper provides further evidence that simple forecasting methods are often to be preferred to sophisticated procedures, a view that I have advanced for some time. Yet, paradoxically, I cannot help feeling that today's results go too far in their assessment of relative accuracies. My reservations about Box–Jenkins arise because the extra complexity of the method is not always justified by the improvement in accuracy, and because it is easy for the method to be misapplied. But I find it hard to believe that Box–Jenkins, if properly applied, can actually be worse than so many of the simple methods. Today's results certainly clash with the results of some previous empirical studies such as Newbold and Granger (1974) and Reid (1975).

Why do empirical studies sometimes give different answers? It may depend on the selected sample of time series, but I suspect it is more likely to depend on the skill of the analysts and on their individual interpretations of what is meant by Method X.

For example, consider Groff (1973) who found exponential smoothing to be as good as Box–Jenkins. If you read his paper you will see that Groff's version of 'Box–Jenkins' bears little resemblance to that originally proposed by Box and Jenkins (1976). So we may disregard Groff's results except insofar as they tell us that the Box–Jenkins procedure may be misinterpreted.

Consider the study of Newbold and Granger (1974) where Box–Jenkins came out better than automatic forecasting procedures. The methods used in the study are clearly defined and I was able to carry out a follow-up study on some of the series which showed (Chatfield, 1978a) that a non-automatic version of the Holt–Winters procedure gives more accurate results than the automatic procedure. Thus in using a phrase like 'The Holt–Winters Forecasting Procedure', it is clearly essential to define precisely what is intended.

One drawback of today's paper is that the forecasting procedures are not described in sufficient detail. For example, the authors' method 11, which I call the (automatic) Holt–Winters procedure, depends heavily on the starting values selected for the mean, trend and seasonals, but we are not told here, or in Makridakis and Wheelwright (1978a), what formulae are used.

I find Appendices A and B unhelpful. I do not understand Appendix B at all. Compare equations (A-46) with (A-49). The descriptions of the methods look confused and incomplete. For example, S_t sometimes denotes a seasonal factor and sometimes without warning a mean level. Formulae (A-10), the formula for b after (A-25), (A-42) and (A-44) look incorrect.

Because the procedures are imprecisely defined, it would be difficult to reproduce the authors' results, even if given access to the series. It therefore appears that the evidence in today's paper is inconclusive as to whether we should believe the results. This being so, the only alternative appears to be to look at the authors' other published work in order to build up a prior probability as to whether their results are likely to be accurate. It is, of course, unusual to comment on other work in this way, but today's empirical study seems to justify such an exception.

As it happens, I have recently (Chatfield, 1978b) been examining a forecasting procedure called 'adaptive filtering' which has been proposed by Professor Makridakis in collaboration with S. C. Wheelwright. In common with a number of other writers (Ekern, 1976; Golder and Settle, 1976; Montgomery and Contreras, 1977) I was unhappy about the method, and I decided to check the claim in Makridakis and Wheelwright (1977) that adaptive filtering gives better forecasts than the Box–Jenkins procedure on the famous airline passengers data. While their comparisons were numerically correct, it turned out that their comparisons were made for a single base period. When forecasts were made for different base periods in the forecast period, it turned out that adaptive filtering was worse on average than both Box–Jenkins and Holt–Winters. Here it is relevant to point out that the forecast comparisons in today's paper are made for a single base period at time $(n - 12)$ whereas the Newbold–Granger study, for example, made comparisons over all the forecast period. I would like to ask the speakers why they chose to make their comparisons in such an unusual way.

My views on today's paper are also influenced by the two books on forecasting co-authored by Professors Makridakis and Wheelwright. As expressed in my review (Chatfield, 1978c) of Wheelwright and Makridakis (1977), I formed the impression that these authors are more at home with simple procedures than with Box–Jenkins, whereas the reverse might be said of Newbold and Granger. Thus I would argue that empirical studies say more about the respective analysts than they do about the methods. I regret to conclude that I have to remain sceptical about today's results, particularly in regard to the conclusions relating to the Box–Jenkins procedure.

often underestimate both these factors: however often he is told that 20 percent

Mr G. STERN (ICL, London): The authors have supplied much interesting information, but I feel that their approach needs modification, otherwise the proposed examination of 1001 series could be a labour of Sisyphus. The basis of their approach is revealed in Section 1, where they contrast judgemental with mathematical forecasts. The same thought is carried on in later sections where they compare forecasting methods as if they were 'treatments' in an experiment, to be applied blindly by an experimenter who is not supposed to take cognizance of anything except the rules for applying the method and the numbers being analysed.

But, to use an Ehrenbergism, we ought to preach what we practise, and I maintain that forecasting is frequently not an expert in forecasting techniques applying a method to data which he knows nothing about, but rather an expert in the data using mathematical methods and judgement to make a prediction. I suggest that the man who successfully forecasts sales of shoes for the next 12 months, month by month, is a shoe expert. He then applies simple methods which seem to have a physical meaning and to be appropriate such as a linear or

logarithmic trend together with seasonal factors. He would then start 'bending':
he may decide to disregard or alter the 1974 figures because of the 3-day week,
and he may alter the final forecast because he believes that there will be a boom
next year. Economic forecasting works this way: Surrey (1971) points out that
one can never go back over an econometric forecast and rerun it to see what the
result would have been had one known what the values of the explanatory
variables were, because the user of an econometric forecast always 'bends'.
Again, I would maintain that Box–Jenkins, properly applied, is really a
sophisticated way of doing this 'bending' and I think this explains at least part of
its success as compared to automatic methods. (I do not accept Groff's contrary
findings as I believe he has not used Box–Jenkins correctly and I suspect this may
apply to the authors' work also). My criticism of Box–Jenkins is that I suspect
that a far simpler system along the lines sketched above would give nearly as good
results and be accessible to more experts in the item being forecast. Chatfield
(1978a) shows that Holt–Winters can be greatly improved by manual
intervention.

The authors quote a study showing that purely judgemental forecasts are less
good than mathematical forecasts, but they do not ask why. I suggest that the
reason is that many time series are subject to high inflation rates and to
multiplicative seasonability. I believe that a purely judgemental forecaster will
often underestimate both these factors: however often he is told that 20 percent
inflation means a doubling over less than 4 years, he still resists accepting the
result, and similarly people tend to feel that seasonal effects are additive when in
fact they are multiplicative and so more important in an inflationary context. A
better test would have been to compare 'judgemental plus trend and seasonals'
with the other methods. I would recommend something of this sort for the
author's future 1001 series tests. I would like to see, for example, if Chatfield's
non-automatic Holt–Winters performs as well as Box–Jenkins. I would like to
see a comparison between, say, the UK government's model of the economy and
a much simpler model with very few variables and manual intervention. And I
think it would be valuable if the authors were to persuade experts in some of the
series to forecast using only, say, trend and seasonal factors with added
judgement, and see if any expert forecaster not expert in the data but using any
methods can do better.

Dr F. H. HANSFORD-MILLER (Inner London Education Authority): I
should like to thank Professor Makridakis, and his co-author, for their excellent
and important paper, which not only impinges on statistical theory, but upon us
as citizens. Are we not all now very much the victims of forecasting? Forecasts are
made, and are used for crucial decisions by politicians in Westminster and
Brussels. They are widely discussed in press and television. In fact we are beset
with forecasting. But this paper questions forecasting, and bearing in mind that
in the 33 years since the last war, in this forecasting age, the United Kingdom has

TABLE D1   The Professions of Westminster MP's Elected in the General Election of October 1970

Profession	Number of MP's	Profession	Number of MP's
Barristers and solicitors	108	Clerical and technical workers	18
Teachers and lecturers	92		
Company directors	80	Mineworkers	17
Journalists, publishers and Public relations officers	58	Underwriters and brokers	17
		Accountants	13
Managers, executives, and administrators	52	Railway and other manual workers	11
Other business workers	50	Doctors and surgeons	9
Engineers	33	Party officials	7
Farmers and landowners	26	Publicans	1
Trade Union officials	19		
		Total	611

suffered a continual economic decline, then I for one give three cheers. In addition, the paper suggests the application of normal scientific experimental design to forecasting, with measures of unbiased testing of forecasts against subsequent reality, for success or failure. A long overdue reform.

With forecasting experts now seen to be fallible it seems to me that this must put a much greater dependence on our administrators – the decision makers – that they should be numerate and able to disentangle the various arguments put to them. This seems a vain hope. 'Is there a scientist in the House?', asked MP Kenneth Warren recently in an article in *Computing Europe* (1978), and he gives the answer 'But nobody, nobody stoops to say he is a 'Scientist'. So what do they say they are? Just what are the numeracy or other qualifications for forecasting decision making of our present Westminster MP's?

Table D1 shows a simplified summary of the October 1974 MP's occupations, as given to *The Times* (1974). The largest group, with 108, are barristers and solicitors, and one would not include numeracy among their first qualifications; next follow teachers and lecturers, who are possibly numerate, but we do not know their subjects; the 33 engineers should known some mathematics, but the far greater number of journalists, farmers, trade union and party officials, mineworkers, and most of the others, seem very questionable as regards being highly numerate. Perhaps we should rely on the publican! For decision making, using complex, and now questionable forecasts, it is a singularly unimpressive list, and I believe we now need a far more highly qualified, and numerate, member, both at Westminster and at Strasbourg, if we are to be properly governed in the future.

Secondly, I think we should at once follow the example set in the paper and, in proper scientific fashion, test forecasts to see how they have turned out. To keep within the political spectrum, and as we are moving into Election Year, 1979, I have taken as an example a test of the main Public opinion polls for the October 1970 General Election. These were Business Decisions, Gallup, Louis Harris, Marplan, NOP, and ORC. Each of these organizations carried out at least two polls during the election period, and I have only time now to consider the final one, that nearest the actual poll, and these took place within the period October 2nd to 9th. I have again gone to *The Times* (1974) for my basic data.

Table D2 shows the Labour, Conservative and Liberal forecasts in terms of percentage votes for each of the six polls, the date of the fieldwork being given in the first column. These forecasts are compared with the General Election Actual Poll, which resulted in 39.3, 35.8 and 18.3 per cent for the three main parties respectively, the difference between the forecast and the actual percentage being the 'Inaccuracy'. The absolute average Inaccuracy of the Labour forecast was 3.17, ranging from 0.7 by Business Decisions to 6.2 by NOP. The Conservative forecasts were closer, with Average Inaccuracy 1.73, and range 0.2, by Gallup, to 4.8, by NOP. In contrast to Labour, and also Liberal, the Tory forecasts were underestimates, except for Gallup, as opposed to overestimates. The Liberal forecasts had Average Inaccuracy of 1.15, with the forecasts within a narrow range of 0.7 (Gallup) to 1.7 (Business Decisions).

So which was the best forecast? Column 6 shows the absolute Inaccuracy total for each, which can be compared with the average of 6.05. In the first place – and the Rank order of accuracy is given in Column 7 – we find Business Decisions, with an Inaccuracy total of only 2.7. Very close behind is Gallup, with a 3.1 error, followed by ORC (5.0), Louis Harris (5.9), Marplan (7.4), and well behind, in sixth place, NOP, with 12.2 Inaccuracy total.

As a result of tonight's paper I hope that appropriate tests, similar to these, will be applied as normal routine to all forecasts, in whatever field, as checks on their accuracy. It would, I feel sure, prove a highly valuable exercise.

Dr D. J. REID (Central Statistical Office and Bank of England): The authors have drawn attention to some apparent inconsistencies between their own results and some work done in the past by Granger and Newbold and by myself, so perhaps I may be allowed a brief comment.

The explanation that I was going to offer has been put far more eloquently than I could have done by Professor Priestley – simply that the authors are comparing different things on the basis of different criteria. It is hardly surprising, therefore, that the results differ.

Where I would differ from Professor Priestley is in his last remark, when he said that it was pointless to try and identify factors which account for forecast accuracy. It may be that by looking for such factors we are looking, albeit partially and inadequately, for evidence about whether the ARMA model fits the data. Insofar as that is what we are doing, perhaps these factors have some value.

I found it of considerable interest that despite the differences in approach of

TABLE D2   The Final, or Pre-Election, Set of Public Opinion Polls for the General Election of October 1970

The forecasting accuracy of Public opinion polls, October 1970 General Election

Date of fieldwork of poll	Public opinion poll	Labour forecast	Conservative forecast	Liberal forecast	Inaccuracy total	Rank order of accuracy
October 2nd	Business Decisions	40.0	35.5	200		
	— Inaccuracy	(+)0.7	(−)0.3	(+)1.7	2.7	1
October 2nd to 5th	NOP	45.5	31.0	19.5		
	— Inaccuracy	(+)6.2	(−)4.8	(+)1.2	12.2	6
October 7th to 8th	Gallup	41.5	36.0	19.0		
	— Inaccuracy	(+)2.2	(+)0.2	(+)0.7	3.1	2
October 8th	Marplan	43.0	33.3	19.5		
	— Inaccuracy	(+)3.7	(−)2.5	(+)1.2	7.4	5
October 8th to 9th	Louis Harris	43.0	34.6	19.3		
	— Inaccuracy	(+)3.7	(−)1.2	(+)1.0	5.9	4
October 9th	ORC	41.8	34.4	19.4		
	— Inaccuracy	(+)2.5	(−)1.4	(+)1.1	5.0	3
October 10th	General Election actual polls	39.3	35.8	18.3		
October 2nd to 9th	Opinion Polls average inaccuracy	3.17	1.73	1.15	6.05	

the authors' work and the work done at Nottingham, nevertheless we come up with the same factors as being important in determining which method should be used. I think we differ in the way in which those factors are interpreted, but the factors themselves are the same. In my own work, which I presented in the form of a decision tree – this has been reproduced rather more accessibly in Kendall (1973) – I listed the following:

1. How long is the series?
2. Are the data seasonal or non-seasonal?
3. Does the random component have high or low variance, compared with the other variability in the series?
4. Are there large peaks, non-stationarities and other discontinuities in the series?
5. Is one trying to predict long or short lead-times?

Almost exactly the same factors are mentioned by the authors in their paper tonight. So perhaps there is more common ground between these studies than appears superficially from the conclusions about which series does best on the respective criteria used.

Professor J. DURBIN (London School of Economics): The discussion so far does not seem to me to have brought out what I feel to be the most interesting point in tonight's paper, namely the spectacular gains in forecasting seasonal series obtained by the authors by projecting seasonally-adjusted data and then reapplying the seasonal factor. I found this most unexpected – as I imagine did many others – because, as the authors say in their paper, they were able to find only one reference where this had been done previously in all their studies of comparative forecasting methods. I had rather taken for granted that what Box and Jenkins claimed was true, namely, that if there is a seasonal series, whether it is a changing seasonal or a constant seasonal, the most effective way of dealing with the seasonality is to take seasonal differences. Yet the authors have obtained much better performance from very simple, unweighted moving average – Method 15 in the Tables – than is given by the standard seasonal Box–Jenkins method.

I believe that is something which would be interesting to see followed up because, to me, it is counter-intuitive. For example, I would suggest to the authors that they make a non-seasonal Box–Jenkins analysis of the seasonally-adjusted series and then reapply the seasonal factor. I find it hard to believe that their Method 15 could not be bettered by doing a Box–Jenkins analysis of the seasonally-adjusted series.

In reference to a point made by Mr Burman, my understanding of the paper was that the authors were assuming a constant seasonal. They state that they tried an X-11 type seasonal adjustment method and had found that it achieved no improvement over an ordinary ratio to moving average method, assuming a constant seasonal. This points to my main worry about this paper which is the extent to which the results are data-dependent. My experience of seasonal time series is that the main problems arise from changing seasonality. One reason for the spectacular success of the X-11 seasonal adjustment procedure, which is used all over the world for adjusting official series, is that it is very much concerned with making provision for changing seasonality. The fact that the authors obtained no differences from the assumption of a constant seasonal, as compared with using X-11, suggests to me that perhaps their series had constant seasonals, and perhaps things would not be quite so good if their methods were tried on series with changing seasonals.

I should like, therefore, to suggest that in their reply to the discussion the authors might give us some more information about these series. I appreciate that it may not be possible to have a complete catalogue of the whole 111 series, but they could be divided into broad categories. For example, the X-11 program contains a test for seasonality; I should like to know how many of these 111 series are classified as non-seasonal, according to the so-called test for stable seasonality. Within each of the two classes, the seasonal and the non-seasonal, I should like to know what is the distribution of the number of observations in the series. This is very relevant for an assessment of the performance of the Box–Jenkins system. I should also like to see the distribution of the numbers of series that are monthly, quarterly, annual, etc., also a classification by the type of series – are they macroeconomic series, microeconomic, sales series and so on? Without some basic information of that kind, we cannot adequately assess the significance of the authors' results.

The following contributions were received in writing, after the meeting.

Mr O. D. ANDERSON (Nottingham): I do not wish to repeat the kind of remarks I made in Anderson (1977), especially as I hold the apparently heretical view (amongst academic statisticians, at any rate) that Spyros Makridakis does considerably more than most of us to improve the level of forecasting *in practice*. I therefore will not cavil with the present paper, except to say that I shall be surprised if it has any real impact on the statistical audience, towards which it appears so bravely aimed.

There are, however, three points on which I must comment, for fear that everyone else may feel they are too obvious to merit mention.

(1) Groff (1973) keeps on getting quoted as evidence that the Box–Jenkins approach is a trifle superfluous. It was mentioned during the Institute of Statisticians Forecasting Conference, in 1976, and again at the recent 1978 Cambridge Time Series Meeting. Unfortunately, Groff did not use the Box–Jenkins methodology – all he did was try out a specific selection of ARIMA models. This is rather like saying, 'estimate some number (say, seven, in fact) by a sophisticated method which could give any integer, but only allow it to choose

multiples of ten; and then compare its hamstrung performances with a less complicated approach, which can just estimate to the nearest 5'. The simpler method evidently has a good chance of doing better than its hobbled competitor.

(2) Makridakis and Hibon should surely have been worried by the implications of their 'Box–Jenkins' modelling. First time round, from 111 series, 83 gave residuals with none of their first 24 serial correlations significant at the 5 percent level. Presumably, then, there were something like 4.8 significant residual correlation values, on average, for the remaining 28 series. This sort of effect was suggested in Anderson (1975, pp. 18–19) and is not too surprising for near perfect identification. However, eventually the authors ended up with 111 models, not one of which had a significant residual correlation amongst its first 24. This has to represent excessive global data mining, and perhaps helps explain why the forecasts were none too good. The moral is that, in the long run, even the perfect analyst should find about 5 percent of his final model residual correlations are one-star significant.

(3) Ending on a more positive note, I believe that it is now easier to distinguish between pairs of similar 'autoregressive' operators, such as $(1 - B)$ and $(1 - \phi B)$, with ϕ near to, but a little less than 1; and that there are usually extra clues about this in the serial correlations. Forthcoming publications, by myself and jointly with Jan de Gooijer (Amsterdam), establish the relevant theory and empirical evidence for such discrimination; and we hope also soon to have some firm indication of gains in forecasting performance, which can then result.

Dr R. T. BAILLIE (University of Aston Management Centre): The authors assert that quantitative forecasting methods can be classified as either 'econometric (explanatory) or time series (mechanistic)'. However, as has been noted by Prothero and Wallis (1976), these are not distinct alternative approaches, but merely different representations of the same structure which are inextricably linked via the notion of the final form of an econometric model. Similarly, multiple time series models may also be reduced to univariate ARIMA representations for each component series; see Chan and Wallis (1978). Many of the comparative papers referred to by the authors compare forecasts from ARIMA models with forecasts from econometric models with rather arbitrary dynamic specifications; hence the conclusions of Naylor *et al.* (1972) for example.

One of the main findings of the authors is that forecasts produced by exponentially smoothing an already seasonally adjusted series compare favourably with those from an ARIMA model. Unless the structure of the series changes between the sample and forecast periods, their finding must be due either to the failure of the ARIMA models to capture the seasonal effects as efficiently as the Census II method, or to their choice of a sub-optimal ARIMA model. As mentioned by the authors their findings differ from the substantial forecasting survey conducted by Newbold and Granger (1974). The authors' case is not helped by their failure to give details of length of series, characteristics of series and their approach in applying Box–Jenkins methodology. For example, did they implement the Box–Cox transform?

It is also unfortunate that the authors did not include in their survey at least some previously analysed series. Indeed, there remains the distinct possibility of another paper appearing to reestablish the superiority of ARIMA models on another data set, to be followed by yet another paper questioning this premise and based on yet another collection of series. Perhaps all such surveys should include some standard series. For example, how does exponential smoothing compare with ARIMA models when applied to the Wolfer sun spot series or the airline data in Box and Jenkins (1976)?

The authors introduce an interesting definition of forecasting accuracy. An alternative approach is to examine the mean squared error and to decompose it in terms of randomness in the model, prediction of exogenous variables and parameter estimation. For short series the latter component is likely to be important. Formal results for regression models with autoregressive errors are given in Baillie (1979) and with ARMA errors in Baillie (1977).

Dr M. BERTRAM (Scicon Computer Services Ltd, Milton Keynes): I should like to add my thanks to the authors for this interesting and provocative paper. With so many methods to choose from it is impossible to consider them all. But I should be interested in the authors' views on the Kalman filtering, mixed-model approach pioneered by Harrison and Stevens (1976).

We have practically applied this approach to a number of economic time series (mainly sales data) with reasonable success. The strength of this approach is that it allows for changes in the model structure of the kind one might expect in economic time series.

Professor D. R. COX (Imperial College, London): The statistical approach, if there is such a thing, includes a large dose of empiricism, and the literature on time series being on the whole theoretical, a very empirical paper is to be welcomed. Yet, like several speakers, I have doubts about several points in the paper. The balance between 10 pages of tables of detailed conclusions, one short paragraph of vague description of the series analysed and nothing on their statistical characteristics makes discussion of the conclusions difficult; I hope the balance can be partly redressed in the authors' reply.

The authors' application of multiple regression in Section 3 seems especially uncritical. Theory and empiricism alike suggest errors of forecasting behaving approximately like $a + b/n$, where n is the number of observations, and to fit a model linear in n is to fit an equation which inevitably has strange properties for large n. Also it seems odd not to express forecasting errors as a fraction of the intrinsic variability of the series.

Dr R. FILDES (Manchester Business School): In a discussion of time series methods of forecasting it is no surprise to find statistical arguments predominate and the 'best' method is determined with respect to some convenient criterion such as expected quadratic loss. Unfortunately, the calculations are only valid as far as the statistical approximations in the argument permit and any declaration on relative model merit is only helpful insofar as the model on which the statistical arguments are based is correct.

Thus claims that the Holt–Winters variant of exponential smoothing as being a special case of Box–Jenkins are beside the point (as well as being technically incorrect). It is apparent from Newbold and Granger's study (1974), and supplemented now by the evidence presented in the paper under discussion, that the *ad hoc* approximations undertaken in initialization, together with the simpler structure of exponential smoothing, sometimes imply that in forecasting applications rather than within sample fitting, exponential smoothing can outperform its more sophisticated alternatives. The fact that simple exponential smoothing is a first-order moving average in first difference is to a large extent irrelevant. Thus we can conclude that the period of fit offers us only limited information, however processed, about forecasting effectiveness. If we adopt Priestley's (1974) categorization of the univariate forecasting problem we find that what is to be determined is 'what class of random processes are we considering?' Only the data can tell us. Unfortunately economic series are not regularly describable as ARIMA processes, despite the apparent optimality of the appropriately identified estimated ARIMA model within that class. Paraphrasing Harrison and Stevens (1976) such well-behaved series seldom occur in practice.

Makridakis and Hibon are to be congratulated in addressing the right problem – establishing which methods work in practice and why. Of course their results do not show us which method to adopt any more than any other study; without a suitably defined population from which to make inferences this is impossible. What they do tell us is that large numbers of economic series are best forecast by simply, robust models, a conclusion my own work wholeheartedly supports (Fildes, 1979), and thoroughly in keeping with the approach of Harrison and Stevens (1976).

Theoretical analysis can only take us so far in the evaluation of a forecasting procedure. In the end what a forecaster needs to know is the likely performance of a particular method in the situation he faces. The authors have taken us some way towards a correct conceptualization of this important practical problem.

Professor ROBIN HOGARTH (INSEAD, France): Forecasting fascinates. The need people have to predict and control the environment guarantees both the interest in and importance of the paper by Makridakis and Hibon. The questions they address are crucial: What are the relative accuracies of different forecasting methods? Under what conditions are methods differentially accurate? The answers clearly depend on the *joint* characteristics of both the series predicted and the method used. Makridakis and Hibon mainly elaborate aspects of the former; my first point concerns the latter.

I believe one useful dimension for characterizing methods is degree of sensitivity of parameters to variations in the data. It used to be thought that sensitivity was desirable. However, recent work in the use of regression methods for repetitive predictions indicates the contrary. Specifically, in situations that Makridakis and Hibon would describe as exhibiting much randomness, simple unit- or equal-weighting procedures will outpredict models based on the 'optimally' estimated least-squares weights. The empirical evidence and theoretical rationale for this statement are provided elsewhere (Dawes and Corrigan, 1974; Einhorn and Hogarth, 1975). However, an additional and, in applied situations, crucial point is that processes generating data rarely meet the assumptions implied by models. There is therefore a need for parameters to be robust against variations in the underlying data generating processes. Indeed, I strongly suspect that the relative success of simple-minded schemes often depends precisely on this factor. Furthermore, it would seem that simple-minded or 'naive' models can, again relative to more complex schemes, simultaneously capitalize on two apparently contradictory features of time-series: first, unpredictable structural changes; and second, the sheer inertia of economic activity which induces high correlations between successive observations. Investigations of the robustness of simple methods are clearly required.

My second point is to support the authors in urging further research into methods for combining forecasts, including human intuition. Once again, simple-minded models, e.g. taking an arithmetic average, seem to be remarkably effective and the investigation of robustness is clearly also relevant here.

Finally, an issue that needs to be faced squarely is whether forecasting based on mechanistic methods is possible beyond short time horizons. As stated above, relative inertia of phenomena across time permits short-term forecasting. However, where are the limits? Interest in forecasting methodology attests to the fact that we need to believe we can make model-based forecasts. However, can we learn to live with the fact that often we cannot? The future is not necessarily 'ours to see'.

Professor COLIN LEWIS (University of Aston Management Centre): Whilst applauding the authors' comprehensive coverage of many of the forecasting models in everyday use, I would point out that the adaptive response rate exponential smoothing model adopted is in practice rarely used; the 'delayed' version generally being preferred because of its lack of response to a single period impulse. In this delayed version the α_t of (A-11) is set equal to $|E_{t-1}/M_{t-1}|$ rather than $|E_t/M_t|$.

This could well explain the poor performance of this model in the authors' analyses.

The AUTHORS replied later, in writing, as follows.

We would first like to thank all discussants of our paper for their comments, the majority of which we have found extremely useful. We understand, of course, that the topics covered in our paper, and the conclusions reached, cannot fail to provoke disagreement from those who believed that their 'favourite' method should have performed better. Furthermore, any empirical study of the sort undertaken cannot but raise many questions, since there is little guidance available to researchers as to how such studies should be done. We understand and accept, therefore, differences in opinions and we hope that further empirical studies will be conducted in a way such that a more standard methodology will become available for this type of research.

The main purpose of our research was not to compare various forecasting methods, but rather to find out the factors that affect *post-sample* forecasting accuracy, as opposed to being concerned with how well a forecasting model fits to a set of available data. Post-sample (i.e. beyond the available data on which the model has been developed) forecasting accuracy is of paramount importance, for practical purposes, since it is the only way to estimate forecasting errors and be able to evaluate when method *A* should take preference over method *B*, for a specific set of data. It is unfortunate that the discussion became concentrated on a comparison between simpler methods and the Box–Jenkins methodology to ARMA models, since this aspect of our paper was only secondary when the study was started. At that time, we believed, as most people involved with forecasting, that that particular question had been settled and that the Box–Jenkins methodology was the most accurate. To our surprise the results of our study showed otherwise; we could indeed have reacted in the same way as the Operational Research Society Forecasting Study Group (see Professor Gilchrist's contribution) and be so horrified by the results as to have 'brushed them under the carpet, the carpet patted down, and no more said about them'. However, we felt that this would be a disservice to the profession and the field of forecasting. Our main concern was that we must 'shed light on the mystery of why, under certain circumstances, simpler methods do as well as or better than sophisticated ones'; similarly, we wanted to be able to know when one method should take preference over another, for a specific set of data, where post-sample accuracy is concerned.

In a recent paper entitled 'Forecasting with econometric methods: folklore versus fact', Armstrong (1978) cites much evidence about 'the tendency of intelligent adults to avoid disconfirming evidence' (pp. 550–551). This is not something new, and Kuhn (1962) has shown its full implications for the evolution (or revolution as he argues) of science. We would like to point out that the empirical evidence which exists at present is extremely weak and conflicting. Most importantly, however, intelligent adults should not judge evidence by different criteria if it is discomforting to their own beliefs. It was suggested by several discussants that Groff's (1973) study should be disregarded as incorrect.

This leaves the studies by Reid (1969) and Newbold and Granger (1974). Some say that these two studies should be counted as one, since both were conducted at Nottingham, at about the same time, without it ever having been made clear whether or not the data used was the same. More importantly, however, Reid's studies compare Box–Jenkins with *Generalized Exponential Smoothing*, which is no longer a method used to any great extent (if at all). Thus the work of Newbold and Granger remains as the only major study which compares Box–Jenkins with the Holt–Winters method. This is very little empirical evidence indeed. But even a close look at the Newbold and Granger study raises the same type of question that we have been asking in our paper. For instance, the percentage of occasions that Box–Jenkins outperforms the Holt–Winters model is 73, 64, 58, 58, 57, 58 and 58 percent, for lead times of 1, 2, 3, 4, 5, 6, 7 and 8 periods. After a lead time of three periods, Box–Jenkins is no better than Holt–Winters but for 8 per cent of the series. How can this be explained? Is the extra accuracy worth the much higher costs involved (in terms of both personal effort and computer cost)? Which is the 42 percent of the series where Holt–Winters outperforms Box–Jenkins? Newbold and Granger do not provide detailed information on accuracy measures other than the percentage of time that method A did better than B, and as Carbone (1978) has shown, this type of accuracy measure can be misleading. Thus, even the Newbold–Granger study does not answer the question as to why, in such a high percentage of series (i.e. 42 percent), exponential smoothing (supposedly a special case of Box–Jenkins) does better than Box–Jenkins.

Armstrong concluded that 'a review of the published empirical evidence yielded little support for ... two important questions. First, do econometric methods provide the most accurate way to obtain short-range forecasts? Second, do complex econometric methods provide more accurate forecasts than simple econometric methods'? Armstrong's conclusions are particularly relevant for time series forecasting because he compared econometric methods with, principally, ARMA models. Secondly, his conclusion that higher complexity does not necessarily lead to greater accuracy is a lesson that time-series forecasters should keep in mind. Even though it is difficult to accept Armstrong's flamboyant style, we have to accept his main conclusions. Econometricians, obviously, cannot be pleased with the implications drawn by Armstrong. Our belief is that time-series forecasting must avoid the mistakes of econometricians: firstly, empirical evidence should not be ignored, but instead used to guide the development of theory; and secondly, it is not necessarily true that more complexity will result in better post-sample forecasting accuracy.

Dr Gilchrist's thoughtful comments have captured the essence of our paper. He raises several important questions and we will try to answer them as best we can. First, we are thankful for his bringing to the attention of forecasters what has been 'under the carpet' concerning the study undertaken by the Operational Research Society Forecasting Study Group. This is obviously a reconfirmation of our point that for *some* series (in particular those which are unstable) random walk (or Naive methods) do as well as, or better than, complex methods. Secondly, he argues, and we agree, that a distinction between seasonal and non-seasonal methods should have been made. We simply refrained from doing so because of the extra tables that would have been required, which would have further increased the size of an already large paper. However, there is a more fundamental reason. Once the data for seasonal series have been deseasonalized (lower part of all tables), then non-seasonal methods can be compared, on an equal basis, with seasonal ones. This provides us with all the comparisons we need. In this respect, methods 1 to 10 in the tables (Original Data: Nonseasonal Methods) are not needed other than to provide us with a yardstick for how much improvement has been achieved by deseasonalization of the data. This information can tell a user, who might not want to go through the trouble of deseasonalizing the data, how much he would lose by not doing so, and was provided only for comparative purposes – no more. Thirdly, Professor Gilchrist is right when he assumes that we only used one-step-ahead forecasts. Unfortunately, the time required to rerun the models for 2, 3, 4, . . ., 12 lead times is enormous: new models must be identified, the differencing might change, the parameters must be re-estimated, etc. The amount of work involved would be multiplied by twelve, and in a huge study of the type we undertook, this is very difficult on practical grounds. However, this deficiency should be corrected in future studies. Fourthly, concerning the accuracy measures, we agree with him that it might have been better if we had not averaged the various errors for forecasting horizons larger than one period ahead. We in fact did this because practitioners care much more about the average errors (e.g. in budgeting) up to a certain month of the year than single error values. We also share his view that 'the dice are very much loaded in favor of ARMA models', and as Chatfield (1978a) has shown, for instance, some subjective modifications to the Holt–Winters automatic version could substantially improve the results. What the effect of this is on the comparisons, we do not know. Finally, we can state that backwards forecasting has little effect on post-sample forecasting accuracy (see Makridakis, 1978). This was one of the first avenues we explored in order to explain the poor performance of ARMA models.

Professor Priestley raises a fundamental question which we believe has extremely important implications for the field of forecasting and statistics in general. To quote him, '*If the series conforms to an* ARMA *model, and the model has been fitted correctly, then the forecast based on this* ARMA *model must, by definition, be optimal.*' This is, obviously, a correct statement, of which every student of statistics is well aware. The statement, however, is based on an assumption, namely the *assumption of constancy*. The least-squares approach – widely used, not only in forecasting, but in the whole field of statistics – fits a model to a set of data. Consequently, *if* the pattern, or the structure, of these data *does not* change, *then* the same model can be used to estimate or forecast beyond the data used (i.e. the post-sample period). Many data series are structurally stable, in which case the assumption of constancy holds. This means that minimizing the mean square error of the sample data also minimizes the mean square error of the post-sample predictions. Unfortunately, however, most of the structurally stable series are in the fields of physical sciences and engineering, and very few in the field of social sciences, business or economic data. It might, therefore, be that ARMA models are more appropriate for engineering applications, where the great majority of series are structurally stable, while exponential smoothing models are better suited to business and economic applications where most series show discontinuities and changes in patterns, and are, therefore, structurally unstable. Obviously, this is only a conjecture at this point, which must be tested in the future, but is, we believe, worth considering. There is a fact that Professor Priestley must accept: empirical evidence is in *disagreement* with his theoretical arguments. Our study is not the only one which has shown that simpler methods do as well as, or better than, sophisticated ones. For instance, Chatfield and Prothero (1973), Groff (1973), Dawes and Corrigan (1974), Geurts and Ibrahim (1975), McWhorter (1975), Narasimham *et al.* (1975) and McCoubrey and McKenzie (1975) have reached similar conclusions; furthermore, our main paper, Armstrong (1978) and Hogarth and Makridakis (1979) cite many more studies which have found that more complex models are no more accurate than simpler ones. Finally, there are also other areas where simple models have been shown to be more robust, and at least as accurate as complex ones (for a review, see Leung, 1978). In theory, this should never have happened, but in practice it does. The reasons are obvious to us: minimizing mean square errors during the model fitting phase does *not* guarantee minimum mean square errors in the post-sample phase simply because real-life data are often structurally unstable. The need, therefore, is for methods which can minimize the post-sample forecasting accuracy (see Fildes and Howell, 1977) and can deal with other types of data than those currently used (e.g. the Airline data) to test statistical methods (Harrison and Stevens, 1976; Fildes, 1979).

We do not understand Professor Priestley's comments saying that we have not used a quadratic loss function. We indeed used Theil's U-coefficient (see Tables 7 and 8). As a matter of fact, ARMA models do even worse than simpler methods when the quadratic loss function is involved. Interestingly enough, however, the mean square error (as measured by the U-statistic) is at its smallest during the model fitting phase, whilst being very large in the post-sample period. Moreover, we did not use a straightforward mean square error measure because the results would not have been comparable with those of *other* studies, since mean square error measures are absolute, depending upon the specific series used.

Professor Priestley makes several comments on our construction of the regression equations. We believe that the procedure used in developing these regression equations is standard in the social sciences. We did not start with more than ten variables and we had close to 1000 observations for each equation of Table 10. Regression results with such a large number of observations cannot be

TABLE D3   Distribution of Number of Data in the 111 Series

Number of data points	Frequency
less than 40	3
40–49	7
50–59	5
60–69	1
70–79	9
80–89	30
90–99	10
100–109	8
110–119	4
120–129	7
130–139	13
140–149	13
more than 150	1
	111

questioned. Furthermore, there is a remarkable consistency in the variables included in the equations – see Tables 9 and 10 – and their values; the chances that such results can be attributed to chance are astronomically low. Concerning the variable of 'number of data points', perhaps it should have been included, and we regret we did not do so. But even if we had included the variable, nothing would have changed.

Mr Burman suggests that we did not present enough information about how the various models were developed and used. This might be so, but we have still provided more details (see Appendix A) than any previous study comparing forecasting accuracies. Furthermore, the book *Interactive Forecasting: Univariate and Multivariate Methods*, by Makridakis and Wheelwright (1978), provides additional information about the methods used and the algorithms employed. Unfortunately, space constraints made it impossible to provide more information about the methods used. But we can assure Mr Burman that we devoted ample time to identifying an appropriate ARMA model (see Appendix A) while using the Box–Jenkins methodology. Concerning number of data points, Table D3 below gives their distribution.

We would like to thank Dr Chatfield for accepting our study as a corroboration of the view that he has advanced for some time; namely that 'simple forecasting methods are often to be preferred to sophisticated procedures'. We can only apologize because our 'results go too far' in the opposite direction. Unfortunately, we had no control over the findings. Furthermore, he argues that some of the results obtained are due to not knowing how to apply the Box–Jenkins methodology properly. This is an easy accusation to make, as Dr Chatfield well knows, after the discussion of his paper with Prothero (1973). Obviously, there is no way of proving or disproving that we properly applied the Box–Jenkins methodology; on the other hand, we have had experience in using it with hundreds of real-life series; moreover, we can say that we did the best we could to apply it as correctly as possible, spending an enormous amount of time in doing so. Dr Chatfield argues that we might not have used the right starting values, for what he calls the Holt–Winters procedure. Had we used more appropriate starting values, however, the Holt–Winters method would have done even better than the Box–Jenkins methodology. We do not, therefore, understand the reason for this type of argument except that it points out that Chatfield (1978) has done some work on the Holt–Winters method.

Finally, Dr Chatfield expresses some personal views about the first author and Professor Wheelwright. (Mrs Hibon has difficulty seeing the relevance of these points to the present paper). He talks about 'prior probabilities'. It might be useful for Dr Chatfield to read some of the psychological literature quoted in the main paper, and he can then learn a little more about biases and how they affect prior probabilities. Specifically, he mentions adaptive filtering. Even though Professors Makridakis and Wheelwright are flattered about the repeated interest of Dr Chatfield (Chatfield and Newbold, 1974; Chatfield, 1978) in this technique, they fail to see its relevance to the present paper. In any event, the answers to the questions he raises on adaptive filtering are as follows: The comparisons made with the Box–Jenkins methodology are based on a single period because this is how the Airline data were forecast by Box–Jenkins (1976, pp. 307–308). If other base periods were taken, no comparisons would have been possible, without the risk of being told that the Box–Jenkins methodology was incorrectly used. (Needless to say, his comment that the comparisons, numerically correct, which show that adaptive filtering – for the base period used – is more accurate than the Box–Jenkins methodology is appreciated). Dr Chatfield also expresses certain further views about adaptive filtering. There is some additional evidence that he might wish to consult. Firstly, there was a 'forecasting tournament' organized by Professor Dave Pack for the Los Angeles ORSA/TIMS meeting of November 1978. There were four series used and adaptive filtering came second, in most measures of accuracy, behind state space forecasting. Box–Jenkins came last (see Carbone, 1978; Granger and McCollister, 1978). Secondly, there have been several papers (see Bretschneider *et al.*, 1978; De Lange and Wojchiechowicz, 1974; Nau and Oliver, 1978) that have established the theoretical correctness of adaptive filtering and its practical usefulness; papers of which he should be aware, due to the great, and repeated, interest he has shown in adaptive filtering.

Another piece of evidence cited by Dr Chatfield (1978) is his review of *Forecasting Methods for Management*. Dr Chatfield should bear in mind that this book was written in 1971 – the part dealing with the Box–Jenkins methodology was not revised when the second edition was published in 1977. A more correct illustration of how at least Professor Makridakis can apply the Box–Jenkins methodology is to be found in Makridakis and Wheelwright

(1978a,b), or in the forthcoming third edition of *Forecasting Methods for Management*, in which a revised chapter for the Box–Jenkins method has been written. Finally, he might want to see some other reviews of *Forecasting Methods for Management* (McR, 1974; Benson, 1975; Brown, 1977) and then he will perhaps understand how prior probabilities can be biased.

We appreciate Mr Stern's suggestions for how to go about replicating our present study with the 1001 series we have already collected. Furthermore, we will propose to him, and to anyone else who would be willing to participate, to use our data. We would also like to see how the improved version of the Holt–Winters method would do, and we hope that Dr Chatfield will also participate. It would indeed be desirable to have access to more information than is available, but unfortunately empirical studies are extremely difficult to conduct and they require countless days of hard work. Moreover, since many difficult decisions have to be made, on many unresolved problems, criticism is easy. Mr Stern criticizes our study for being 'applied blindly' without subjective adjustments. First of all, we never pretended to have done otherwise. Secondly, it is not at all clear from the psychological literature dealing with human judgement that subjective interventions will improve the results. As unlikely and unintuitive this statement may seem, it is accepted as a well-established fact amongst psychologists. Reading Slovic (1972), Dawes and Corrigan (1974) or Slovic *et al.* (1977) will easily establish the belief of psychologists concerning human judgement, as compared to quantitative models.

We cannot disagree with Dr Hansford-Miller, and we would like to thank him for enlarging the scope of our study by relating it to many additional implications.

We greatly appreciate Dr Reid's comment 'perhaps there is more common ground between these studies [ours and previous ones] than appears superficially from the conclusions about which series does best on the respective criteria used'. Our main purpose was not that of comparing various forecasting methods and we therefore hope that some logical explanations will be found. Our results have been checked and rechecked, and leave us in no doubt as to their correctness. Thus, we might not have used the best starting values for the Holt–Winters model, or applied the best procedure for other methods, but given the way the various methods were used (see Appendix A), the results are correct and in close agreement with programs other than ours, with which we have checked the numerical answers. As a most critical point, we believe that the forecasting accuracy of various methods differs significantly depending upon the characteristics of the series (point No. 3 in Dr Reid's reply). Unfortunately, we know very little about this particular factor, as well as the others mentioned by Dr Reid. In our opinion, the challenge is to quantify and measure the influence of these factors; hence our present study, which is unique in trying to identify and measure such criteria as those mentioned by Dr Reid.

We would like to thank Professor Durbin for the insight of his comments, and for reemphasizing that no tests concerning seasonally-adjusted data (except involving a single series) have been reported in the literature. We think it is a good idea to apply Box–Jenkins to seasonally-adjusted data and we will try to follow his advice. We indeed hope that the results of our study are data-dependent, but still maintain that we must know which types of data enable some methods to perform better than others. Table D3 gives the distribution of the number of data points. About 80 percent of the series were monthly, 12 percent quarterly and 8 percent annual. About one-third of the series were macroeconomic and two-thirds were taken from business sources – some of them originating from industry. About two-thirds of the series were seasonal. It is difficult to assess the ending date of the series (we did not record this information), but we would estimate that about half of them ended between 1974 and 1976, thus being influenced by the 1974–75 recession.

Most of the points made by Dr Baillie have already been dealt with – we will not therefore repeat them. His comment about the structure of the series changing is correct. Many of our series were not structurally stable; but this should not surprise anyone, since real economic and business series are often structurally unstable. Dr Baillie regrets that we did not include in our survey 'at least some previously analysed series'. We regret it too, since we wrote to all those who had carried out previous surveys, but we received no series – not even a single reply. We indeed, however, included the famous Airline data whose MAPE, for a number of methods, is shown in Table D4. It should be noted that 90 data points were used for model fitting. Thus the 12 forecasts refer to periods 91 to 102.

TABLE D4   Mean Average Percentage Error (MAPE) of the Airline Data
(errors refer to periods 91–102)

Forecasting method	Model fitting	Forecasting horizons							
		1	2	3	4	5	6	9	12
Naive 1	8.56	9.44	8.55	7.48	11.17	16.54	17.48	16.99	14.76
Single exponential smoothing	8.56	9.94	8.55	7.48	11.17	16.54	17.48	16.99	14.76
Brown's linear exponential smoothing	9.59	3.13	4.72	13.09	24.97	38.49	45.49	60.28	67.41
Holt's linear exponential smoothing	8.60	8.94	7.72	7.62	12.07	18.18	19.68	20.54	18.09
Linear trend	10.72	25.77	24.76	20.63	16.07	16.14	14.09	11.34	11.42
Harrison's harmonic smoothing	3.64	3.46	2.67	2.01	2.11	2.07	2.26	3.13	3.3
Winter's exponential smoothing	3.72	1.26	0.85	2.15	3.81	4.75	5.75	7.59	7.15
Adaptive filtering	3.60	2.36	2.48	1.76	1.50	1.48	1.49	1.64	1.87
Box–Jenkins	3.58	1.32	1.10	1.60	2.45	2.64	3.11	4.05	4.16
Naive 2	3.02	1.75	0.93	1.00	1.35	1.24	1.14	1.34	2.66
Exponential smoothing (seasonal adjustment)	3.00	2.30	1.18	1.17	1.47	1.34	1.22	1.40	2.70
Brown's exponential smoothing (seasonal adjustment)	3.00	2.82	1.65	2.10	3.02	3.53	4.09	5.94	5.95
Holt's exponential smoothing (seasonal adjustment)	3.10	1.86	1.46	2.03	2.91	3.31	3.74	5.08	4.78
Linear trend (seasonal adjustment)	4.19	11.30	10.14	9.21	8.29	7.90	7.56	6.61	7.35

Mr Anderson is very generous in the opinion he expresses about Spyros Makridakis. Concerning his remarks made in his 1977 paper, there is no point in arguing here since there has been a reply (Makridakis, 1978). However, we would like to point out to Mr Anderson that it is a little premature to talk now about the 'real impact' that our study will have. Any good forecaster must know that prejudgements of this type have little value, and add nothing to the discussion. Concerning his specific points, our answers are as follows: (1) We do not want to make a value judgement of whether or not Groff's study was correctly conducted; in the same way that we do not do so for Reid's (1969) or Newbold and Granger's (1974) studies. What we fail to see, however, is why nobody questions Newbold and Granger, for instance, on whether or not they applied the various methods correctly (or is it, as we have already argued, because their conclusions are in agreement with preconceived notions?). They gave no information as to how the different methods were used, the algorithms employed, the starting values, etc. How can it be known, then, that what they did was correct? (This is not to suggest that it was not). (2) We believe that Mr Anderson did not read carefully the description of how we used the Box–Jenkins methodology. Thus, when he says that 'The moral is that, in the long run, even the perfect analyst should find about 5 per cent of his final model residual correlations are one-star significant', he is testifying exactly to what we describe in Appendix A, where we state: 'There were rarely autocorrelations [of the residuals] outside the 95 percent confidence limits . . .'. If this is not what is represented in the above statement by Anderson, then perhaps we have been let down by our knowledge of English, which after all is not our native language. (3) We fail to see the connection of this point with our paper, and as before we will let the acid test of time judge the importance of Anderson's positive note.

We thank Dr Bertram for his suggestion, and we do hope that Bayesian Forecasting will be included in the next comparative study. Unfortunately, when the present study was conducted, no computer programs were available to us, hence we were unable to include the approach of Harrison and Stevens.

We hope that our reply has already answered the several questions raised by Professor Cox; there is no point in repeating them here. Concerning forecasting errors and how they behave, we can only say that n – for the data we used – did not have a statistically significant effect on forecasting accuracy, but for a very slight number of exceptions. Yes, indeed we do 'express forecasting errors as a fraction of the intrinsic variability of the series'; this is precisely what most of the independent variables are (e.g. the mean absolute percentage change in randomness).

We are in complete agreement with Professor Fildes' well thought-out comments. His point that 'the period of fit offers us only limited information, however processed, about forecasting effectiveness' needs to be framed and put before anyone working in forecasting because, somehow, training and habit lead all of us to ignore this fact; namely, that with real-life data model fitting and post-sample forecasting are not necessarily the same.

TABLE D5

Types of time series data

Time interval between successive observations	Micro-data Total firm I	Micro-data Major divisions II	Micro-data Below major divisions III	Industry	Macro-data GNP or its major components I	Macro-data Below GNP or its major components II	Demographic	Total
Yearly	17	30	11	29	28	30	30	175
Quarterly Ending date							40	200
July 1969–December 1970 / July 1974–December 1975	3	3	3	3	15	23		70
Ending dates other than								
July 1969–December 1970 / July 1974–December 1975	4	15	13	13	34	31		130
Quarterly Ending date							76	626
July 1969–December 1970 / July 1974–December 1975	4	23	33	15	20	30		163
Ending dates other than								
July 1969–December 1970 / July 1974–December 1975	12	49	90	165	48	61		463
Total	40	120	150	225	145	175	146	1001

We would like to thank our colleague, Robin Hogarth, for his comments, with which we wholeheartedly agree. Most importantly, however, we would like to thank him for making us aware of the huge psychological literature dealing with human judgement and its comparison with quantitative models. Somehow, in getting to know this literature, as well as looking at additional comparisons between time-series and other methods, one cannot but summarize the findings as 'there is some bad news and some good news for time series forecasters': Well, the bad news is that simpler time-series methods do as well as or better than more sophisticated ones, whilst the good news is that time-series methods (for the short term) do at least as well as all alternatives (including human judgement) for which comparisons are available.

Table D5 shows 1001 real-life time series, collected on a quota sampling. They come from countries all over the world, and have various ending dates, as well as the different characteristics listed as the headings of Table D5. We believe it is imperative to put an end to the argument of what method(s) is (are) better than others and under which circumstances. We have, therefore, decided to make these data available to all researchers. (A tape containing these data can be obtained, at cost, from Tim Davidson at Applied Decision Systems, Temple, Barker & Sloane, Inc., 33 Hayden Avenue, Lexington, MA 02163, USA). We hope others will contribute to enlarge this data bank and provide a realistic basis on which time-series forecasting methods can be tested. In addition, we have asked recognized experts in the various forecasting methods to forecast each of the 1001 series. Doing so will introduce an indisputable level of objectivity, and eliminate any accusations that the different methods have not been properly used. Up to now, we have received positive replies from:

1. Cameron–Mehra for State space forecasting;
2. Carbone for Adaptive estimation procedure;
3. Fildes–Stevens for Bayesian forecasting;
4. Johnson–Montgomery for the various Exponential smoothing methods;
5. Reilly for the automatic Box–Jenkins methodology;
6. Wheelwright for Adaptive filtering.

We hope that the remaining invited persons will join [eight more persons have joined at the time of correcting the proofs] this 'forecasting competition', and that the final results will be available to the profession not later than the end of 1980.

ADDITIONAL REFERENCES

Abdrabbo, N. A. and Priestley, M. B. (1967). On the prediction of non-stationary processes. *J. R. Statist. Soc.* B, **29**, 570–585.
Anderson, O. D. (1975). *Time Series Analysis and Forecasting: The Box–Jenkins Approach*. London: Butterworth.

Anderson, O. D. (1977). A commentary on '*A Survey of Time Series*'. *Int. Statist. Rev.*, **45**, 273–297.

Armstrong, J. C. (1978). Forecasting with econometric methods: folklore versus fact. *J. Business*, **S1**, 549–600.

Baillie, R. T. (1977). Asymptotic prediction mean square error for dynamic single equation models. Working paper No. 107, Aston Management Centre.

Baillie, R. T. (1979). The asymptotic mean squared error of multistep prediction from the regression model with autoregressive errors. *J. Amer. Statist. Ass.*, **74**, 175–184.

Benson, P. H. (1975). Book review of *Forecasting Methods for Management. Interfaces, J. Marketing Res.*, **XII**, 492.

Box, G. E. P. and Jenkins, G. M. (1976). *Times Series Analysis, Forecasting and Control.* San Francisco: Holden-Day.

Bretschneider, S. *et al.* (1979). An adaptive approach to time series forecasting. *Decision Sci.* (in press).

Brown, W. J. (1977). Book review of *Forecasting Methods for Management. Business Econ.*, May 1977, 62–63.

Carbone, R. (1978). On the performance of the adaptive methods in the L.A. competition. Paper presented at the TIMS/ORSA National Conference, Los Angeles.

Chan, W.-Y. T. and Wallis, K. F. (1978). Multiple time series modelling: another look at the mink–muskrat interaction. *Appl. Statist.*, **27**, 168–175.

Chatfield, C. (1978a). The Holt–Winters forecasting procedure. *Appl. Statist.*, **27**, 264–279.

Chatfield, C. (1978b). Adaptive filtering: a critical assessment. *J. Oper. Res. Soc.*, **29**, 891–896.

Chatfield, C. (1978c). Book review of *Forecasting Methods for Management. J. R. Statist. Soc.* A, **141**, 113–114.

Computing Europe (1978). Article entitled 'In Parliament—is there a scientist in the House?' 6, No. 23 (issue of June 8th), p. 14.

De Lange, P. W. and Wojchiechowicz, K. (1974). Application of adaptive filtering forecasting to the gold bullion price. *Business Manag.*, August 1974, 3–6.

Einhorn, H. J. and Hogarth, R. M. (1975). Unit weighting schemes for decision making. *Organiz. Behav. and Human Perform.*, **13**, 171–192.

Ekern, S. (1976). Forecasting with adaptive filtering. *Oper. Res. Q.*, **27**, 705–715.

Fildes, R. (1979). Quantitative forecasting—the state of the art. *J. Oper. Res. Soc.*, **30** (in press).

Fildes, R. A. and Howell, S. (1979). On selecting a forecasting model. *TIMS Studies in Manag.*, **12** (in press).

Golder, E. R. and Settle, J. G. (1976). On adaptive filtering. *Oper. Res. Q.*, **27**, 857–868.

Granger, C. W. J. and McCollister, G. (1978). Comparison of forecasts of selected series by adaptive, Box–Jenkins and state space methods. Paper presented in the TIMS/ORSA National Conference, Los Angeles.

Harrison, P. J. and Stevens, C. F. (1976). Bayesian forecasting (with Discussion). *J. R. Statist. Soc.* B, **38**, 205, 247.

Hogarth, R. M. and Makridakis, S. (1979). Forecasting and planning: an assessment and some suggestions, INSEAD Working paper 185.

Kendall, M. G. (1973). *Time Series.* London: Griffin.

Kuhn, T. S. (1962). *The Structure of Scientific Revolution.* Chicago: University of Chicago Press.

Leung, P. (1978). Sensitivity analysis of the effect of variations in the form and parameters of a multivariate utility model: a survey. *Behav. Sci.*, **23**, 478–485.

McR, M. J. (1974). Book review of *Forecasting Methods for Management. Interfaces, J. Marketing Res.*, **XI**, 67–69.

Makridakis, S. (1978). Time series analysis and forecasting: an update and evaluation. *Int. Statist. Rev.*, **46**, 255–278.

Makridakis, S. and Wheelwright, S. C. (1977). Adaptive filtering: an integrated autoregressive-moving average filter for time series forecasting. *Oper. Res. Q.*, **28**, 425–437.

Makridakis, S. and Wheelwright, S. C. (1978a). *Interactive Forecasting: Univariate and Multivariate Methods*. San Francisco: Holden Day.

Makridakis, S. and Wheelwright, S. C. (1978b). *Forecasting: Methods and Applications*. New York: Wiley.

Montgomery, D. C. and Contreras, L. E. (1977). A note on forecasting with adaptive filtering. *Oper. Res. Q.*, **28**, 87–91.

Nau, R. F. and Oliver, R. M. (1978). Adaptive filtering revisited. University of California Working paper.

Newbold, P. and Granger, C. W. J. (1974). Experience with forecasting univariate time series and the combination of forecasts (with Discussion). *J. R. Statist. Soc.* A, **137**, 131–165.

Priestley, M. B. (1974). Discussion contribution to the paper by Newbold and Granger. *J. R. Statist. Soc.* A, **137**, 152–153.

Prothero, D. L. and Wallis, K. F. (1976). Modelling macroeconomic time series (with Discussion). *J. R. Statist. Soc.* A, **139**, 468–500.

Reid, D. J. (1975). A review of short-term projection techniques. In *Practical Aspects of Forecasting* (H. A. Gordon, ed.), pp. 8–25. London: Operational Research Society.

Slovic, P., Fischhoff, B. and Lichtenstein, S. (1977). Behavioral decision theory. *Ann. Rev.Psychol.*, **28**, 1–39.

Surrey, M. J. C. (1971). *The Analysis and Forecasting of the British Economy*. Cambridge: Cambridge University Press. (N.I.E.S.R. Occasional papers XXV.) *The Times* (1974). *The Times Guide to the House of Commons, October*, 1974. London: Times Books.

Wheelwright, S. C. and Makridakis, S. (1977). *Forecasting Methods for Management*. New York: Wiley.

CHAPTER 4

The Accuracy of Extrapolation (Time Series) Methods: Results of a Forecasting Competition

S. Makridakis
INSEAD, Fontainebleau, France

A. Andersen
University of Sydney, Australia

R. Carbone
Université Laval, Quebec, Canada

R. Fildes
Manchester Business School, Manchester, England

M. Hibon
INSEAD, Fontainebleau, France

R. Lewandowski
Marketing Systems, Essen, Germany

J. Newton and E. Parzen
Texas A & M University, Texas, U.S.A.

R. Winkler
Indiana University, Bloomington, U.S.A.

ABSTRACT

In the last few decades many methods have become available for forecasting. As always, when alternatives exist, choices need to be made so that an appropriate forecasting method can be selected and used for the specific situation being considered. This paper reports the results of a forecasting competition that provides information to facilitate such choice. Seven experts in each of the 24 methods forecasted up to 1001 series for six up to eighteen time horizons. The results of the competition are presented in this paper whose purpose is to provide empirical evidence about *differences* found to exist among the various extrapolative (time series) methods used in the competition.

Reproduced from the *Journal of Forecasting*, **1**, 111–153 (1982).

Forecasting is an essential activity both at the personal and organizational level. Forecasts can be obtained by:

(a)   purely judgemental approaches;
(b)   causal or explanatory (e.g. econometric or regression) methods;
(c)   extrapolative (time series) methods; and
(d)   any combination of the above.

Furthermore, as there are many approaches or methods available within (a), (b), (c), there is considerable choice for selecting a single approach, method, or combination procedure to predict future events. The implications of making the right choice are extremely important both from a theoretical standpoint and in practical terms. In many situations even small improvements in forecasting accuracy can provide considerable savings.

It is important to understand that there is no such thing as the best approach or method as there is no such thing as the best car or best hi-fi system. Cars or hi-fis differ among themselves and are bought by people who have different needs and budgets. What is important, therefore, is not to look for 'winners' or 'losers', but rather to understand how various forecasting approaches and methods differ from each other and how information can be provided so that forecasting users can be able to make rational choices for their situation.

Empirical studies play an important role in better understanding the pros and cons of the various forecasting approaches or methods (they can be thought of as comparable to the tests conducted by consumer protection agencies when they measure the characteristics of various products).

In forecasting, accuracy is a major, although not the only factor (see note by Carbone in this issue of the *Journal of Forecasting*) that has been dealt with in the forecasting literature by empirical or experimental studies. Summaries of the results of published empirical studies dealing with accuracy can be found in Armstrong (1978), Makridakis and Hibon (1979), and Slovic (1972). The general conclusions from these three papers are: (a) Judgemental approaches are not necessarily more accurate than objective methods; (b) Causal or explanatory methods are not necessarily more accurate than extrapolative methods; and (c) More complex or statistically sophisticated methods are not necessarily more accurate than simpler methods.

The present paper is another empirical study concerned mainly with the *post-sample* forecasting accuracy of extrapolative (time series) methods. The study was organized as a 'forecasting competition' in which expert participants analysed and forecasted many real life time series.

This paper extends and enlarges the study by Makridakis and Hibon (1979). The major differences between the present and the previous study owe their origins to suggestions made during a discussion of the previous study at a meeting of the Royal Statistical Society (see Makridakis and Hibon, 1979) and in private communications. The differences are the following:

1. The number of time series used was increased from 111 to 1001 (because of time constraints, not all methods used all 1001 series).
2. Several additional methods were considered and, in some cases, different versions of the same method were compared.
3. Instead of a single person running all methods, experts in each field analysed and forecasted the time series.
4. The type of series (macro, micro, industry, demographic), time intervals between successive observations (monthly, quarterly, yearly) and the number of observations were recorded and used (see Table 1).
5. The time horizon of forecasting was increased (18 periods for monthly data, 8 for quarterly and 6 for yearly).
6. Initial values for exponential smoothing methods were obtained by 'back-forecasting'—a procedure common in the Box–Jenkins method.
7. Additional accuracy measures were obtained (notably mean square errors, average rankings and medians).

The paper is organized as follows: first, an estimate of the time needed and computer cost incurred for each method will be given; second, the data used are briefly described; third, summary measures of overall accuracy are given; fourth, the effects of sampling errors are discussed: what would have happened had another set of series been selected?; fifth, differences among the various methods will be presented; sixth, the conditions under which various methods are better than others are discussed. Finally, an evaluation of the results and some general conclusions are presented. There will also be two appendices describing the accuracy measures and the methods used.

TIME AND COST OF RUNNING THE VARIOUS METHODS

According to statements by the participants of the competition, the Box–Jenkins methodology (ARMA models) required the most time (on the average more than one hour per series). This time included looking at the graph of each series, its autocorrelation and partial autocorrelation functions, identifying an appropriate model, estimating its parameters and doing diagnostic checking on the residual autocorrelations. The method of Bayesian forecasting required about five minutes of personal time to decide on the model to be used and get the program started. Apart from that, the method was run mechanically.

All other methods were run on a completely automatic basis. That is, the various data series were put in the computer, and forecasts were obtained with no human interference. This means that the model selection (if needed) and parameter estimation were done automatically and that the forecasts were *not* modified afterwards through any kind of human intervention. All results can, therefore, be exactly replicated by passing the data through the program.

THE DATA

The 1001 time series were selected on a quota basis. Although the sample is not random, in a statistical sense, efforts were made to select series covering a wide spectrum of possibilities. This included different sources of statistical data and different starting/ending dates. There were also data from firms, industries and nations. Table 1 shows the major classifications of the series. All accuracy measures were computed, using these classifications. Unfortunately, the output is many thousands of pages long and can only be reported in this paper in a summary form. However, a computer tape containing the original series and the forecasts of each method, together with the programs used for the evaluation, can be obtained by writing to A. Andersen, R. Carbone or S. Makridakis (whoever is geographically closest), because a major ground rule for this competition has been that all of the results could be *replicated* by anyone interested in doing so. Also, interested readers can write to S. Makridakis to obtain more or all of the results, or they may wish to wait until a book (Makridakis *et al.*, 1983) describing the methods and the study in detail is published.

Running 1001 time series is a formidable and time-consuming task. It was

TABLE 1

| Time interval between successive observations | Types of time series data | | | | | | | |
| | Micro-data | | | | Macro-data | | | |
	Total firm	Major divisions	Below major divisions	Industry	GNP or its major components	Below GNP or its major components	Demographic	Total
Yearly	16	29	12	35	30	29	30	181
Quarterly	5	21	16	18	45	59	39	203
Monthly	10	89	104	183	64	92	75	617
Subtotal	31	139	132	236	139	180	144	1001
TOTAL	302			236	319		144	1001

decided, therefore, by the organizer of this competition, to allow some of the participants to run 111 series only. These 111 series were selected through a random systematic sample. The series in this sample were every ninth entry starting with series 4 (a randomly selected starting point): 4, 13, 22, 31, ..., 994. These 111, as well as the remaining 890 series, are different from the 111 series used in the study reported in JRSS (Makridakis and Hibon, 1979). The Box–Jenkins, Lewandowski, and Parzen methodologies utilized the same systematic sample of 111 series, whereas the rest employed all series. The various tables are, therefore, presented in terms of both the 111 series for all methods and the 1001 series for all methods except the three above-mentioned methods.

SUMMARY MEASURES OF OVERALL ACCURACY

What are the most appropriate accuracy measures to describe the results of this competition? The answer obviously depends upon the situation involved and the person making the choice. It was decided, therefore, to utilize many important accuracy measures. Interestingly enough, the performance of the various methods differs – sometimes considerably – depending upon the accuracy measure (criterion) being used.

Five summary accuracy measures are reported in this paper: Mean Average Percentage Error (MAPE), Mean Square Error (MSE), Average Ranking (AR), Medians of absolute percentage errors (Md), and Percentage Better (PB).

Table 2(a) shows the MAPE for each method for all 1001 series, whereas Table 2(b) shows the MAPE for the 111 series.

Tables 3(a) and 3(b) show the MSE for each method for the 1001 and 111 series respectively.

It should be noted that a few series whose MAPE were more than 1000% were excluded (this is why not all methods in the various tables of MAPE and MSE have the same n(max) value – see Tables 2(a), 2(b), 3(a) and 3(b)).

Tables 4(a) and 4(b) show the AR for each method for all and the 111 series.

Tables 5(a) and 5(b) show the Md for each method for all the 111 series.

Finally, Tables 6(a), 6(b), 7(a), 7(b), 8, 9(a), and 9(b) show the percentage of times that methods Naive 1, Naive 2, Box–Jenkins and Winters' exponential smoothing are better than the other methods (these four methods were chosen because the same results were reported in the JRSS paper).

The accuracy measures reported in Tables 2 to 9 are overall averages. A breakdown of most of these measures also exists for each of the major categories (and often subcategories) shown in Table 1. Unfortunately, space restrictions make it impossible to report them in this paper. Findings concerning some subcategories will be given below. Some general conclusions will be provided in a later section of this paper. It is believed, however, that the best way to understand the results is to consult the various tables carefully.

TABLE 2(a)   Average MAPE: All Data (1001)

METHODS	MODEL FITTING	Forecasting Horizons										Average of Forecasting Horizons						n(max)
		1	2	3	4	5	6	8	12	15	18	1–4	1–6	1–8	1–12	1–15	1–18	
NAIVE 1	14.2	11.9	16.8	17.0	17.7	22.4	25.0	24.7	17.1	24.9	32.1	15.9	18.5	19.6	20.0	20.8	21.9	1001
Mov.Averag	12.5	13.0	18.1	17.6	18.4	22.8	25.0	23.2	17.8	23.3	31.3	16.8	19.1	19.8	19.8	20.4	21.4	1001
Single EXP	13.7	11.2	16.5	16.2	17.1	21.6	24.1	22.6	17.5	23.3	31.2	15.2	17.8	18.7	19.0	19.7	20.8	998
ARR EXP	15.4	12.4	18.9	17.5	18.0	21.7	24.6	21.7	18.7	24.2	30.8	16.7	18.8	19.2	19.4	20.2	21.2	1001
Holt EXP	13.2	11.4	15.9	16.6	18.6	23.7	26.8	30.1	26.5	39.5	56.7	15.6	18.8	21.0	22.8	25.2	28.3	1001
Brown EXP	13.8	11.4	15.2	17.6	18.8	24.2	27.8	31.4	29.3	43.9	58.9	15.9	19.3	21.7	23.8	26.6	30.0	998
Quad.EXP	13.3	12.8	18.0	21.1	23.1	31.9	38.3	52.8	56.0	91.3	149.4	18.8	24.2	29.3	34.3	41.5	52.4	1001
Regression	19.7	17.6	21.8	22.1	21.0	25.8	26.7	28.0	30.8	51.9	75.7	20.7	22.5	23.4	26.1	29.1	33.6	997
NAIVE2	9.6	9.1	11.3	13.3	14.6	18.4	19.9	19.1	17.7	21.9	26.3	12.1	14.4	15.2	15.7	16.4	17.4	1001
D Mov.Avrg	8.4	11.5	14.9	14.0	14.1	21.5	22.3	20.6	17.8	20.6	29.4	12.4	14.9	16.1	18.1	18.6	19.6	1001
D Sing EXP	9.5	8.6	11.6	13.2	14.1	17.7	19.5	17.9	16.9	21.4	26.1	11.9	14.1	14.8	15.3	16.0	16.9	998
D ARR EXP	10.6	9.4	13.5	14.0	15.3	18.1	20.2	18.0	17.1	21.4	26.0	13.1	15.1	15.6	15.9	16.5	17.4	1001
D Holt EXP	8.8	8.7	11.0	13.3	15.0	18.7	21.1	24.8	23.1	33.7	48.3	12.1	14.7	16.7	18.0	20.2	22.9	998
D brownEXP	9.0	8.7	10.9	13.8	15.0	18.7	21.0	24.5	23.0	30.8	43.7	12.1	14.7	16.6	18.0	19.6	21.9	1001
D Quad.EXP	9.3	9.8	12.7	16.6	18.8	25.7	31.0	45.1	40.7	64.0	108.3	14.5	19.1	23.7	26.9	31.2	38.5	1001
D Regress	15.6	15.5	16.9	19.1	18.3	21.9	23.0	24.3	29.7	49.1	70.7	17.4	19.4	20.0	22.6	25.5	29.8	997
WINTERS	9.3	8.7	11.9	13.4	14.9	19.0	21.5	24.3	23.0	32.8	47.0	11.9	14.7	16.5	18.1	19.8	22.4	997
Autom. AEP	9.9	9.1	11.9	13.4	13.7	19.8	22.3	20.3	19.3	24.8	28.8	12.0	14.4	15.5	17.6	17.5	18.8	1001
Bayesian F	13.7	11.2	12.8	14.5	16.2	16.7	19.7	22.6	18.6	24.2	30.8	13.7	16.1	17.2	16.3	18.3	19.3	998
Combining A	8.6	8.1	10.4	12.1	13.3	17.6	19.2	24.2	24.2	24.2	30.3	11.0	13.3	14.5	17.6	18.5	19.9	997
Combining B	9.0	8.5	11.1	12.8	13.8	17.6	19.2	18.9	18.4	23.3	30.3	11.6	13.8	14.8	15.6	16.5	17.8	1001
Average	11.8	10.9	14.4	15.8	16.8	21.2	23.8	25.4	23.7	34.1	48.1	14.5	17.2	18.7	20.0	21.7	24.3	

TABLE 2(b)   Average MAPE: All Data (111)

METHODS	MODEL FITTING	Forecasting Horizons										Average of Forecasting Horizons						n(max)
		1	2	3	4	5	6	8	12	15	18	1-4	1-6	1-8	1-12	1-15	1-18	
NAIVE 1	14.4	13.2	17.3	20.1	18.6	22.4	23.5	27.0	14.5	31.9	34.9	17.3	19.2	20.7	19.9	20.9	22.3	111
Mov.Averag	12.8	14.1	16.9	19.1	18.9	21.8	23.6	23.9	16.3	28.7	31.9	17.3	19.1	20.1	18.9	19.7	20.8	111
Single EXP	13.2	12.2	14.8	17.4	17.6	20.3	22.5	22.7	16.1	28.8	32.5	15.5	17.5	18.5	17.8	18.8	20.1	111
ARR EXP	15.1	13.0	17.1	18.4	18.3	20.7	22.8	22.4	16.1	29.6	32.2	16.7	18.4	19.2	18.3	19.3	20.5	111
Holt EXP	13.6	12.2	13.9	17.6	19.2	23.1	24.9	31.2	22.6	40.4	40.3	15.7	18.5	21.1	21.3	23.4	25.1	111
Brown EXP	13.6	13.0	15.1	18.6	19.5	25.2	27.1	35.0	28.0	54.0	59.6	16.5	19.7	22.8	23.6	26.8	30.3	111
Quad.EXP	13.9	13.2	16.1	21.3	23.2	30.3	34.1	51.5	49.0	103.0	106.0	18.6	23.1	28.4	31.7	40.4	47.7	111
Regression	16.6	17.9	19.9	21.1	21.2	23.2	25.0	26.2	26.1	49.5	60.2	20.0	21.4	22.5	22.9	25.4	29.5	110
NAIVE2	9.1	8.5	11.4	13.9	15.4	16.0	17.4	17.8	14.5	31.2	30.8	12.3	13.8	14.9	14.9	16.4	17.8	111
D Mov.Avrg	8.1	10.7	13.6	17.8	19.4	22.0	23.1	22.7	15.7	28.3	34.0	15.4	17.8	19.0	18.4	19.1	20.6	111
D Sing EXP	8.6	7.8	10.8	13.1	14.5	15.7	17.2	16.5	13.6	29.3	30.1	11.6	13.2	14.0	14.0	15.8	16.8	111
D ARR EXP	9.8	8.3	12.1	14.0	16.4	16.7	18.1	16.5	13.7	28.6	29.3	12.9	14.4	15.1	14.7	15.8	17.1	111
D Holt EXP	8.6	7.9	10.5	13.3	15.1	17.3	19.3	23.1	16.5	35.6	35.2	11.7	13.8	16.1	17.0	18.0	19.7	111
D brownEXP	8.3	8.5	10.8	13.3	14.5	17.3	19.3	19.0	19.0	43.1	45.4	11.7	13.9	16.2	17.0	19.5	22.3	111
D Quad.EXP	8.4	8.8	11.8	15.0	16.9	21.9	24.1	35.7	29.7	56.1	63.6	13.1	16.4	20.3	22.2	25.9	30.2	111
D Regress	12.0	12.5	14.9	17.2	18.4	19.7	21.0	21.0	23.4	46.5	57.3	15.7	17.3	18.2	18.8	21.3	25.6	110
WINTERS	9.3	9.2	10.5	13.2	15.5	18.7	18.8	23.3	15.9	33.4	34.5	12.5	14.1	16.3	16.4	17.8	19.5	111
Autom. AEP	10.8	9.8	11.3	13.7	15.1	15.9	18.8	23.3	16.2	30.2	33.9	12.5	14.3	16.3	16.2	17.4	19.0	111
Bavesian F	11.3	10.3	12.8	13.6	14.4	15.2	17.1	19.2	14.2	27.5	30.6	12.8	14.1	15.2	15.0	16.1	17.6	111
Combining A	8.1	7.9	9.9	11.2	13.5	15.4	16.4	20.1	15.5	32.1	33.3	10.8	12.6	14.4	14.7	15.9	17.7	111
Combining B	8.2	8.2	10.1	11.8	14.7	16.4	17.1	18.9	16.4	26.2	31.4	11.2	12.8	14.4	14.7	16.2	17.7	111
Box-Jenkins	7.4	10.3	10.7	11.4	14.5	16.4	17.1	18.9	16.4	26.2	34.2	11.7	13.4	14.8	15.1	16.3	18.0	111
Lewandowski	12.3	11.6	12.9	12.9	15.3	14.3	14.7	16.0	13.7	33.0	28.6	13.5	14.7	15.5	15.6	17.2	18.6	111
Parzen	8.9	10.6	10.7	10.7	13.5	14.3	14.7	16.0	13.7	22.5	26.5	11.4	12.4	13.3	13.4	15.4	15.4	111
Average	10.7	10.8	13.7	15.5	16.8	19.3	20.8	24.0	19.7	37.5	40.7	14.1	16.1	17.8	18.0	19.9	22.1	

ARR   = Adaptive Response Rate
Mov.   = Moving
Quad.   = Quadratic
EXP.   = Exponential Smoothing
Sing.   = Single
D   = Deseasonalized
WINTERS = Holt–Winters Exponential Smoothing

TABLE 3(a) Average MSE: All Data (1001)

METHODS	MODEL FITTING	1	2	4	6	8	12	18	1-4	1-12	1-18	n(max)
NAIVE 1	.1340E+11	.9317E+11	.1476E+12	.2083E+12	.9360E+12	.3051E+09	.1574E+10	.1576E+10	.1904E+12	.2031E+12	.1491E+12	1001
Mov.Averg	.1226E+11	.2701E+12	.3245E+12	.3354E+12	.1151E+13	.2022E+09	.1345E+10	.1372E+10	.3504E+12	.2998E+12	.2199E+12	1001
Single EXP	.1267E+11	.8553E+11	.1387E+12	.1984E+12	.9145E+12	.1251E+09	.9177E+09	.9292E+09	.1804E+12	.1955E+12	.1433E+12	998
ARR EXP	.1911E+11	.8594E+11	.1326E+12	.1959E+12	.9168E+12	.1196E+09	.8087E+09	.8289E+09	.1756E+12	.1928E+12	.1414E+12	1001
Holt EXP	.1178E+11	.5117E+11	.7670E+11	.9468E+11	.5395E+12	.3248E+09	.2005E+10	.2559E+10	.9991E+11	.1094E+12	.8067E+11	998
Brown EXP	.1384E+11	.4004E+11	.6033E+11	.9878E+11	.4698E+12	.3134E+09	.1969E+10	.2514E+10	.8805E+11	.9755E+11	.7202E+11	1001
Quad.EXP	.1431E+11	.3242E+11	.4706E+11	.1365E+12	.3829E+12	.1255E+10	.2980E+10	.4806E+10	.8645E+11	.9237E+11	.6867E+11	1000
Regression	.2168E+11	.7996E+11	.1249E+12	.1564E+12	.6928E+12	.1958E+09	.1713E+10	.2045E+10	.1535E+12	.1540E+12	.1132E+12	997
NAIVE2	.1338E+11	.9316E+11	.1476E+12	.2083E+12	.9360E+12	.3565E+09	.1574E+10	.1549E+10	.1904E+12	.2031E+12	.1491E+12	1001
D Mov.Avrg	.1224E+11	.2701E+12	.3245E+12	.3354E+12	.1151E+13	.3338E+09	.1308E+10	.1289E+10	.3504E+12	.2998E+12	.2198E+12	1001
D Sing EXP	.1264E+11	.8551E+11	.1387E+12	.1984E+12	.9145E+12	.2496E+09	.8968E+09	.8711E+09	.1804E+12	.1954E+12	.1433E+12	998
D ARR EXP	.1908E+11	.8586E+11	.1325E+12	.1958E+12	.9167E+12	.2001E+09	.6971E+09	.6851E+09	.1756E+12	.1928E+12	.1413E+12	1001
D Holt EXP	.1175E+11	.5113E+11	.7668E+11	.9468E+11	.5394E+12	.5644E+09	.1947E+10	.2408E+10	.9988E+11	.1094E+12	.8065E+11	998
D brownEXP	.1381E+11	.4001E+11	.6032E+11	.9876E+11	.4698E+12	.4874E+09	.1908E+10	.2309E+10	.8803E+11	.9754E+11	.7198E+11	1001
D Quad.EXP	.1426E+11	.3233E+11	.4698E+11	.1364E+12	.3825E+12	.1514E+10	.2803E+10	.4409E+10	.8632E+11	.9225E+11	.6850E+11	1001
D Regress	.2165E+11	.7962E+11	.1249E+12	.1564E+12	.6927E+12	.3994E+09	.1652E+10	.1933E+10	.1535E+12	.1540E+12	.1132E+12	997
WINTERS	.1175E+11	.5113E+11	.7668E+11	.9467E+11	.5394E+12	.4847E+09	.1927E+10	.2393E+10	.9988E+11	.1094E+12	.8064E+11	998
Autom. AEP	.9105E+10	.4893E+11	.6687E+11	.7060E+11	.4544E+12	.1141E+10	.3308E+10	.4006E+10	.8562E+11	.9134E+11	.6740E+11	1001
Bavesian F	.2345E+11	.2173E+11	.2898E+11	.1047E+12	.2488E+12	.3286E+09	.1708E+10	.2228E+10	.5945E+11	.6198E+11	.4593E+11	997
Combining A	.9389E+10	.5759E+11	.8462E+11	.1061E+12	.5981E+12	.4191E+09	.1655E+10	.1906E+10	.1114E+12	.1219E+12	.8967E+11	1001
Combining R	.1027E+11	.8525E+11	.1379E+12	.1977E+12	.9118E+12	.2319E+09	.8898E+09	.8693E+09	.1798E+12	.1950E+12	.1430E+12	1001
Average	.1437E+11	.8287E+11	.1190E+12	.1630E+12	.7027E+12	.4548E+09	.1695E+10	.2071E+10	.1517E+12	.1556E+12	.1144E+12	

e.g. .1340E + 11 = 1340 × 10^{11}

TABLE 3(b)   Average MSE: All Data (111)

METHODS	MODEL FITTING	1	2	4	6	8	12	18	1-4	1-12	1-18	n(max)
NAIVE 1	.4510E+08	.3049E+08	.7776E+08	.4657E+09	.4657E+09	.1891E+09	.4620E+08	.2742E+08	.2044E+08	.2164E+09	.1736E+09	111
Mov.Averag	.3792E+08	.2297E+09	.1976E+09	.5622E+09	.5409E+09	.1116E+08	.4710E+08	.2492E+08	.3377E+08	.2810E+09	.2171E+09	111
Single Exp	.3921E+08	.4161E+08	.8897E+08	.5170E+09	.4983E+09	.1158E+08	.5808E+08	.2284E+08	.2327E+08	.2325E+09	.1821E+09	111
ARR EXP	.5109E+08	.5570E+08	.2011E+09	.7274E+09	.6816E+09	.1306E+08	.4610E+08	.3026E+08	.3316E+08	.2991E+09	.2312E+09	111
Holt EXP	.3316E+08	.2576E+08	.3781E+08	.2193E+09	.1000E+09	.1389E+08	.7728E+08	.2206E+08	.1011E+09	.1062E+09	.9572E+08	111
Brown EXP	.3838E+08	.2738E+08	.5240E+08	.2373E+09	.1223E+09	.1989E+08	.5073E+08	.2483E+08	.1131E+09	.1109E+09	.9773E+08	111
Quad.EXP	.4191E+08	.2635E+08	.4378E+08	.2190E+09	.1099E+09	.5745E+08	.9348E+08	.1292E+08	.1024E+09	.1215E+09	.1324E+09	111
Regression	.3253E+08	.3345E+08	.5241E+08	.2294E+09	.1641E+09	.1403E+08	.8243E+08	.2103E+08	.1127E+09	.1129E+09	.9505E+08	110
NAIVE2	.2594E+08	.2970E+08	.6000E+08	.4559E+09	.4489E+09	.2569E+08	.4620E+08	.7348E+08	.1954E+09	.1996E+09	.1637E+09	111
D Mov.Avrg	.2178E+08	.2316E+08	.1857E+09	.5691E+09	.5324E+09	.8741E+08	.6792E+08	.1171E+09	.3360E+08	.3051E+09	.2473E+09	111
D Sing Exp	.2186E+08	.3890E+08	.7564E+08	.5117E+09	.4810E+09	.1625E+08	.3250E+08	.6251E+08	.2258E+08	.2155E+09	.1714E+09	111
D ARR EXP	.2849E+08	.5199E+08	.1793E+09	.7197E+09	.6550E+09	.1090E+08	.2428E+08	.5661E+08	.3209E+08	.2776E+09	.2145E+09	111
D Holt EXP	.1501E+08	.2301E+08	.2525E+08	.2140E+09	.9188E+08	.2535E+08	.5790E+08	.8529E+08	.9464E+08	.9053E+08	.8773E+08	111
D brownExp	.1693E+08	.2619E+08	.3161E+08	.2360E+09	.1039E+09	.4356E+08	.8036E+08	.1143E+09	.1055E+09	.1073E+09	.1085E+09	111
D Quad.EXP	.1783E+08	.2435E+08	.2738E+08	.2123E+09	.9759E+08	.7443E+08	.9877E+08	.1551E+09	.9447E+08	.1081E+09	.1177E+09	111
D Regress	.1454E+08	.2640E+08	.3315E+08	.2238E+09	.1455E+09	.1241E+08	.4071E+08	.6628E+08	.1035E+08	.9149E+08	.8066E+08	110
WINTERS	.1567E+08	.2158E+08	.2538E+08	.2128E+09	.9347E+08	.2842E+08	.8271E+08	.1004E+09	.9443E+08	.9478E+08	.9593E+08	111
Autom. AEP	.2377E+08	.6811E+08	.4696E+08	.4011E+09	.1999E+09	.3214E+08	.6256E+08	.7469E+08	.1958E+09	.1763E+09	.1543E+09	111
Bayesian F	.6023E+08	.2641E+08	.2549E+08	.1328E+09	.9407E+08	.2626E+08	.3494E+08	.5723E+08	.6660E+08	.6401E+09	.6383E+09	111
Combining A.	.1994E+08	.3431E+08	.5011E+08	.3501E+08	.2079E+09	.2286E+08	.4810E+08	.7083E+08	.1561E+08	.1409E+09	.1213E+09	111
Combining B.	.2057E+08	.3107E+08	.3422E+08	.3403E+09	.1446E+09	.1409E+08	.2657E+08	.5626E+08	.1504E+08	.1250E+09	.1032E+09	111
Box-Jenkins	N.A	.5293E+08	.3422E+08	.2491E+09	.1325E+09	.3433E+08	.7849E+08	.1172E+09	.1233E+09	.1156E+09	.1129E+09	111
Lewandowski	.4079E+08	.2015E+08	.1771E+09	.8285E+08	.9147E+08	.1941E+08	.4196E+08	.5307E+08	.7456E+08	.7621E+08	.7024E+08	111
Parzen	.2231E+08	.7054E+08	.1522E+08	.1105E+08	.8625E+08	.2806E+08	.6992E+08	.1136E+08	.9494E+08	.9351E+08	.9097E+08	111
Average	.2850E+08	.5199E+08	.6793E+08	.3416E+09	.2620E+09	.2757E+08	.5814E+08	.6985E+08	.1653E+09	.1568E+09	.1345E+09	

e.g. .4510E+08 = .4510 × 10⁸

TABLE (4a) Average Ranking: All Data (1001)

METHODS	MODEL FITTING	Forecasting Horizons																		AVERAGE OF ALL FORECASTS	n(max)
		1	2	3	4	5	6	7	8	9	10	11	12	13	14	15	16	17	18		
NAIVE 1	15.8	11.9	12.4	12.3	11.7	12.2	12.0	11.9	11.6	11.6	11.2	11.1	10.0	11.1	11.5	11.1	11.5	11.3	11.2	11.62	1001
Mov.Averaj	13.3	11.8	12.3	11.9	11.9	11.8	11.5	11.0	11.0	10.8	10.8	10.9	10.9	10.7	11.1	10.4	11.0	10.6	10.6	11.28	1001
Single Exp	12.9	11.9	12.2	11.9	11.9	11.9	11.6	11.6	10.8	10.4	10.5	10.6	10.6	10.6	10.9	10.5	10.8	10.6	10.6	11.18	1001
ARR Exp	18.3	12.8	14.0	12.4	13.0	12.1	12.4	10.8	11.2	10.7	11.0	10.7	11.5	11.2	11.6	11.1	11.2	10.9	10.9	11.82	1001
Holt Exp	10.5	10.9	10.9	11.0	10.7	10.9	11.0	12.1	11.4	11.7	12.0	12.0	11.3	11.9	12.4	12.0	11.7	11.6	11.8	11.41	1001
Brown Exp	12.4	10.8	10.9	10.9	10.8	11.2	11.8	11.8	11.8	12.0	11.9	12.0	11.9	12.6	13.0	12.1	12.1	12.3	12.6	11.68	1001
Quad.Exp	13.8	11.8	12.0	12.5	12.1	12.6	13.1	14.1	14.2	13.8	13.8	13.8	11.8	15.5	15.5	15.1	15.0	15.3	15.7	13.68	1001
Regression	15.6	14.2	13.4	12.8	12.2	11.6	11.4	11.6	11.5	11.9	12.3	12.4	11.8	12.0	12.3	11.3	11.2	11.0	11.1	12.08	1001
NAIVE2	11.7	10.4	10.5	10.6	10.8	11.0	10.6	11.5	11.1	10.5	10.8	10.3	10.0	9.8	9.5	10.2	11.0	10.1	9.9	10.36	1001
D Mov.Avrg	8.1	11.1	10.4	10.6	12.3	12.1	11.6	11.5	11.1	10.5	10.8	10.8	10.8	11.0	10.8	10.8	11.0	10.9	10.8	11.34	1001
D Sing Exp	7.6	10.3	10.4	10.6	10.7	10.8	10.5	9.8	9.7	9.6	9.6	9.7	9.8	9.3	9.0	9.7	9.5	9.5	9.4	10.00	1001
D ARR Exp	13.6	11.4	12.4	11.6	12.0	11.5	11.5	10.4	10.5	10.7	10.5	10.4	10.5	10.8	9.7	10.3	10.1	10.0	10.0	10.87	1001
D Holt Exp	4.8	9.4	8.9	9.3	9.7	9.7	9.9	10.5	10.5	10.6	10.5	10.7	10.7	11.0	10.7	11.0	10.6	10.7	10.7	10.09	1001
D brownEXP	6.5	9.4	9.0	9.5	9.7	9.9	10.0	10.6	10.6	11.4	10.8	10.6	10.9	11.0	10.7	11.0	11.0	11.3	11.3	10.29	1001
D Quad.EXP	8.3	10.2	10.2	12.1	11.4	10.8	11.9	12.8	13.1	12.9	13.0	13.0	13.7	13.7	13.5	14.1	14.2	14.4	14.6	12.44	1001
D Regress	12.3	13.3	12.0	10.3	9.6	9.7	9.8	10.1	10.4	10.8	11.4	11.4	11.2	10.3	10.4	10.2	10.2	10.5	10.5	11.21	1001
WINTERS	7.2	9.4	9.0	9.3	9.7	10.0	10.0	10.4	10.7	10.7	10.5	10.5	10.4	10.9	9.9	10.3	10.3	10.3	10.3	9.96	1001
Autom. AEP	9.1	9.8	9.8	10.2	10.3	10.1	10.0	10.4	10.7	10.3	10.8	10.6	10.6	10.9	11.0	10.5	10.7	10.6	10.7	10.32	1001
Bayesian F	15.6	11.0	11.0	10.1	9.2	10.3	10.1	9.3	10.3	9.4	10.3	10.2	10.7	10.5	10.2	10.4	10.7	10.4	10.2	10.38	1001
Combining A	6.7	9.0	8.8	8.9	10.1	9.2	9.4	9.3	9.3	9.4	9.4	9.2	9.4	9.1	8.8	9.3	9.1	9.5	9.1	9.17	1001
Combining B	7.5	9.8	10.0	10.0	11.0	11.0	10.1	9.8	9.8	9.7	9.6	9.7	9.8	9.5	9.1	9.6	9.4	9.6	9.6	9.80	1001
Average	11.0	11.0	11.0	11.0	11.0	11.0	11.0	11.0	11.0	11.0	11.0	11.0	11.0	11.0	11.0	11.0	11.0	11.0	11.0	11.00	

Table 4(b) Average Ranking: All Data (111)

METHODS	MODEL FITTING	Forecasting Horizons																		AVERAGE OF ALL FORECASTS	n(max)
		1	2	3	4	5	6	7	8	9	10	11	12	13	14	15	16	17	18		
NAIVE 1	17.4	13.5	14.5	15.2	13.8	14.9	14.0	14.4	14.1	14.2	12.5	13.3	11.1	12.5	14.6	14.0	13.4	13.6	13.2	13.83	111
Mov.Averag	15.5	13.4	14.0	14.2	13.5	14.5	13.8	13.7	12.8	12.1	12.6	12.4	12.7	12.0	12.9	12.7	12.0	11.9	12.1	13.09	111
Single Exp	14.8	13.4	13.7	14.1	14.4	14.2	14.1	12.8	12.7	12.1	12.6	12.4	12.7	12.4	12.4	13.1	12.4	12.8	12.8	13.20	111
ARR Exp	20.3	14.1	15.2	15.1	14.9	14.6	15.1	12.7	13.5	12.7	13.3	12.8	13.7	12.9	13.3	13.4	13.7	13.7	13.4	13.95	111
Holt Exp	11.9	12.0	12.8	13.0	12.7	13.6	12.6	14.0	13.4	13.6	13.7	13.7	13.4	12.8	14.4	14.3	13.1	13.4	12.6	13.25	111
Brown EXP	13.8	13.1	13.2	12.9	12.5	13.9	12.7	13.5	13.5	13.3	13.1	12.8	13.4	13.4	13.7	14.8	13.7	13.6	13.2	13.30	111
Quad.EXP	15.1	13.7	14.0	14.5	13.4	14.9	13.8	15.3	16.3	15.7	14.8	15.8	16.0	17.0	17.3	17.1	16.9	16.1	16.4	15.27	111
Regression	16.9	16.7	15.4	15.5	15.2	14.6	15.2	14.3	16.3	14.2	14.6	14.8	14.6	13.7	14.7	17.1	16.9	16.1	12.7	14.61	111
NAIVE2	11.8	11.5	12.1	12.7	13.1	12.3	12.6	12.4	12.1	14.2	14.6	14.8	14.6	13.7	14.7	12.9	12.3	12.0	12.6	12.32	111
D Mov.Avrg	9.3	12.8	13.6	14.7	14.9	11.6	14.5	14.1	13.9	13.8	13.7	14.4	12.9	13.3	12.3	12.2	13.8	14.4	14.2	13.86	111
D Sing Exp	8.4	10.9	12.2	12.7	12.5	13.3	12.7	10.9	11.7	11.0	11.2	11.4	11.9	11.8	10.5	11.6	11.8	11.5	11.9	11.57	111
D ARR EXP	14.5	12.7	13.8	13.4	14.2	13.3	13.9	11.6	11.7	12.1	12.0	11.5	11.9	11.8	12.6	12.6	11.8	12.4	12.9	12.72	111
D Holt Exp	4.7	10.0	10.0	10.3	10.6	10.2	10.9	12.4	11.6	12.1	11.7	11.5	12.0	10.9	11.5	12.1	11.6	11.8	11.8	11.15	111
D brownEXP	6.3	10.8	11.4	11.1	10.7	11.8	11.7	13.8	11.8	11.9	12.0	12.6	12.5	12.5	12.4	12.2	12.9	11.8	12.8	11.47	111
D Quad.EXP	8.0	11.4	11.4	11.1	11.0	11.8	13.8	14.2	14.2	14.7	14.0	15.1	15.1	15.2	14.4	15.2	15.8	15.3	14.9	13.23	111
D Regress	12.5	14.7	13.4	14.4	14.1	12.7	13.8	12.6	12.6	12.0	12.3	12.4	12.7	11.6	12.0	12.5	11.5	12.7	11.9	12.94	111
WINTERS	7.6	11.9	11.5	12.5	11.3	10.6	10.3	11.3	11.6	11.7	12.1	12.6	12.0	11.6	11.1	10.5	12.5	11.3	10.8	11.26	111
Autom. AEP	10.3	11.5	11.5	12.0	11.3	11.6	12.1	11.4	11.4	12.8	12.3	11.7	12.1	12.1	11.1	11.0	11.8	11.6	12.7	11.77	111
Bavesian F	17.1	12.7	12.2	11.5	11.1	10.8	11.9	12.3	13.4	11.8	11.9	11.6	12.3	10.4	12.4	11.7	12.6	12.0	11.8	11.90	111
Combining A	7.3	10.1	10.3	9.9	10.6	11.3	10.5	11.3	10.8	10.0	10.4	10.4	10.1	10.3	9.5	10.8	10.5	10.2	11.0	10.40	111
Combining B	7.5	10.8	11.6	10.5	10.8	11.1	10.8	11.0	11.1	11.6	10.7	10.4	12.1	12.1	12.1	10.8	11.6	11.6	11.6	11.30	111
BoxJenkins	N.A	12.4	11.3	10.5	11.2	10.5	10.8	11.0	10.8	12.7	12.7	12.6	12.4	13.2	12.6	10.7	11.9	11.8	12.1	11.53	111
Lewandowski	15.6	13.5	12.4	11.4	11.6	10.5	10.7	10.4	10.9	11.5	10.4	9.9	10.9	10.7	10.0	10.3	9.5	9.5	8.8	10.87	111
Parzen	9.4	12.5	10.7	10.9	10.6	11.2	10.8	10.5	10.8	11.9	12.0	10.8	11.0	12.7	11.5	10.1	10.9	10.9	12.1	11.22	111
Average	11.5	12.5	12.5	12.5	12.5	12.5	12.5	12.5	12.5	12.5	12.5	12.5	12.5	12.5	12.5	12.5	12.5	12.5	12.5	12.50	

TABLE 5(a) Median MAPE: All Data (1001)

METHODS	MODEL FITTING	FORECASTING HORIZONS 1	2	3	4	5	6	8	12	15	18	CUMULATIVE FORECASTING HORIZONS 1-4	1-6	1-8	1-12	1-15	1-18	N(MAX)
NAIVE 1	5.9	6.0	8.8	9.1	10.8	12.0	13.3	12.2	10.4	13.9	15.7	8.5	9.6	10.1	10.5	11.0	11.6	1001
MOV.AVERAG	5.6	6.3	8.9	9.0	10.7	11.7	13.1	11.4	11.8	13.4	15.4	8.5	9.6	9.9	10.3	10.8	11.3	1001
SINGLE EXP	5.7	6.0	8.2	8.6	10.5	11.2	12.5	11.1	11.3	13.4	14.9	8.2	9.3	9.7	10.0	10.6	11.0	1001
ARR EXP	6.5	6.7	9.7	8.5	10.2	11.2	12.7	11.2	12.2	13.4	14.7	8.8	9.6	9.9	10.3	10.9	11.3	1001
HOLT EXP	5.3	6.1	7.2	7.1	8.9	9.5	11.4	11.5	11.3	14.2	17.2	7.2	8.2	8.9	9.7	10.3	10.9	1001
BROWN EXP	5.5	5.7	6.9	7.6	8.8	10.0	11.8	12.0	14.6	15.5	19.1	7.2	8.3	9.1	10.2	11.0	11.6	1001
QUAD.EXP	5.5	6.3	7.8	8.9	9.8	12.0	14.3	16.2	19.7	25.8	29.9	8.2	9.6	10.7	12.0	13.3	14.6	1001
REGRESSION	7.6	9.0	9.9	9.6	10.8	11.0	12.2	11.7	12.6	13.1	14.1	9.8	10.4	10.6	11.1	11.7	11.9	1001
NAIVE2	3.8	4.8	6.6	6.9	8.8	9.9	10.8	10.4	10.4	12.7	12.6	6.6	7.6	8.2	8.6	9.1	9.5	1001
D MOV.AVRG	3.5	5.8	7.9	9.5	11.6	12.3	13.8	11.5	11.5	14.1	16.4	8.3	9.6	10.2	10.6	11.0	11.4	1001
D SING EXP	3.6	4.7	6.2	6.8	8.2	9.0	10.5	10.0	10.3	12.1	12.5	6.3	7.3	7.6	8.3	8.8	9.2	1001
D ARR EXP	4.1	5.3	7.3	7.1	8.6	9.6	10.9	10.0	10.6	12.6	12.9	6.9	7.8	8.3	8.8	9.2	9.5	1001
D HOLT EXP	3.3	4.5	5.3	5.6	7.3	8.1	9.2	9.8	9.9	12.2	13.6	5.5	6.4	7.0	7.7	8.3	8.8	1001
D BRO*NEXP	3.4	4.4	5.1	6.2	7.3	8.5	9.6	10.0	11.1	12.7	15.9	5.6	6.5	7.2	8.0	8.8	9.4	1001
D QUAD.EXP	3.4	4.5	5.7	6.9	8.3	10.0	11.7	13.9	15.8	20.4	24.8	6.3	7.5	8.6	9.6	10.6	11.8	1001
D REGRESS	5.6	7.8	7.9	9.1	9.1	9.8	11.2	10.6	10.9	12.6	11.9	8.5	9.0	9.3	9.8	10.1	10.3	1001
WINTERS	3.7	4.4	5.4	5.5	6.6	7.9	9.6	9.7	10.1	11.4	12.2	5.5	6.4	6.9	7.5	8.0	8.5	1001
AUTOM. AEP	4.0	4.7	5.8	6.6	7.2	8.5	9.7	9.6	10.5	12.4	14.1	6.1	6.9	7.5	8.2	8.8	9.3	1001
BAYESIAN F	5.0	5.1	5.5	6.4	7.8	8.5	10.0	10.2	10.1	12.5	13.1	6.1	7.0	7.6	8.2	8.7	9.1	1001
COMBINING A	3.3	4.2	5.4	5.9	7.1	8.0	9.7	9.3	9.5	11.8	12.6	5.6	6.4	6.9	7.5	8.0	8.4	1001
COMBINING B	3.4	4.6	6.0	6.6	7.7	8.8	10.1	10.1	10.4	12.3	12.8	6.0	7.0	7.6	8.1	8.6	9.1	1001

TABLE 5(b)   Median APE: All Data (111)

| METHODS | MODEL FITTING | FORECASTING HORIZONS | | | | | | | | | | CUMULATIVE FORECASTING HORIZONS | | | | | | N(MAX) |
		1	2	3	4	5	6	8	12	15	18	1-4	1-6	1-8	1-12	1-15	1-18	
NAIVE 1																		111
MOV.AVERAG																		111
SINGLE EXP																		111
ARR EXP																		111
HOLT EXP																		111
BROWN EXP																		111
QUAD.EXP																		111
REGRESSION																		111
NAIVE2																		111
DD MOV.AVRG																		111
DD SING EXP																		111
DD ARR EXP																		111
DD HOLT EXP																		111
DD BROWNEXP																		111
DD QUAD.EXP																		111
DD REGRESS																		111
WINTERS																		111
AUTOM. AEP																		111
BAYESIAN F																		111
COMBINING A																		111
COMBINING B																		111
BOX-JENKINS																		111
LEWANDOWSKI																		111
PARZEN																		111

TABLE 6(a) Percentage of Time that the Naive 1 Method is Better than Other Methods ($n = 1001$)

METHODS	MODEL FITTING	Forecasting Horizons 1	2	3	4	5	6	7	8	9	10	11	12	13	14	15	16	17	18	AVERAGE OF ALL FORECASTS	n(max)
Mov.Averag	30.5	48.8	48.0	47.2	49.7	46.7	46.5	42.3	44.9	43.7	45.0	46.8	54.7	47.1	45.8	45.9	45.8	44.7	46.8	46.76	
Single EXP	22.1	52.3	49.8	46.6	52.4	49.5	48.4	39.1	43.2	40.4	41.2	43.9	52.8	46.5	45.5	46.1	43.9	44.0	44.7	46.61	
ARR EXP	65.4	55.1	60.5	52.0	59.2	50.6	54.0	41.8	47.4	43.8	47.8	50.8	58.8	52.3	50.9	50.4	48.3	47.5	48.9	51.63	
Holt EXP	16.4	43.7	40.9	43.6	45.4	43.3	46.3	50.9	49.3	51.9	55.1	53.0	56.6	55.8	56.6	56.7	53.2	53.3	55.3	49.55	
Brown EXP	26.5	43.5	42.3	43.9	46.1	45.0	47.9	51.7	51.7	52.0	53.0	53.8	59.3	58.7	57.1	55.3	53.2	55.1	57.1	50.57	
Quad.EXP	32.6	48.2	47.1	50.9	50.2	52.0	54.4	61.0	61.3	60.3	61.9	64.5	68.7	72.1	71.6	70.8	67.9	69.9	73.3	59.62	
Regression	50.8	62.6	54.6	52.0	50.1	44.2	44.9	48.0	45.9	50.6	54.1	56.2	57.7	54.5	54.5	51.2	48.1	45.5	48.0	51.17	
NAIVE2	25.4	42.3	46.1	41.0	47.4	44.7	43.5	42.6	45.4	44.7	44.2	46.2	49.9	43.8	43.4	38.6	42.6	42.0	42.0	43.46	
D Mov.Avrg	19.5	46.5	46.1	49.4	54.2	49.1	48.5	46.2	49.5	47.6	46.8	49.4	57.7	43.8	43.4	49.0	47.5	49.5	49.3	48.57	
D Sing EXP	14.7	45.4	42.4	43.1	49.3	46.1	45.2	38.7	42.0	40.3	40.3	42.4	49.2	39.8	36.7	42.3	41.1	39.8	40.1	42.87	
D ARR EXP	42.5	50.4	54.2	48.8	56.4	48.8	51.1	42.5	47.1	45.1	47.5	47.0	53.8	48.1	41.2	46.0	45.9	43.3	43.3	48.05	
D Holt EXP	8.9	32.8	35.2	37.8	38.9	37.8	39.6	44.9	47.1	47.5	48.6	48.0	53.0	48.1	46.0	46.0	45.9	48.1	48.0	43.00	
D brownEXP	13.5	36.8	35.0	37.8	40.3	40.2	41.5	47.2	46.2	46.5	48.9	50.1	55.4	50.6	46.4	52.0	48.6	50.6	51.5	44.67	
D Quad.EXP	16.8	40.8	39.6	44.3	45.1	47.5	48.8	54.8	56.0	55.9	58.3	58.9	65.0	60.8	58.0	63.4	61.3	63.9	64.8	53.02	
D Regress	39.7	57.8	47.3	49.1	47.4	42.1	42.2	46.6	47.0	50.4	52.7	50.6	53.8	46.5	45.7	45.7	44.6	47.8	44.7	47.99	
WINTERS	10.0	38.5	34.4	36.6	40.0	37.3	39.4	43.4	45.4	47.8	48.5	48.5	52.8	49.3	43.3	47.0	44.6	46.8	46.4	42.85	
Autom. AEP	10.3	36.8	34.6	38.4	39.1	38.7	41.6	43.9	44.0	46.3	48.3	49.4	54.0	49.9	47.0	45.9	46.8	46.2	45.9	43.23	
Bayesian F	43.2	44.7	39.7	41.0	42.4	45.1	42.1	41.6	45.0	43.6	48.3	44.6	51.4	46.5	43.1	44.7	45.7	45.2	44.6	43.72	
Combining A	9.5	36.9	31.3	31.7	36.2	35.4	35.1	40.2	41.7	39.5	43.6	43.9	47.3	41.7	36.6	41.5	38.7	41.5	41.7	38.40	
Combining B	19.3	41.3	38.6	39.3	44.3	42.4	41.3	40.0	43.2	42.3	42.6	43.1	49.6	42.3	38.9	41.3	40.7	40.8	42.8	41.80	
Average	25.9	45.5	43.0	43.5	46.6	44.1	45.2	45.4	47.2	47.0	48.8	49.6	55.1	49.6	47.3	49.6	47.7	48.3	48.9	46.88	

TABLE 6(b)   Percentage of Time that the Naive 1 Method is Better than Other Methods ($n = 111$)

METHODS	MODEL FITTING	1	2	3	4	5	6	7	8	9	10	11	12	13	14	15	16	17	18	AVERAGE OF ALL FORECASTS	n(max)
								Forecasting Horizons													
Mov.Averag	31.5	50.5	45.0	47.7	49.1	50.5	50.5	46.2	44.0	41.2	48.5	41.2	62.5	49.3	41.2	44.9	40.4	36.0	44.1	46.66	
Single EXP	23.4	53.2	45.9	45.9	54.5	49.1	50.9	36.4	40.1	33.8	41.9	36.8	49.3	50.7	33.1	43.4	43.3	34.6	41.2	44.50	
ARR EXP	68.5	55.4	55.9	52.3	56.8	50.9	57.7	38.5	42.9	41.9	50.0	50.0	64.7	55.9	47.1	51.5	51.5	54.4	57.4	52.06	
Holt EXP	16.2	43.2	39.6	43.2	46.8	43.2	45.9	48.4	45.1	51.5	58.8	54.4	66.2	51.5	50.0	52.9	50.0	52.9	55.9	48.82	
Brown EXP	23.4	46.8	41.4	39.6	43.2	42.3	43.2	46.2	45.1	38.2	48.5	42.6	64.7	55.9	44.1	57.4	51.5	51.5	45.6	46.34	
Quad.EXP	30.6	49.5	45.9	48.6	44.1	54.1	47.7	51.6	57.1	58.8	57.4	60.3	67.6	67.6	66.2	66.2	65.2	63.2	61.8	55.82	
Regression	48.6	64.0	55.9	50.5	51.4	45.0	53.2	47.3	44.0	52.9	54.4	54.4	72.1	55.9	48.5	51.5	47.1	44.1	48.5	52.23	
NAIVE2	24.8	40.5	41.4	39.2	46.8	38.7	42.8	42.9	45.1	39.7	44.9	48.5	51.1	48.5	39.7	51.5	43.4	44.4	47.8	43.36	
D Mov.Avrg	20.7	44.6	46.8	48.2	55.9	52.7	53.2	48.9	49.5	50.0	50.7	52.9	63.2	47.8	40.4	41.2	52.9	51.5	47.6	50.07	
D Sing EXP	13.5	43.2	44.1	35.6	49.1	42.3	45.0	35.2	38.5	35.3	41.2	39.0	48.5	44.1	32.4	33.1	40.4	36.0	42.6	40.71	
D ARR EXP	45.9	49.1	55.9	43.2	57.7	50.5	35.7	38.5	38.5	39.7	45.6	54.4	50.0	44.1	44.1	41.1	45.6	48.5	47.1	47.41	
D Holt EXP	6.3	35.1	33.3	27.0	38.7	32.4	37.8	44.0	45.1	44.1	52.9	44.1	54.4	44.1	44.1	41.2	44.1	48.5	45.6	40.58	
D brownEXP	10.8	37.8	36.9	29.7	36.9	28.8	38.7	41.8	40.7	41.2	44.1	45.6	50.0	51.5	38.2	41.2	44.1	44.1	45.6	39.92	
D Quad.EXP	15.3	42.3	38.7	34.2	36.9	37.8	41.4	49.5	53.8	52.9	53.4	52.9	63.2	58.8	52.9	54.4	57.4	55.9	54.4	47.77	
D Regress	36.0	55.9	45.0	46.8	45.9	41.4	46.8	47.3	45.6	54.4	54.4	50.0	54.4	44.1	41.2	45.6	47.1	52.9	46.5	47.32	
WINTERS	9.0	41.4	34.2	29.7	39.6	32.4	36.0	41.8	44.0	41.2	58.8	50.0	55.9	48.5	35.3	33.8	47.1	45.6	44.1	41.10	
Autom. AEP	9.9	40.5	30.6	36.0	39.6	36.0	39.6	37.4	37.3	42.6	47.1	47.1	54.4	48.5	35.3	29.4	39.7	39.7	41.2	39.79	
Bayesian F	43.2	44.1	40.5	36.0	36.0	36.9	39.6	42.9	47.3	42.6	50.0	42.6	51.5	42.6	44.1	50.0	44.1	47.1	47.1	42.67	
Combining A	10.8	36.0	29.7	21.6	32.4	29.7	31.5	38.5	40.7	32.4	44.1	41.2	41.2	41.2	25.0	35.3	32.4	39.7	41.2	34.49	
Combining B	16.2	41.4	38.7	25.7	43.2	36.0	38.7	37.4	39.6	39.7	44.1	42.6	51.5	43.4	32.4	34.6	36.0	36.8	43.4	38.84	
Box-Jenkins	N.A	41.4	31.5	31.5	39.6	29.7	36.9	37.4	37.4	36.8	51.5	55.9	58.8	42.6	42.6	44.1	44.1	44.1	43.4	40.71	
Lewandowski	42.3	48.6	39.6	34.2	35.1	28.4	36.9	35.2	35.2	36.8	44.1	39.0	45.6	39.7	32.4	35.3	33.8	30.9	33.8	36.91	
Parzen	9.9	45.9	29.7	36.0	36.0	30.6	35.1	33.0	31.9	44.1	52.9	42.6	48.5	52.9	36.8	36.8	45.5	41.2	44.1	39.20	
Average	28.6	45.7	41.2	38.4	44.2	39.8	43.4	41.8	43.0	42.7	49.8	46.7	56.1	50.1	41.0	43.4	46.0	45.2	46.8	44.23	

TABLE 7(a)   Percentage of Time that the Naive 2 Method is Better than Other Methods (n = 1001)

METHODS	MODEL FITTING	\[Forecasting Horizons\] 1	2	3	4	5	6	7	8	9	10	11	12	13	14	15	16	17	18	AVERAGE OF ALL FORECASTS	n(max)
NAIVE 1	74.6	57.7	59.5	59.0	54.5	55.3	56.5	57.4	54.6	55.3	55.8	53.8	50.1	57.5	61.4	55.5	57.4	56.6	58.0	56.54	
Mov.Averag	66.3	58.0	60.1	56.5	56.0	55.3	54.6	54.3	52.0	51.1	54.4	52.3	54.8	56.5	58.9	51.9	55.3	54.5	55.5	55.35	
Single EXP	61.2	61.1	61.4	58.1	58.4	57.3	56.3	51.7	52.1	49.6	52.0	51.1	52.9	54.1	59.2	50.6	54.3	52.7	53.8	55.40	
ARR EXP	89.2	63.0	70.2	62.0	65.4	58.4	61.9	53.6	55.1	51.3	56.3	57.2	58.9	57.9	61.5	56.7	57.8	55.8	55.7	59.46	
Holt EXP	53.9	53.3	50.2	51.0	49.1	48.6	50.8	58.0	53.3	51.3	58.3	58.5	56.6	59.6	63.5	59.2	59.0	57.9	60.0	54.91	
Brown EXP	59.9	51.2	50.9	50.9	49.3	51.0	53.7	59.6	56.5	58.3	58.0	57.7	59.3	61.4	65.8	59.6	60.0	59.8	62.4	56.06	
Quad.EXP	63.3	53.9	54.6	57.3	55.0	57.4	60.5	67.3	65.5	63.5	65.2	66.5	68.7	74.1	75.2	72.1	71.2	72.6	76.0	63.93	
Regression	75.7	68.8	62.4	58.9	54.4	51.0	51.2	55.0	52.7	58.0	57.7	59.5	57.7	60.5	61.8	54.5	54.5	53.8	54.5	57.09	
D Mov.Avrg	31.0	54.8	56.3	57.3	58.4	55.8	56.4	57.1	53.2	53.4	53.4	54.7	57.8	52.8	56.2	56.2	55.2	55.2	53.6	55.43	
D Sing EXP	22.3	52.0	51.4	53.1	51.4	51.5	52.8	46.5	46.2	45.6	45.9	46.5	49.3	47.1	49.3	46.1	47.9	47.0	47.6	49.21	
D ARR EXP	66.7	56.0	61.6	58.1	59.3	54.1	58.5	50.7	51.6	51.0	51.8	51.1	53.9	53.6	53.1	52.4	51.3	50.4	52.1	54.49	
D Holt EXP	13.7	46.0	41.5	42.3	43.1	42.7	46.6	50.2	50.2	51.4	51.8	51.4	52.0	54.0	54.2	52.4	52.4	54.0	54.0	48.39	
D brownEXP	22.1	44.1	42.8	44.9	44.1	43.7	47.2	51.1	50.5	49.9	53.0	51.9	55.4	55.9	55.9	54.8	55.4	54.1	56.9	49.60	
D Quad.EXP	30.5	47.3	47.2	51.0	50.0	52.4	55.4	60.0	60.6	59.8	60.8	61.8	65.0	66.5	65.6	67.4	68.7	71.2	71.3	58.53	
D Regress	59.1	65.5	56.9	55.1	51.1	46.3	48.8	51.7	50.6	53.0	53.8	53.0	53.8	53.2	52.6	52.2	49.3	51.2	50.1	52.77	
WINTERS	28.1	44.7	42.3	43.0	43.2	41.8	46.1	48.4	48.8	51.1	52.5	51.2	52.8	53.2	52.0	52.2	52.7	52.5	53.3	48.07	
Autom. AEP	45.2	45.7	45.7	46.3	44.1	46.3	47.0	51.3	49.6	51.8	54.6	52.7	54.0	57.4	57.9	52.7	54.3	52.4	55.1	50.16	
Bayesian F	78.2	50.4	45.5	47.3	46.4	45.3	47.0	47.4	49.5	48.9	48.9	46.2	51.4	51.5	52.7	49.1	50.9	49.3	48.9	48.40	
Combining A	19.4	41.9	38.3	39.2	40.0	39.7	42.5	44.3	43.3	45.1	46.8	46.2	47.3	47.5	46.7	45.9	44.1	45.2	46.5	43.28	
Combining B	29.0	47.7	46.5	47.6	46.2	46.2	48.2	46.7	46.8	46.8	46.0	47.5	49.6	47.5	49.3	48.3	45.9	45.5	48.5	47.20	
Average	49.5	53.2	52.3	52.0	51.0	50.0	52.1	53.1	52.1	52.5	53.8	53.6	55.1	56.0	57.0	54.2	54.9	54.5	55.7	53.21	

TABLE 7(b) Percentage of Time that the Naive 2 Method is Better than Other Methods ($n = 111$)

METHODS	MODEL FITING	1	2	3	4	5	6	7	8	9	10	11	12	13	14	15	16	17	18	AVERAGE OF ALL FORECASTS	n(max)
NAIVE 1	75.2	59.5	58.6	60.8	53.2	61.3	57.5	57.1	54.9	60.3	55.1	55.9	48.5	51.5	60.3	58.1	52.2	56.6	52.2	56.64	
Mov.Averg	72.1	64.0	58.1	56.8	53.2	65.3	58.6	61.0	54.4	54.4	56.6	45.6	61.0	50.7	54.4	52.9	45.6	51.5	49.3	55.96	
Single EXP	64.0	62.2	59.9	57.2	58.1	63.3	55.4	53.8	51.1	49.3	50.7	46.3	47.8	52.9	51.5	47.8	44.9	47.8	45.6	53.66	
ARR EXP	93.7	67.6	68.5	66.2	63.1	64.0	60.4	52.7	58.2	52.9	58.8	54.4	64.7	57.4	63.2	60.3	57.4	58.8	57.4	60.96	
Holt EXP	55.9	55.0	49.5	51.4	51.4	56.8	51.4	62.6	56.0	55.9	60.3	58.8	66.2	52.9	57.4	58.8	57.4	58.8	51.5	55.69	
Brown EXP	60.4	55.9	54.1	54.1	45.9	56.8	48.6	56.6	54.9	58.8	51.5	50.0	64.7	52.9	55.9	58.8	57.4	51.5	52.9	54.19	
Quad.EXP	64.9	59.5	55.0	58.6	53.2	62.2	54.1	61.5	59.3	63.2	57.4	63.2	67.6	70.6	72.1	58.8	57.4	60.3	64.7	62.11	
Regression	76.6	73.6	59.0	63.1	59.5	60.4	59.5	61.5	59.3	61.8	57.4	61.8	61.8	54.4	50.0	58.8	55.9	60.3	54.4	61.91	
D Mov.Avrg	32.4	53.6	59.9	59.0	59.5	52.7	54.1	43.4	42.9	39.7	41.9	44.9	47.1	44.1	36.8	32.4	41.9	44.9	43.4	58.08	
D Sing EXP	23.4	50.9	53.6	47.3	49.1	52.7	54.1	43.4	47.3	39.0	47.1	44.9	54.4	39.7	50.0	45.6	32.4	44.9	43.4	46.04	
D ARR EXP	67.6	57.7	62.2	56.8	58.0	55.0	56.8	44.5	47.3	47.1	47.1	54.4	52.9	48.5	54.4	47.1	48.5	47.1	51.5	52.26	
D Holt EXP	11.7	40.5	42.3	39.6	37.8	42.3	39.6	52.7	47.3	47.1	45.6	44.1	52.9	39.7	48.5	45.6	45.6	45.6	47.1	44.11	
D brownEXP	16.2	43.2	45.9	41.4	37.8	39.6	45.0	44.0	46.2	41.2	41.2	54.4	63.2	48.5	47.1	47.1	58.8	52.9	52.9	45.03	
D Quad.EXP	27.0	47.7	48.6	41.4	45.0	45.0	55.9	54.9	53.8	41.2	54.4	54.4	63.2	63.2	63.2	64.7	67.6	64.7	64.7	53.21	
D Regress	53.2	64.9	55.9	57.7	52.3	48.6	54.1	49.5	54.9	45.6	48.5	51.5	54.4	39.7	47.1	44.1	47.1	51.5	50.0	51.77	
WINTERS	28.8	45.9	39.6	36.9	41.4	40.5	38.7	41.8	48.4	48.5	55.4	52.9	55.9	47.1	50.0	42.6	52.9	48.5	41.2	45.03	
Autom. AEP	46.8	45.0	45.0	41.4	43.2	43.2	45.0	46.2	50.5	48.5	55.9	52.9	51.5	51.5	52.9	51.5	45.6	48.5	47.1	47.71	
Bavesian F	85.6	54.1	48.6	45.9	40.5	41.4	43.2	46.2	54.9	42.6	47.1	48.5	51.5	35.3	51.5	44.1	52.9	48.5	45.6	46.40	
Combining A	22.5	37.8	39.6	32.4	31.5	36.0	37.8	38.5	40.7	36.8	38.2	41.2	41.2	42.6	29.4	36.8	41.2	42.6	36.8	37.57	
Combining B	24.3	51.4	48.6	39.6	41.4	43.2	44.6	42.1	46.7	45.2	42.6	51.5	51.5	44.1	41.9	37.5	39.0	46.3	42.6	44.54	
Box-Jenkins	N.A	53.2	43.2	38.7	40.5	38.7	45.0	45.1	46.2	41.2	51.5	55.9	57.0	44.1	51.5	39.7	50.0	47.1	51.5	46.66	
Lewandowski	73.0	58.6	50.5	44.1	42.3	39.6	37.8	41.8	38.2	38.2	44.1	36.8	45.6	44.1	36.8	38.2	38.2	41.2	32.4	42.80	
Parzen	47.7	57.3	43.2	42.3	38.7	45.9	44.1	40.7	42.9	48.5	45.6	48.5	48.5	50.0	45.6	39.7	48.5	47.1	50.0	45.35	
Average	53.2	54.5	51.9	49.2	47.3	50.7	49.4	50.6	51.7	49.1	50.7	50.8	55.8	50.6	51.2	48.4	51.1	52.0	49.6	50.77	

TABLE 8 Percentage of Time that the Box–Jenkins Method is Better than Other Methods ($n = 111$)

METHODS	MODEL FITTING	Forecasting Horizons																		AVERAGE OF ALL FORECASTS	n(max)
		1	2	3	4	5	6	7	8	9	10	11	12	13	14	15	16	17	18		
NAIVE 1	N.A	58.6	68.5	68.5	60.4	70.3	64.0	62.6	62.6	63.2	48.5	44.1	42.6	41.2	57.4	63.2	55.9	55.9	55.9	59.29	
Mov.Averaj	N.A	56.8	61.0	63.1	62.2	72.1	67.6	62.6	65.9	58.8	51.5	47.1	52.9	41.2	58.8	60.3	52.9	51.5	54.4	59.23	
Single EXP	N.A	55.0	59.5	64.9	64.0	66.7	66.7	60.4	62.6	55.9	51.5	50.0	51.5	41.2	57.4	61.8	54.4	57.4	54.4	58.51	
ARR EXP	N.A	57.7	63.1	66.7	64.3	65.8	69.4	59.1	58.8	58.8	50.0	45.6	51.5	47.1	51.5	61.8	60.3	57.4	51.5	59.10	
Holt EXP	N.A	47.7	55.9	54.1	52.3	56.8	53.2	57.1	58.2	58.8	51.5	58.8	54.4	55.9	55.9	61.8	54.4	54.4	45.6	54.12	
Brown EXP	N.A	50.5	59.9	59.5	53.2	60.4	53.2	59.3	58.2	60.3	50.0	55.9	54.4	48.5	50.0	58.8	51.5	54.4	51.5	55.27	
Quad.EXP	N.A	52.3	63.1	65.8	55.9	66.7	62.2	68.1	68.1	61.8	58.8	63.2	66.2	60.3	72.1	70.6	69.1	60.3	69.1	63.68	
Regression	N.A	64.0	68.5	72.1	65.8	67.6	66.7	63.7	67.0	61.8	58.8	66.2	61.8	55.9	61.8	61.8	69.1	51.5	50.0	63.15	
NAIVE2	N.A	46.8	56.8	61.3	59.5	61.3	55.0	54.9	53.8	58.8	48.5	44.1	42.6	42.6	48.5	60.3	50.0	52.9	48.5	53.34	
D Mov.Avrj	N.A	55.0	61.3	69.4	65.8	71.2	69.4	65.9	65.9	60.3	51.5	58.8	54.4	47.1	52.9	55.9	60.3	64.7	54.4	61.32	
D Sing EXP	N.A	45.0	52.3	62.2	56.8	60.4	59.5	50.5	53.8	55.9	41.2	45.6	54.4	44.1	44.1	50.0	50.0	52.9	51.5	52.29	
D ARR EXP	N.A	49.5	57.7	62.2	59.5	64.9	60.4	49.5	58.2	52.9	42.6	41.2	47.1	41.2	45.6	57.4	48.5	54.4	55.9	53.80	
D Holt EXP	N.A	39.6	45.5	45.0	47.7	45.0	45.0	53.8	47.3	55.9	51.5	50.0	48.5	45.6	38.2	61.8	52.9	54.4	52.9	48.33	
D brownEXP	N.A	44.1	43.7	49.5	46.8	47.7	52.3	52.7	56.0	47.1	52.9	54.4	54.4	51.5	47.1	55.9	54.4	55.9	45.6	50.29	
D Quad.EXP	N.A	43.2	49.5	55.9	48.6	56.8	54.1	57.1	61.5	61.8	57.4	55.5	58.8	58.8	54.4	66.2	64.7	61.8	60.3	56.15	
D Regress	N.A	59.5	55.0	65.8	64.0	56.8	61.3	58.2	57.1	51.5	52.9	55.9	51.5	50.0	47.1	60.3	47.1	52.9	52.9	56.41	
WINTERS	N.A	50.5	44.1	48.6	45.0	45.9	46.8	51.6	46.2	50.0	39.7	50.0	48.5	45.6	36.8	50.0	52.9	50.0	41.2	46.92	
Autom. AEP	N.A	45.9	54.1	50.5	50.0	52.3	54.9	54.9	58.8	48.5	48.5	48.5	45.6	44.1	51.5	54.4	47.1	50.0	61.8	52.03	
Bayesian F	N.A	51.4	54.1	56.8	54.1	51.4	54.1	54.9	60.4	52.9	51.5	45.6	47.1	39.7	41.2	48.5	50.0	42.6	45.6	50.92	
Combining A	N.A	39.6	45.9	48.6	48.6	54.1	53.2	53.8	49.5	52.9	39.7	39.7	45.6	39.7	38.2	55.9	48.5	44.1	45.6	47.25	
Combining R	N.A	41.4	49.5	50.5	55.9	49.5	52.3	49.5	51.6	55.9	36.8	42.6	48.5	44.1	44.1	55.9	50.0	54.4	51.5	49.28	
Lewandowski	N.A	56.8	55.9	61.3	55.0	52.3	50.5	39.6	45.1	52.9	44.1	39.7	45.6	47.1	36.8	47.1	45.6	36.8	38.2	48.43	
Parzen	N.A	49.5	45.0	49.5	45.9	54.1	50.5	50.5	48.4	52.9	48.5	39.7	42.6	57.4	48.5	45.6	39.7	45.6	50.0	48.23	
Average	N.A	50.5	55.2	58.9	55.7	58.8	57.3	56.3	57.2	56.5	49.0	49.7	50.5	46.9	49.6	57.8	52.7	52.9	51.8	54.23	

TABLE 9(a)   Percentage of Time that the Winters Method is Better than Other Methods ($n = 1001$)

| METHODS | MODEL FITTING | forecasting horizons | | | | | | | | | | | | | | | | | | AVERAGE OF ALL FORECASTS | n(max) |

METHODS	MODEL FITTING	1	2	3	4	5	6	7	8	9	10	11	12	13	14	15	16	17	18	AVERAGE OF ALL FORECASTS	n(max)
NAIVE 1	90.0	61.5	65.6	63.4	60.0	62.7	60.6	56.6	54.6	52.2	51.5	51.5	47.2	50.7	56.7	53.0	55.4	53.2	53.6	57.15	
Mov.Averag	84.5	61.6	66.5	63.0	61.5	59.6	58.7	53.9	53.9	50.1	50.2	51.2	50.6	50.1	54.5	49.9	53.0	49.6	48.5	55.96	
Single EXP	81.9	62.3	63.8	61.4	60.0	60.3	58.1	52.1	52.5	48.9	48.8	50.1	49.1	49.4	53.5	50.9	52.7	51.4	50.7	55.31	
ARR EXP	97.3	64.2	70.2	64.0	63.7	60.6	59.7	51.1	52.8	49.1	51.1	52.2	53.6	53.3	56.4	53.3	51.4	52.5	49.6	57.22	
Holt EXP	72.9	56.7	59.2	57.9	54.2	55.1	55.8	57.6	54.5	55.7	58.8	52.4	59.2	56.5	61.8	58.6	55.5	54.9	58.4	56.70	
Brown EXP	83.9	56.9	56.3	56.3	55.6	55.5	57.5	58.5	55.7	55.9	57.7	59.3	61.3	64.7	60.3	60.3	60.5	59.5	63.5	58.31	
Quad.EXP	88.0	61.2	63.6	63.2	60.9	63.2	66.5	69.4	65.0	65.8	66.3	66.8	70.2	71.5	74.4	73.3	72.1	74.1	76.8	67.78	
Regression	86.5	70.4	71.6	66.7	65.8	60.1	60.3	56.5	57.4	55.8	60.3	59.0	55.6	58.5	63.5	56.9	55.4	54.6	55.9	60.86	
NAIVE2	71.9	55.3	57.7	57.0	56.8	58.2	53.9	51.1	51.2	48.9	47.5	48.8	47.2	46.8	48.0	47.8	47.3	47.5	46.7	51.93	
D Mov.Avrg	50.4	60.4	64.8	63.0	62.2	61.8	57.7	56.1	53.0	51.1	50.4	51.5	50.4	48.0	50.7	51.5	53.2	50.9	50.1	55.98	
D Sing EXP	47.3	54.4	57.5	56.0	55.8	53.6	49.0	47.8	44.4	44.4	46.0	45.2	48.0	45.4	44.7	45.1	53.8	46.5	45.5	50.23	
D ARR EXP	79.3	59.0	63.9	61.2	59.2	58.2	56.5	52.6	49.9	48.0	49.1	48.6	49.3	47.6	48.9	49.3	47.5	48.0	47.5	53.64	
D Holt EXP	28.6	49.5	50.2	50.0	50.0	50.0	49.8	51.0	50.4	48.7	50.6	49.4	51.5	53.7	52.4	52.9	50.2	51.9	51.9	50.58	
D BrownEXP	43.0	51.1	51.1	50.8	52.8	52.8	51.3	53.0	51.1	52.5	55.9	55.9	53.1	55.9	55.1	54.6	55.4	55.4	57.1	52.65	
D Quad.EXP	52.4	54.0	57.1	57.9	58.7	59.3	60.7	65.4	64.6	62.7	64.2	64.2	67.7	69.0	67.4	70.0	69.5	71.3	72.8	63.25	
D Regress	68.2	68.5	63.4	63.4	59.1	57.2	57.1	55.9	54.9	54.0	56.7	53.4	56.2	53.0	55.9	54.6	51.5	53.0	50.2	57.39	
Autom. AEP	69.0	60.8	54.5	55.1	50.5	49.2	49.2	51.6	48.6	48.6	51.4	53.4	50.4	52.5	53.0	52.7	52.7	53.2	53.8	51.74	
Bayesian F	98.2	57.9	52.1	53.7	49.8	51.1	52.4	52.4	52.0	47.8	49.6	50.1	52.8	53.8	54.8	52.5	53.8	51.5	51.5	52.59	
Combining A	41.8	50.3	51.7	50.7	49.8	48.8	50.8	44.0	42.6	40.4	42.1	40.8	44.6	44.4	44.0	44.7	43.9	42.5	43.1	46.24	
Combining B	44.4	51.1	55.5	53.8	52.6	52.9	52.4	47.3	46.2	43.6	43.8	44.7	46.8	45.7	44.2	46.0	43.9	46.5	46.0	48.79	
Average	69.0	57.9	60.0	58.5	56.9	56.7	56.2	54.3	53.2	51.1	52.5	52.7	52.8	53.5	55.3	53.8	53.4	53.4	53.7	55.21	

TABLE 9(b) Percentage of Time that the Winters Method is Better than Other Methods ($n = 111$)

METHODS	MODEL FITTING	Forecasting Horizons 1	2	3	4	5	6	7	8	9	10	11	12	13	14	15	16	17	18	AVERAGE OF ALL FORECASTS	n(max)
NAIVE 1	91.0	58.6	65.8	70.3	60.4	67.6	64.0	59.2	56.0	58.8	41.2	50.0	44.1	51.5	64.7	66.2	52.9	54.4	55.9	58.90	
Mov.Averag	91.0	55.0	65.4	64.9	60.4	64.0	63.1	52.7	53.8	48.5	47.1	47.1	50.0	51.5	55.9	61.8	48.5	50.0	50.0	56.15	
Single EXP	86.5	57.7	61.3	63.1	63.1	63.1	62.2	53.8	51.6	48.5	47.1	45.6	47.1	52.9	52.9	60.3	50.0	61.8	51.5	56.22	
ARR EXP	98.2	56.8	64.9	69.4	64.7	60.4	62.2	49.5	50.5	54.4	51.5	45.6	54.4	60.3	51.5	60.3	50.0	63.2	52.9	57.66	
Holt EXP	73.4	51.8	61.7	59.0	57.2	60.8	60.8	61.0	55.5	58.1	55.1	56.6	49.3	61.8	64.0	69.1	57.8	55.1	58.1	57.26	
Brown EXP	82.9	56.8	62.2	60.4	61.3	68.5	64.0	60.4	57.1	61.8	58.8	52.9	57.4	61.8	64.7	75.0	70.6	67.6	61.8	61.13	
Quad.EXP	89.3	57.7	63.1	61.3	60.4	67.5	65.8	70.3	72.5	72.1	61.8	72.1	70.6	70.6	72.1	61.8	54.4	67.6	76.5	67.34	
Regression	87.4	72.1	69.4	65.8	66.7	71.1	71.2	58.2	60.4	58.8	64.7	60.3	58.8	58.8	67.6	61.8	47.1	55.9	52.9	63.48	
NAIVE2	71.2	54.1	60.4	63.1	58.5	59.5	61.3	58.2	51.6	51.5	45.6	47.1	44.1	52.9	50.0	50.0	50.0	51.5	58.9	54.97	
D Mov.Avrg	54.1	56.8	64.9	66.7	65.8	65.8	65.8	59.3	58.2	52.9	55.9	55.9	50.0	57.4	51.5	55.9	44.1	50.0	60.3	59.49	
D Sing EXP	50.5	44.1	55.9	55.9	54.1	56.8	56.8	46.2	46.2	44.1	48.5	41.2	48.5	52.9	44.1	48.5	44.1	52.9	50.0	50.85	
D ARR EXP	79.3	51.4	60.4	64.0	60.4	60.4	64.0	56.0	46.2	52.9	52.9	48.5	50.0	57.4	48.5	58.8	47.1	52.9	61.8	55.96	
D Holt EXP	26.6	37.4	47.3	50.0	40.1	48.2	50.9	57.7	53.3	52.2	47.8	49.3	50.7	41.9	52.2	53.7	43.4	52.2	53.7	48.63	
D BrownEXP	35.1	45.0	50.5	47.6	45.9	53.7	56.8	59.3	61.5	55.5	55.9	54.4	52.9	61.8	61.8	63.2	51.5	54.4	61.8	53.99	
D Quad.Exp	46.8	43.2	55.0	66.7	50.5	53.2	61.5	61.5	61.5	66.2	64.7	60.3	69.1	69.1	69.1	73.5	69.1	70.6	69.1	59.42	
D Regress	62.2	63.1	62.2	66.7	65.8	57.7	64.9	57.1	54.9	52.9	54.4	54.4	57.4	50.0	58.8	61.8	45.6	63.2	55.9	58.97	
Autor.AEP	71.2	47.7	55.0	58.6	50.5	55.9	50.5	51.6	50.5	52.9	47.1	41.2	50.0	47.1	55.9	51.5	47.1	52.9	61.8	51.77	
Bavesian F	99.1	53.2	55.0	56.8	48.6	45.9	55.0	57.1	62.6	51.5	44.1	44.1	52.9	39.7	57.4	54.4	51.5	54.4	54.4	52.36	
Combining A	40.5	41.4	52.3	45.9	47.7	50.5	54.1	49.5	47.3	39.7	45.6	35.3	44.1	44.1	41.2	55.9	45.6	42.6	52.9	46.86	
Combining B	45.0	42.3	53.2	54.1	54.1	53.2	57.7	49.5	47.3	45.6	44.1	42.6	51.5	57.4	44.1	50.0	48.5	52.9	55.9	50.52	
BoxJenkins	N.A	49.5	55.9	51.4	55.0	54.1	51.4	42.9	53.8	50.0	60.3	50.0	51.5	54.4	63.2	50.0	47.1	50.0	58.8	53.08	
Lewandowski	90.1	60.4	61.3	52.3	49.5	51.4	51.4	49.5	45.1	51.5	36.8	38.2	45.6	42.6	38.2	48.5	38.2	42.6	42.6	47.84	
Parzen	64.0	49.5	51.4	53.2	47.7	55.0	53.2	49.5	47.3	51.5	55.9	42.6	50.0	51.5	57.4	48.5	45.6	52.9	54.4	50.98	
Average	71.1	52.4	58.9	58.9	56.1	58.0	59.4	55.1	53.8	53.4	51.6	49.4	52.2	53.8	55.9	58.7	50.1	55.4	57.2	55.38	

EFFECTS OF SAMPLING

Can the results of this study be generalized? Surprisingly, not much is known about the sampling distribution of actual post-sample forecasting errors. Furthermore, not much is known about the relative desirability of different accuracy measures for the purpose of comparing various forecasting methods.

The five accuracy measures reported in this study (i.e. MAPE, MSE, AR, Md and PB) are not exhaustive. Average Percentage Errors (APE), Mean Absolute Deviations (MAD), Mean Root Square Errors (MRSE), and other accuracy measures could have been used (APE and MAD have been computed but are not reported in this paper).

Having to report the accuracy measures for both the 1001 and 111 series is a disadvantage because it increases the length of the paper and the time and effort required to read it. The advantage, however, is that the reader can examine how each of the five accuracy measures differs between the 111 and all 1001 series for the 21 methods that are reported on both the 111 *and* 1001 series. Although the 111 series is only a part of the 1001, much can be learned by looking at the (a) and (b) parts of Tables 2 to 9 and seeing how the various accuracy measures vary among the (a) and (b) parts.

In general, the MSE fluctuates much more than the other measures, whereas Md, PB and AR fluctuate the least with MAPE somewhere in between. For instance, the overall average MSE of the Automatic AEP method is one of the best for all 1001 series and one of the worst for the 111 series. On the other hand, the other four measures are more consistent between the (a) and (b) parts of the tables.

In order to obtain a more precise idea of sampling variations, Table 10 shows the behaviour of five measures for nine systematic samples from the 1001 series for a single method, chosen arbitrarily: Holt–Winters exponential smoothing. It is not difficult to see that the variations in the results from the nine different samples are relatively smaller for MAPE than for MSE while the average rankings and the percentage better measures seem to fluctuate the least.

Would the results of the systematic samples for the Holt–Winters method vary more if other data were used? To deal with this type of question, Table 11 compares the percentage of times the Box–Jenkins methodology was better than other methods used, both in the present study and in that reported in JRSS. (The entries for Table 11 have been taken from Table 7 of the present study and Table 6 on p. 108 of the JRSS paper.) The results do vary, as can be expected, but for most methods they are similar, in particular for the overall average.

SOME GENERAL OBSERVATIONS

The performance of various methods differs considerably sometimes, depending upon the accuracy measure (criterion) being used. Parzen and Holt–Winters are

TABLE 10

AVERAGE MAPE: ALL DATA (n=111)

METHODS	MODEL FITTING	1	2	3	4	5	6	8	12	15	18	1-4	1-6	1-8	1-12	1-15	1-18	n(max)
							Forecasting Horizons						Average of Forecasting Horizons					
Winter 1	9.3	9.0	13.2	11.6	13.7	17.8	23.2	24.0	12.4	16.9	20.7	11.9	14.8	16.5	16.0	16.1	16.5	110
Winter 2	9.3	9.2	10.5	13.4	15.5	17.5	18.7	23.3	15.9	33.4	34.5	12.1	14.1	16.3	16.4	17.8	19.5	111
Winter 3	8.4	8.8	10.5	14.4	12.0	15.9	24.5	21.3	23.5	27.4	34.6	11.4	14.3	15.6	16.7	17.5	18.7	111
Winter 4	12.7	8.6	11.8	12.6	16.5	18.4	22.3	24.5	21.3	28.6	35.9	12.4	15.0	17.3	20.9	22.3	23.7	111
Winter 5	9.9	9.3	11.4	16.4	17.0	22.7	22.0	26.5	17.1	22.2	39.4	13.6	16.5	17.2	16.9	17.6	19.9	110
Winter 6	8.4	8.0	10.0	12.7	16.6	18.8	19.6	26.5	28.8	37.6	62.4	11.8	14.3	16.4	19.1	21.8	25.4	111
Winter 7	7.6	6.9	7.8	10.7	12.8	14.4	16.3	21.6	33.9	33.9	67.1	9.5	11.6	12.4	14.6	16.8	21.1	110
Winter 8	8.9	7.4	9.8	12.3	12.0	25.0	25.4	37.6	32.5	54.3	86.4	10.4	15.3	19.1	22.1	24.9	30.7	111
Winter 9	9.3	10.3	13.5	14.8	17.1	19.8	21.0	18.8	27.5	38.6	41.3	13.9	16.1	16.7	19.0	21.8	24.0	111
Average	9.3	8.7	10.9	13.2	14.8	18.9	21.4	24.0	22.7	32.5	46.9	11.9	14.7	16.4	18.0	19.6	22.2	

AVERAGE MSE: ALL DATA (n=111)

METHODS	MODEL FITTING	1	2	3	4	5	6	8	12	15	18	n(max)
					Forecasting Horizons							
Winter 1	.6965E+08	.1308E+09	.1545E+09	.1328E+10	.4857E+10	.6196E+08	.7221E+07	.1790E+08	.4660E+09	.7891E+09	.5805E+09	110
Winter 2	.1567E+08	.2158E+08	.2538E+08	.2128E+08	.9347E+08	.2842E+08	.8271E+08	.1004E+09	.9443E+09	.9478E+08	.9593E+08	111
Winter 3	.2266E+11	.4191E+12	.6340E+12	.6227E+12	.4665E+13	.1536E+10	.3300E+09	.1215E+09	.8086E+12	.9095E+12	.6667E+12	111
Winter 4	.1685E+11	.1620E+11	.3461E+11	.2198E+10	.2994E+11	.2972E+09	.4083E+08	.1008E+08	.1329E+11	.8644E+10	.6340E+10	111
Winter 5	.1398E+10	.1242E+11	.1835E+10	.5398E+12	.1016E+11	.1496E+11	.7458E+07	.7050E+08	.4367E+10	.3416E+10	.2512E+10	110
Winter 6	.3906E+11	.3016E+10	.2187E+10	.1961E+12	.1208E+12	.8029E+09	.1250E+10	.2285E+10	.5337E+10	.3818E+11	.2847E+11	111
Winter 7	.1908E+11	.1215E+11	.1158E+11	.1047E+11	.7350E+11	.4576E+11	.6269E+07	.3138E+08	.9038E+10	.1226E+11	.8980E+10	110
Winter 8	.6696E+10	.9842E+10	.4911E+10	.1212E+11	.5011E+11	.1761E+08	.1381E+08	.7739E+07	.7952E+10	.1066E+11	.7816E+10	111
Winter 9	.1047E+08	.2636E+08	.7233E+08	.3985E+08	.2921E+08	.8246E+08	.9330E+08	.3955E+08	.3409E+08	.3409E+08	.4136E+08	111
Average	.1176E+11	.5108E+11	.7660E+11	.9429E+11	.5384E+12	.4853E+09	.1944E+09	.3042E+09	.9969E+11	.1093E+12	.8017E+11	

e.g. .6965E+08 = $.6965 \times 10^8$

TABLE 10—*cont.*

AVERAGE RANKING : ALL DATA (111)

METHODS	MODEL FITTING	1	2	3	4	5	6	7	8	9	10	11	12	13	14	15	16	17	18	AVERAGE OF ALL FORECASTS	n(max)
Winter 1	5.2	5.2	5.3	5.1	5.1	5.1	5.1	5.1	5.0	4.8	4.9	5.2	5.2	4.9	5.1	5.1	5.0	4.8	5.0	5.06	111
Winter 2	5.0	5.2	5.0	4.8	4.8	5.0	4.8	5.0	5.0	4.8	4.9	5.2	4.9	5.0	4.9	5.0	5.0	4.8	5.1	4.99	111
Winter 3	4.6	4.7	4.8	4.8	4.6	4.7	4.9	4.7	4.7	4.4	4.3	4.1	4.4	4.2	4.4	4.4	4.3	4.3	4.6	4.55	111
Winter 4	4.9	4.7	4.8	4.9	4.9	4.9	4.9	5.1	5.1	5.0	5.1	5.2	5.1	4.8	4.8	4.8	4.7	5.0	4.9	4.95	111
Winter 5	5.2	5.1	5.0	5.1	5.1	5.1	5.1	4.9	5.1	5.2	5.1	4.8	4.7	4.8	4.9	4.8	4.7	5.0	4.7	4.98	111
Winter 6	5.5	5.3	5.4	5.5	5.4	5.6	5.5	5.5	5.7	5.8	6.1	5.9	5.7	5.8	5.9	6.1	5.9	5.9	6.0	5.67	111
Winter 7	5.2	5.3	5.4	5.5	5.5	5.6	5.6	5.6	5.8	5.3	5.2	5.0	5.2	5.0	5.2	5.2	5.1	5.1	5.0	5.23	111
Winter 8	4.7	4.8	4.7	5.1	5.3	4.5	4.5	4.6	4.9	4.9	4.7	5.0	5.2	5.0	4.6	4.8	5.1	5.1	4.9	4.70	111
Winter 9	4.7	4.8	4.7	4.9	4.9	4.8	4.7	4.9	4.7	4.9	5.1	5.0	4.9	5.1	5.1	4.8	5.0	5.1	4.9	4.88	111
Average	5.0	5.0	5.0	5.0	5.0	5.0	5.0	5.0	5.0	5.0	5.0	5.0	5.0	5.0	5.0	5.0	5.0	5.0	5.0	5.00	

AVERAGE MEDIANS : ALL DATA (N=111)

METHODS	MODEL FITTING	FORECASTING HORIZONS														CUMULATIVE FORECASTING HORIZONS						N(MAX)
		1	2	3	4	5	6	7	8	9	10	11	12	15	18	1-4	1-6	1-8	1-12	1-15	1-18	
WINTER 1	0.0	4.5	5.6	6.6	8.1	8.3	9.5	10.7	8.6	9.5	10.7	13.2	13.2	6.3	7.7	7.9	9.1	8.2	10.8	12.3		111
WINTER 2	0.0	4.6	6.8	6.7	8.8	9.2	11.3	14.3	11.0	10.3	14.8	13.0	13.0	7.6	7.0	8.0	9.0	9.1	10.5	13.2		111
WINTER 3	0.0	4.4	6.4	6.9	7.5	7.2	8.1	8.3	10.0	8.7	8.0	10.3	10.3	5.1	6.6	6.1	6.4	7.3	8.0	10.3		111
WINTER 4	0.0	4.1	4.8	5.8	7.0	6.6	9.1	10.0	10.0	9.4	10.0	9.5	10.5	6.0	8.4	7.5	7.0	8.5	9.8	10.7		111
WINTER 5	0.0	4.2	6.0	6.0	6.2	8.1	8.6	11.6	11.5	9.4	11.6	11.5	10.5	5.0	6.1	6.4	6.9	6.5	7.6	10.3		111
WINTER 6	0.0	4.4	6.3	6.9	6.6	6.5	9.4	17.6	17.6	11.9	17.6	18.6	18.0	6.0	6.4	6.4	6.0	6.4	6.9	18.0		111
WINTER 7	0.0	4.5	5.1	6.5	7.3	9.3	11.9	12.9	12.9	12.0	11.8	18.0	18.0	6.9	7.2	7.0	7.2	6.9	7.6	19.8		111
WINTER 8	0.0	3.6	4.3	5.7	6.6	7.4	8.5	10.2	8.2	8.5	10.2	11.8	9.8	6.1	6.1	7.4	6.6	7.4	7.7	9.8		111
WINTER 9	0.0	3.6	4.3	5.7	6.6	7.4	8.5	10.2	8.2	8.5	10.2	11.8	9.8	6.1	6.1	7.4	6.6	7.4	7.7	9.8		111

PERCENTAGE OF TIME THAT METHOD Winter 1 IS BETTER THAN OTHER METHODS (n=111)

METHODS	MODEL FITTING	1	2	3	4	5	6	7	8	9	10	11	12	13	14	15	16	17	18	AVERAGE OF ALL FORECASTS	n(max)
Winter 2	45.0	46.8	40.5	47.7	48.6	46.8	47.7	51.6	57.1	54.3	51.4	51.4	45.7	55.8	48.6	50.0	52.9	52.9	50.0	49.55	
Winter 3	43.2	46.8	46.8	42.3	41.4	45.0	49.5	45.1	47.3	42.8	45.7	38.4	42.8	42.8	42.8	41.3	42.8	47.1	50.0	44.68	
Winter 4	43.2	43.2	47.7	48.6	51.4	45.0	54.1	49.5	54.3	54.3	54.3	49.5	48.6	48.6	47.1	45.7	54.3	51.4	51.4	48.90	
Winter 5	51.4	47.7	53.2	46.8	48.6	51.4	54.1	51.6	51.6	53.6	42.0	43.5	42.0	50.7	47.8	49.3	44.9	49.3	43.5	48.83	
Winter 6	53.2	48.6	47.7	59.5	52.3	50.5	55.9	52.7	54.9	60.4	66.7	52.2	50.7	53.6	56.5	50.7	56.5	62.3	56.5	54.81	
Winter 7	50.5	47.7	49.5	53.2	49.5	50.5	55.1	45.1	55.1	55.1	52.2	55.1	50.7	53.6	46.4	50.7	55.1	52.2	55.1	50.91	
Winter 8	39.6	43.2	45.0	44.1	40.5	46.8	42.3	44.6	45.1	52.2	43.5	43.5	47.8	49.3	46.4	44.9	52.2	50.7	47.8	45.84	
Winter 9	49.5	47.7	41.4	50.5	52.3	47.7	46.8	50.0	48.9	50.7	55.1	50.7	49.3	53.6	55.1	47.8	56.5	55.1	49.3	50.00	
Average	47.0	46.5	46.5	49.1	48.2	49.1	49.7	49.7	49.9	53.0	51.4	47.6	47.2	51.2	48.8	48.3	51.9	52.6	50.5	49.19	

TABLE 11   Percentage of Times that the Box–Jenkins Method was Better than Methods Listed for the Study Reported in JRSS and the Present Competition

Forecasting Method	Forecasting Horizons [1]																Simple Average	
	1		2		3		4		5		6		9		12			
	JRSS	Now	JRSS	Now	JRSS	Now	JRSS	Now	JRSS	Now	JRSS	Now	JRSS	Now	JRSS	Now	JRSS	Now
1. Naive 1	59.1	58.6	63.6	68.5	57.3	68.5	63.6	60.4	64.5	70.3	59.1	64.0	63.6	63.2	67.3	42.6	62.26	62.01
2. Single Mov. Average	49.1	64.9	56.4	66.7	53.6	71.2	60.9	68.5	64.5	66.7	61.8	64.9	65.5	50.0	66.4	54.4	59.77	63.41
3. Single Exp. Smoothing	53.6	55.0	57.3	59.5	55.5	64.9	60.0	64.0	62.7	66.7	62.7	66.7	65.5	55.9	66.4	51.5	60.46	60.52
4. Adapt.	49.1	57.7	56.4	63.1	56.4	66.7	60.0	64.0	62.7	65.8	64.5	69.4	62.7	58.8	67.3	51.5	59.88	62.12
5. Linear Mov. Average	NA	NA	NA	NA	NA	NA	NA	NA	NA	NA	NA	NA	NA	NA	NA	NA	NA	NA
6. Brown's	42.7	50.5	51.8	59.9	50.0	59.5	53.6	53.2	59.1	60.4	59.1	53.2	67.3	60.3	70.9	54.4	56.81	56.42
7. Holt's	50.0	47.7	55.5	55.9	50.0	54.1	54.5	52.3	62.7	56.8	61.8	53.2	62.7	58.8	66.4	51.5	57.95	53.79
8. Brown's Quadr.	48.2	52.3	51.8	63.1	50.9	65.8	57.3	55.9	61.8	66.7	63.6	62.2	65.5	61.8	73.6	66.2	59.09	61.75
9. Linear Trend	60.9	64.0	61.8	68.5	63.6	72.1	65.5	65.8	69.1	67.6	68.2	66.7	71.8	61.8	72.7	61.8	66.7	66.04
10. Harrison's	NA	NA	NA	NA	NA	NA	NA	NA	NA	NA	NA	NA	NA	NA	NA	NA	NA	NA
11. Winters'	47.3	50.5	49.1	44.1	51.8	48.6	50.0	45.0	50.0	45.9	50.9	46.8	46.4	50.0	45.5	48.5	48.88	47.43
12. Adaptive Filt. [2]	49.1	45.9	54.5	52.3	52.7	54.1	55.5	50.5	56.4	55.0	56.4	52.3	58.2	58.8	51.8	45.6	54.33	51.81
13. Naive 2	53.6	46.8	46.4	56.8	44.5	61.3	42.7	59.5	48.2	61.3	41.8	55.0	38.2	58.8	39.1	42.6	44.31	55.26
14. Single Mov. Average	46.4	70.3	42.7	70.3	43.6	72.1	39.1	67.6	42.7	70.3	39.1	65.8	37.3	58.8	38.2	42.6	41.14	64.73
15. Single Exp. Smoothing	50.0	45.0	42.7	52.3	42.7	62.2	38.2	56.8	43.6	60.4	39.1	59.5	41.8	55.9	40.0	48.5	42.26	55.07
16. Adaptive Resp.	44.5	49.5	46.4	57.7	42.7	62.2	40.9	59.5	47.3	64.9	48.2	60.4	44.5	52.9	41.8	47.1	44.54	56.77
17. Linear Mov. Average	NA	NA															NA	NA
18. Brown's Linear	42.7	44.1	48.2	43.7	44.5	49.5	44.5	46.8	49.1	47.7	47.3	52.3	49.1	47.1	46.4	54.4	46.48	48.2
19. Holt's Linear	41.8	39.6	43.6	45.5	42.7	45.0	40.0	47.7	46.4	45.9	44.5	45.0	46.4	55.9	45.5	48.5	43.86	46.64
20. Brown's Quadrat.	41.8	43.2	44.5	49.5	44.5	55.9	45.5	48.6	47.3	56.8	47.3	54.1	50.9	61.8	59.1	58.8	47.61	53.59
21. Linear Trend	63.6	59.5	59.1	55.0	62.7	65.8	60.0	64.0	59.1	56.8	57.3	61.3	53.6	51.5	50.0	51.5	58.18	58.17

[1]. No comparisons for model fitting were made because no model fitting forecasts were provided for Box-Jenkins in the present study.

[2]. The method of A.E.P. Filtering used in the current study is somewhat different (see description of A.E.P.)

two methods which exhibit a higher degree of consistency among most of the five accuracy measures than the remaining methods.

Differences among methods were influenced by differences in the type of series used and the length of the forecasting horizon. These differences are discussed next mainly within the subset of the 111 series.

Effects of the type of series

The relative forecasting accuracy of the various methods was affected significantly by (a) the yearly, quarterly or monthly nature of data; (b) the micro, macro classification; and (c) whether the data were seasonal or not. Thus, while some methods (e.g. deseasonalised single exponential smoothing) perform well for monthly data, they may do badly for, say, yearly data. Tables 12, 13, 14, 15, 16, 17 and 18 show the MAPE for yearly, quarterly, monthly, micro, macro, non-seasonal and seasonal data. Tables 19, 20, 21, 22, 23, 24 and 25 do so for the average rankings, whereas Tables 26, 27, 28, 29, 30, 31, and 32 do so for the medians.

It is to be expected that methods which do not take trend into account will not do as well as methods which do for data subject to substantial trends (e.g. yearly). Single exponential smoothing does not do very well therefore, whereas Holt or Holt–Winters (for yearly data the two are equivalent) and Lewandowski do the best. Single exponential smoothing is progressively worse as the time horizon increases, precisely because it does not take trend into account. Bayesian forecasting and the Box–Jenkins method do about the same as single exponential smoothing (the reason could be that the trend is over-extended in the forecasting). For monthly data, deseasonalized single exponential smoothing does relatively better than Holt–Winters, Automatic AEP, Bayesian forecasting, Box–Jenkins and Lewandowski.

The most striking differences are between micro and macro data (see Tables 15, 16, 22 and 23). In micro data the simple methods do much better than the statistically sophisticated methodologies, which, in turn, are at their best with macro data. For instance, the overall MAPE for Lewandowski is 13.7% for micro and 18.2% for macro, whereas that of Parzen is 18.4% for micro and 11.2% for macro. Even for the small number of series in each category (33 micro and 35 macro) these differences are significant.

Finally, it is interesting to note that for seasonal data, deseasonalized single and adaptive response rate exponential smoothing, deseasonalized regression, Bayesian forecasting and Parzen do about the same as far as overall MAPE is concerned. For non-seasonal data the MAPEs are much more spread out as sophisticated methods do relatively better than with seasonal data. Furthermore, the differences in overall average ranking for the various methods are even more pronounced for non-seasonal data, whereas they (excluding non-seasonal methods) are very small for seasonal data.

TABLE 12   Average MAPE: Yearly Data (20)

METHODS	MODEL FITTING	Forecasting Horizons										Average of Forecasting Horizons						n(max)
		1	2	3	4	5	6	8	12	15	18	1-4	1-6	1-8	1-12	1-15	1-18	
NAIVE 1	10.9	6.8	9.7	16.6	21.1	23.8	24.8	0.0	0.0	0.0	0.0	13.6	17.1	17.1	17.1	17.1	17.1	20
Mov.Averag	10.7	8.6	10.9	17.7	21.9	24.7	26.0	0.0	0.0	0.0	0.0	14.8	18.3	18.3	18.3	18.3	18.3	20
Single EXP	11.4	6.2	9.1	16.3	21.0	23.6	25.4	0.0	0.0	0.0	0.0	13.1	16.9	16.9	16.9	16.9	16.9	20
ARR EXP	13.4	7.8	13.7	17.7	24.4	25.3	29.3	0.0	0.0	0.0	0.0	15.9	19.7	19.7	19.7	19.7	19.7	20
Holt EXP	12.9	5.6	7.2	11.9	16.2	19.0	16.5	0.0	0.0	0.0	0.0	10.2	12.7	12.7	12.7	12.7	12.7	20
Brown EXP	10.8	6.7	8.2	12.0	16.5	19.8	16.4	0.0	0.0	0.0	0.0	10.8	13.3	13.3	13.3	13.3	13.3	20
Quad.EXP	10.6	7.0	8.6	11.8	16.0	20.7	17.4	0.0	0.0	0.0	0.0	10.9	13.6	13.6	13.6	13.6	13.6	20
Regression	8.5	6.9	7.8	14.9	18.4	20.0	20.6	0.0	0.0	0.0	0.0	12.0	14.8	14.8	14.8	14.8	14.8	19
NAIVE2	10.9	6.8	9.7	16.6	21.1	23.8	24.8	0.0	0.0	0.0	0.0	13.6	17.1	17.1	17.1	17.1	17.1	20
D Mov.Avrg	10.7	8.6	10.9	17.7	21.9	24.7	26.0	0.0	0.0	0.0	0.0	14.8	18.3	18.3	18.3	18.3	18.3	20
D Sing EXP	11.4	6.2	9.1	16.3	21.0	23.6	25.4	0.0	0.0	0.0	0.0	13.1	16.9	16.9	16.9	16.9	16.9	20
D ARR EXP	13.4	7.8	13.7	17.7	24.4	25.3	29.3	0.0	0.0	0.0	0.0	15.9	19.7	19.7	19.7	19.7	19.7	20
D Holt EXP	12.9	5.6	8.3	12.0	16.5	19.0	16.5	0.0	0.0	0.0	0.0	10.8	12.7	12.7	12.7	12.7	12.7	20
D brownEXP	10.8	6.7	8.6	11.8	16.0	19.8	16.4	0.0	0.0	0.0	0.0	10.8	13.3	13.3	13.3	13.3	13.3	20
D Quad.EXP	10.6	7.0	8.6	11.9	18.4	20.7	20.6	0.0	0.0	0.0	0.0	12.0	13.6	13.6	13.6	13.6	13.6	20
D Regress	8.6	6.9	7.2	14.9	16.2	20.0	16.5	0.0	0.0	0.0	0.0	10.2	14.8	14.8	14.8	14.8	14.8	19
WINTERS	12.9	5.6	7.2	11.9	17.8	19.0	19.1	0.0	0.0	0.0	0.0	11.9	12.7	12.7	12.7	12.7	12.7	19
Autom. AEP	8.7	7.1	8.8	14.1	18.0	21.8	20.6	0.0	0.0	0.0	0.0	11.9	14.8	14.8	14.8	14.8	14.8	20
Bavesian F.	20.2	12.2	12.6	14.9	17.4	20.6	17.8	0.0	0.0	0.0	0.0	11.4	16.5	16.5	16.5	16.5	16.5	20
Combining A	8.4	5.7	7.7	12.5	17.4	20.0	20.1	0.0	0.0	0.0	0.0	10.8	13.5	13.5	13.5	13.5	13.5	20
Combining B	9.0	6.3	8.3	13.7	17.5	19.7	20.1	0.0	0.0	0.0	0.0	11.5	14.3	14.3	14.3	14.3	14.3	20
Box-Jenkins	N.A	7.2	10.8	13.7	18.6	23.2	22.3	0.0	0.0	0.0	0.0	12.6	16.0	16.0	16.0	16.0	16.0	20
Lewandowski	20.1	7.3	8.3	14.7	13.8	16.8	15.1	0.0	0.0	0.0	0.0	12.7	12.7	12.7	12.7	12.7	12.7	20
Parzen	9.6	7.6	7.7	12.8	16.0	20.5	18.0	0.0	0.0	0.0	0.0	11.0	13.8	13.8	13.8	13.8	13.8	20
Average	11.1	7.1	9.2	14.4	18.6	21.5	20.9	0.0	0.0	0.0	0.0	12.3	15.3	15.3	15.3	15.3	15.3	

TABLE 13   Average MAPE: Quarterly Data (23)

METHODS	MODEL FITTING	Forecasting Horizons										Average of Forecasting Horizons						n(max)
		1	2	3	4	5	6	8	12	15	18	1-4	1-6	1-8	1-12	1-15	1-18	
NAIVE 1	11.3	8.5	11.2	19.2	21.5	25.6	25.2	23.3	0.0	0.0	0.0	15.1	18.5	20.9	20.9	20.9	20.9	23
Mov.Averag	9.3	16.3	18.9	24.3	28.5	32.9	32.6	31.0	0.0	0.0	0.0	22.0	25.6	28.1	28.1	28.1	28.1	23
Single Exp	9.6	10.0	11.7	16.8	21.3	24.6	24.2	23.2	0.0	0.0	0.0	14.9	18.3	20.6	20.6	20.6	20.6	23
ARR Exp	11.4	13.6	16.7	19.3	22.9	26.2	24.6	24.1	0.0	0.0	0.0	18.1	20.6	22.4	22.4	22.4	22.4	23
Holt Exp	9.1	10.0	10.3	21.1	25.9	34.3	37.3	41.9	0.0	0.0	0.0	16.8	23.1	29.0	29.0	29.0	29.0	23
Brown Exp	9.3	11.3	10.0	19.6	26.6	36.1	39.3	44.9	0.0	0.0	0.0	16.9	23.8	30.1	30.1	30.1	30.1	23
Quad.Exp	9.4	12.3	13.3	26.2	36.7	49.2	58.0	79.3	0.0	0.0	0.0	22.1	32.6	43.7	43.7	43.7	43.7	23
Regression	12.4	20.1	20.2	22.2	25.9	31.0	25.2	23.9	0.0	0.0	0.0	22.1	24.1	25.5	25.5	25.5	25.5	23
NAIVE2	9.0	7.6	12.0	15.8	21.5	23.1	22.3	23.3	0.0	0.0	0.0	14.2	16.9	19.0	19.0	19.0	19.0	23
D Mov.Avrg	7.7	14.4	18.3	23.2	27.4	30.8	31.3	29.5	0.0	0.0	0.0	20.8	24.2	26.5	26.5	26.5	26.5	23
D Sing Exp	7.7	9.0	12.0	14.4	20.5	21.0	21.9	22.6	0.0	0.0	0.0	14.0	16.5	18.5	18.5	18.5	18.5	23
D ARR Exp	9.6	12.3	16.8	18.2	25.0	25.3	24.3	26.0	0.0	0.0	0.0	18.1	20.7	22.2	22.2	22.2	22.2	23
D Holt Exp	7.2	9.2	10.4	17.1	25.1	30.3	32.2	39.2	0.0	0.0	0.0	14.5	19.3	24.0	24.0	24.0	24.0	23
D brownExp	7.3	10.0	10.4	15.1	22.5	27.1	30.5	36.5	0.0	0.0	0.0	15.1	21.6	24.8	24.8	24.8	24.8	23
D Quad.Exp	7.6	11.1	12.5	21.1	32.0	39.2	46.0	66.6	0.0	0.0	0.0	19.2	27.0	35.6	35.6	35.6	35.6	23
D Regres	11.5	18.1	21.0	21.6	26.3	28.6	32.0	40.3	0.0	0.0	0.0	22.0	23.5	24.8	24.8	24.8	24.8	23
WINTERS	7.3	8.9	9.1	17.1	25.6	32.6	34.7	40.2	0.0	0.0	0.0	15.2	21.3	26.4	26.4	26.4	26.4	23
Autom. AEP	16.5	12.7	18.6	15.4	22.4	27.2	28.8	28.8	0.0	0.0	0.0	17.3	20.9	25.9	25.9	25.9	25.9	23
Bavesian F	10.4	8.3	8.6	20.4	24.7	27.8	26.4	31.0	0.0	0.0	0.0	15.4	19.5	24.6	24.6	24.6	24.6	23
Combining A	8.8	8.0	8.0	11.7	19.4	24.4	26.1	28.8	0.0	0.0	0.0	11.8	16.3	20.7	20.7	20.7	20.7	23
Combining B	10.3	8.5	10.1	13.9	23.6	26.7	27.7	33.5	0.0	0.0	0.0	14.0	18.4	22.4	22.4	22.4	22.4	23
Box-Jenkins	N.A	7.6	8.2	13.9	21.3	26.1	26.1	25.4	0.0	0.0	0.0	12.7	17.2	20.1	20.1	20.1	20.1	23
Lewandovski	9.8	12.5	14.1	14.2	21.8	24.8	22.8	26.9	0.0	0.0	0.0	15.7	18.4	20.6	20.6	20.6	20.6	23
Parzen	7.7	6.8	7.6	12.0	16.5	21.1	20.4	21.0	0.0	0.0	0.0	10.7	14.1	16.7	16.7	16.7	16.7	23
Average	9.2	11.1	12.9	18.1	24.4	23.0	29.0	33.6	0.0	0.0	0.0	16.6	20.9	24.8	24.8	24.8	24.8	

TABLE 14   Average MAPE: Monthly Data (68)

METHODS	MODEL FITTING	Forecasting Horizons										Average of Forecasting Horizons						n(max)
		1	2	3	4	5	6	8	12	15	18	1-4	1-6	1-8	1-12	1-15	1-18	
NAIVE 1	16.5	16.7	21.6	21.4	16.9	21.0	22.6	28.3	14.5	31.9	34.9	19.1	20.0	21.4	20.0	21.4	23.0	68
Mov.Averag	14.5	15.0	18.0	17.7	14.8	17.2	19.9	21.4	16.3	28.7	31.9	16.4	17.1	17.7	16.9	18.3	19.9	68
Single EXP	15.0	14.7	17.6	17.9	15.4	17.5	20.8	22.5	16.1	28.8	32.5	16.4	17.3	18.2	17.2	18.7	20.3	68
ARR EXP	16.8	14.4	18.2	18.4	14.9	17.5	20.3	21.9	16.1	29.6	32.2	16.5	17.3	18.0	17.2	18.7	20.3	68
Holt EXP	15.3	14.9	17.1	18.1	17.8	20.5	23.3	27.5	22.6	40.4	40.3	17.0	18.6	20.3	20.8	23.7	25.8	68
Brown EXP	15.8	15.4	18.8	20.2	20.8	23.1	26.2	31.7	28.0	54.0	59.6	18.1	20.3	22.5	23.6	27.8	32.0	68
Quad.EXP	16.4	15.4	19.2	23.4	20.4	26.8	31.0	42.1	49.0	103.0	106.0	19.7	22.8	26.5	31.6	43.0	51.7	68
Regression	20.3	20.2	23.1	22.5	20.4	21.4	26.2	27.0	26.1	49.5	60.2	21.6	22.3	23.2	23.4	26.6	31.4	68
NAIVE2	8.6	9.2	11.7	12.4	11.7	12.6	13.5	16.0	14.5	28.3	30.8	11.3	11.8	13.0	13.7	15.8	17.7	68
D Mov.Avrg	7.5	10.1	12.8	16.0	11.7	18.2	18.2	20.4	15.7	28.3	34.0	13.0	15.4	16.6	16.6	17.8	20.0	68
D Sing EXP	8.0	7.9	10.9	11.7	10.6	11.6	13.2	14.4	13.6	29.3	30.1	10.9	11.0	12.0	12.6	14.5	16.5	68
D ARR EXP	8.8	8.2	10.5	11.5	11.4	11.3	12.7	13.3	13.7	28.6	29.3	10.3	11.8	11.6	12.8	14.2	16.1	68
D Holt EXP	7.9	7.9	11.5	13.0	11.2	13.3	16.4	19.6	19.0	35.6	35.2	10.2	12.3	14.8	15.5	17.2	19.5	68
D brownEXP	8.0	8.2	11.6	13.8	11.6	14.8	18.7	25.3	29.7	43.1	45.4	11.1	13.7	14.6	16.0	19.5	22.9	68
D Quad.EXP	8.0	8.6	12.5	16.0	12.1	16.3	16.3	18.7	29.7	56.1	63.6	11.7	15.0	16.6	20.4	25.7	31.0	68
D Regress	13.1	12.2	12.2	12.5	11.8	11.9	14.8	19.6	17.0	46.5	57.3	14.7	15.2	18.1	18.1	21.4	26.7	68
WINTERS	9.0	10.3	12.0	12.0	11.8	11.2	13.4	17.5	15.9	33.4	34.5	11.7	12.2	14.6	14.6	16.8	19.1	68
Autom. AEP	9.5	11.2	12.8	13.0	11.9	11.2	12.8	17.6	16.2	30.2	33.9	12.2	12.0	14.2	14.2	16.1	18.4	68
Bayesian F	12.3	12.8	10.9	10.8	10.4	10.4	12.4	16.0	16.1	27.5	30.6	12.2	12.7	14.2	13.1	14.5	16.6	68
Combining A	7.8	8.4	11.1	11.8	11.1	10.2	11.5	15.6	14.2	32.4	33.3	10.4	11.0	13.1	13.0	15.3	17.6	68
Combining B	7.5	8.6	10.7	10.6	10.8	11.0	12.5	15.6	15.5	31.3	31.4	10.2	11.3	13.0	13.8	15.3	17.4	68
Box-Jenkins	N.A	12.1	11.5	9.9	11.1	11.0	12.6	16.7	16.4	26.4	34.2	11.1	12.7	13.8	13.8	15.6	17.9	68
Lewandowski	10.8	12.6	13.6	14.6	13.5	13.5	13.6	16.2	17.0	33.0	28.6	13.6	14.1	14.9	14.9	17.1	18.9	68
Parzen	9.0	12.7	12.6	9.6	11.7	10.2	11.8	14.3	13.7	22.5	26.5	11.7	11.4	12.6	12.6	13.9	15.4	68
Average	11.1	11.8	14.4	15.0	13.7	15.3	17.7	20.7	19.2	37.5	40.7	13.7	14.7	16.0	16.8	19.5	22.3	68

TABLE 15   Average MAPE: Micro Data (33)

METHODS	MODEL FITTING	Forecasting Horizons										Average of Forecasting Horizons						n(max)
		1	2	3	4	5	6	8	12	15	18	1-4	1-6	1-8	1-12	1-15	1-18	
NAIVE 1	19.2	17.7	25.4	31.3	25.9	33.4	25.0	17.8	19.7	38.9	31.3	25.1	26.5	25.6	24.5	25.7	26.8	33
Mov.Averag	18.1	15.0	19.9	24.7	21.6	28.3	20.7	10.8	22.0	31.1	24.5	20.3	21.7	20.7	19.8	20.5	21.4	33
Single EXP	18.4	14.8	19.0	24.0	22.7	28.5	21.6	10.7	21.1	30.1	25.9	20.4	21.8	20.6	19.8	20.5	21.5	33
ARR EXP	20.3	15.0	20.3	24.9	21.2	23.0	20.7	9.8	21.0	28.2	22.9	20.4	21.5	20.2	19.2	19.6	20.4	33
Holt EXP	20.1	15.8	20.2	26.6	22.6	32.0	22.1	15.8	22.0	41.9	33.1	21.3	23.2	23.0	22.3	23.6	25.5	33
Brown EXP	19.5	18.1	22.9	27.6	22.6	35.5	24.0	16.9	25.6	49.5	43.0	22.8	25.1	24.8	24.3	26.6	29.5	33
Quad.EXP	19.9	18.5	23.6	31.0	25.6	39.1	28.7	29.3	41.5	82.2	85.5	24.7	27.7	28.5	30.5	36.3	43.1	32
Regression	20.3	17.7	20.1	25.5	21.1	27.3	19.6	12.4	18.9	28.5	17.8	21.1	21.9	21.2	19.7	19.8	20.2	33
NAIVE2	14.3	12.4	19.7	17.9	17.0	22.9	19.4	14.9	19.7	28.8	26.3	16.8	18.2	18.5	18.5	19.6	20.4	33
D Mov.Avrg	13.0	12.6	17.6	20.2	18.1	25.6	22.2	17.4	17.4	29.2	32.0	17.1	19.4	20.2	20.1	21.0	22.2	33
D Sing EXP	13.6	11.8	17.6	15.8	15.1	19.9	18.0	11.5	17.4	24.7	25.3	15.1	16.4	16.5	16.3	17.2	18.0	33
D ARR EXP	15.2	11.6	17.4	15.1	14.9	19.5	17.5	10.9	18.3	23.3	23.6	14.8	16.0	16.0	16.0	16.7	17.3	33
D Holt EXP	15.1	12.8	18.8	18.9	15.4	23.0	18.4	15.0	17.6	33.3	34.0	16.4	17.9	19.0	18.6	19.3	21.7	33
D brownExp	13.9	14.2	19.4	18.0	15.4	25.1	21.7	18.4	22.7	41.7	46.0	17.1	19.2	20.4	21.0	23.4	26.3	32
D Quad.EXP	14.2	14.8	21.5	20.0	16.0	29.8	26.6	29.6	39.4	69.4	82.7	18.1	21.5	23.8	27.0	32.5	39.0	32
D Regress	15.6	11.9	17.1	17.8	15.9	20.8	19.1	15.0	17.6	29.3	34.5	15.7	16.6	17.0	18.9	20.0	21.7	33
WINTERS	15.8	14.9	19.7	16.7	16.3	23.0	19.1	12.2	13.2	29.2	30.1	17.4	18.6	18.1	18.4	20.0	20.9	33
Autom. AEP	14.0	18.0	19.9	17.1	16.5	20.5	18.8	18.2	21.8	36.4	37.4	16.3	17.4	18.1	18.4	19.6	20.9	33
Bavesian F	18.8	13.7	18.2	17.1	15.0	21.4	18.0	18.0	22.6	29.6	30.7	16.3	17.4	18.4	18.9	20.7	22.7	33
Combining A	12.3	13.3	16.6	16.6	14.8	16.6	17.0	10.3	16.1	24.2	22.4	16.1	16.9	17.4	17.2	18.3	19.8	33
Combining B	11.6	12.9	16.9	13.5	14.8	16.6	13.9	10.9	13.6	25.9	28.7	14.5	14.8	14.9	15.0	16.0	16.6	33
Box-Jenkins	N/A	17.1	18.8	14.1	17.4	22.7	14.5	12.6	18.6	24.2	13.7	17.1	17.6	17.7	17.7	18.9	20.2	33
Lewandowski	20.3	13.2	14.4	12.1	11.6	13.9	14.1	10.9	12.6	17.8	13.7	12.8	13.1	13.5	13.7	13.8	13.7	33
Parzen	14.8	16.9	21.4	14.1	16.7	18.1	14.1	11.9	17.2	26.2	27.5	17.8	17.2	17.0	16.8	17.6	18.4	33
Average	15.8	14.9	19.6	20.1	18.1	24.8	19.5	15.0	20.9	34.3	33.4	18.2	19.5	19.7	19.6	21.0	22.7	

TABLE 16 Average MAPE: Macro Data (35)

METHODS	MODEL FITTING	Forecasting Horizons										Average of Forecasting Horizons						n(max)
		1	2	3	4	5	6	8	12	15	18	1-4	1-6	1-8	1-12	1-15	1-18	
NAIVE 1	7.7	6.8	10.5	10.5	12.8	14.9	15.4	16.2	13.1	32.1	33.8	10.1	11.8	12.8	13.0	14.5	15.9	35
Mov.Averag	6.2	8.5	12.4	11.8	15.8	17.3	18.6	16.6	14.9	31.5	33.4	12.1	14.1	14.7	14.3	15.5	16.7	35
Single EXP	6.5	5.9	8.0	9.5	13.2	14.6	16.1	13.7	15.6	32.0	33.8	9.6	11.5	12.1	12.4	13.9	15.3	35
ARR EXP	8.1	5.5	11.0	10.1	14.6	14.6	17.8	15.9	17.3	36.7	38.5	10.3	12.3	13.1	13.5	15.3	17.1	35
Holt EXP	5.8	5.2	8.0	6.7	10.0	11.9	13.3	14.0	16.6	32.3	34.0	7.5	9.2	10.3	10.9	12.7	14.3	35
Brown EXP	6.2	5.5	8.5	8.1	11.8	14.2	15.8	19.0	30.2	64.8	79.6	8.5	10.7	12.4	14.9	19.4	24.3	35
Quad.EXP	6.3	4.9	7.8	7.0	10.9	13.4	15.5	21.4	24.9	60.7	86.2	7.6	9.9	12.1	13.6	17.8	23.3	35
Regression	8.8	9.8	12.8	11.4	14.4	15.0	15.8	16.1	16.1	29.5	30.4	12.1	13.2	13.8	14.0	15.3	16.3	35
NAIVE2	4.9	4.1	7.4	11.4	13.7	13.2	16.1	14.0	13.1	52.2	39.2	9.2	11.0	11.7	12.5	15.3	17.4	35
D Mov.Avrg	4.4	4.2	10.1	11.4	10.8	13.2	16.4	13.9	17.7	34.7	34.0	12.2	11.4	15.0	12.9	16.0	17.6	35
D Sing EXP	4.4	4.3	7.6	9.5	11.4	13.6	16.4	14.0	12.9	34.0	39.1	9.3	11.1	11.8	12.5	15.3	17.4	35
D ARR EXP	5.5	4.8	8.9	11.1	15.0	13.3	17.3	14.2	13.0	52.0	37.3	10.0	11.8	12.3	12.8	15.3	17.2	35
D Holt EXP	3.6	3.4	5.8	8.4	10.9	10.2	13.6	16.0	12.7	48.8	45.8	7.0	8.7	10.1	11.9	15.2	18.0	35
D brownEXP	3.8	3.6	5.8	8.5	11.2	10.9	14.0	15.9	15.6	71.4	53.9	8.9	8.9	10.3	11.9	16.3	19.6	35
D Quad.EXP	3.9	3.5	5.9	8.5	11.2	10.9	15.1	20.3	12.1	56.5	42.8	7.3	9.2	11.3	11.9	15.2	17.6	35
D Regress	7.1	9.3	9.1	10.9	13.1	13.1	15.1	13.7	13.1	34.9	25.1	10.6	11.9	12.2	12.4	14.0	14.9	35
WINTERS	4.1	5.2	5.2	9.0	11.8	11.1	13.6	15.7	14.6	68.7	54.4	7.8	9.3	10.8	12.9	15.7	19.4	35
Autom. AEP	9.8	5.9	6.0	8.3	10.8	10.8	11.1	14.2	14.2	48.6	37.4	7.8	8.8	9.8	11.2	13.8	16.2	35
Bayesian F	8.6	6.2	7.7	9.4	10.6	12.3	13.5	12.2	13.2	31.5	22.7	8.5	8.9	9.8	10.6	12.1	12.9	35
Combining A	4.7	4.8	6.8	10.1	12.1	11.2	13.4	12.5	12.4	57.9	43.0	7.5	8.9	9.0	11.1	14.3	16.8	35
Combining B	5.8	6.0	6.8	8.8	11.1	10.3	12.4	16.0	16.7	50.9	38.6	8.5	9.7	10.4	11.6	14.1	16.1	35
Box-Jenkins	7.8	6.6	6.7	6.5	10.5	11.8	16.0	14.4	14.4	36.3	35.5	7.2	8.6	10.0	11.5	13.5	15.3	35
Lewandowski	6.3	4.9	10.9	11.5	14.0	14.0	15.8	15.8	16.1	61.0	36.6	10.7	11.8	12.0	12.0	16.4	18.2	35
Parzen	4.0	4.6	4.6	5.5	9.0	9.6	9.2	10.8	10.9	26.5	25.3	6.0	7.1	7.9	8.4	9.9	11.2	35
Average	5.7	5.7	8.1	9.5	12.5	12.7	14.9	15.3	15.3	46.5	40.8	8.9	10.6	11.5	12.3	14.9	17.0	35

TABLE 17   Average MAPE: Seasonal Data (60)

METHODS	MODEL FITTING	Forecasting Horizons										Average of Forecasting Horizons						n(max)
		1	2	3	4	5	6	8	12	15	18	1-4	1-6	1-8	1-12	1-15	1-18	
NAIVE 1	17.8	16.1	20.7	21.6	19.3	22.1	24.8	32.2	13.0	30.7	31.2	19.5	20.8	22.9	21.3	22.2	22.9	60
Mov.Averag	15.2	14.7	17.5	17.3	17.2	18.5	22.4	25.1	15.6	27.0	27.8	16.7	17.9	19.1	18.0	18.9	19.5	60
Single EXP	15.7	14.1	17.2	17.8	19.1	19.1	23.1	26.3	15.6	27.1	28.7	16.7	18.2	19.6	18.2	19.3	20.0	60
ARR EXP	17.5	14.7	19.0	18.1	16.1	18.2	21.7	25.0	15.4	28.3	28.8	17.0	18.0	19.2	18.2	19.2	20.0	60
Holt EXP	16.0	14.5	16.1	18.3	20.6	21.9	26.2	31.8	23.8	43.3	40.3	17.4	19.6	22.0	22.3	25.0	26.8	60
Brown EXP	16.7	14.8	17.5	19.7	21.5	25.1	29.3	36.6	27.5	53.3	54.7	18.4	21.5	24.3	24.7	28.1	31.1	60
Quad.EXP	17.2	14.3	17.3	22.9	24.7	28.4	34.6	47.8	46.3	103.1	93.7	19.8	23.7	28.4	31.7	41.9	48.5	60
Regression	18.6	20.8	21.7	18.7	18.1	18.6	21.7	23.0	16.1	27.5	26.5	19.8	19.9	20.6	19.4	20.3	20.6	60
NAIVE2	7.9	7.3	9.9	10.1	13.5	11.4	13.3	18.3	16.1	29.8	25.7	10.2	10.9	12.6	13.3	14.9	16.0	60
D Mov.Avrg	6.7	8.4	11.3	14.9	18.2	18.8	21.3	23.3	14.8	26.5	30.5	13.2	15.5	17.3	17.2	17.9	19.3	60
D Sing EXP	7.1	6.1	9.7	9.8	12.0	10.6	13.2	16.9	12.3	27.8	25.5	9.4	10.2	11.8	12.3	13.8	14.9	60
D ARR EXP	7.8	6.9	10.3	9.8	13.0	10.9	13.0	16.0	12.2	37.0	24.9	9.9	10.6	11.9	12.3	13.7	14.7	60
D Holt EXP	6.9	6.6	10.8	10.2	13.1	11.2	15.1	19.7	15.7	37.0	33.6	9.9	11.0	13.1	14.2	16.5	18.4	60
D BrownEXP	7.1	6.4	9.5	9.9	13.0	10.5	14.8	19.6	15.4	38.7	35.6	9.5	10.6	12.7	14.0	16.6	18.8	60
D Quad.EXP	7.0	6.4	9.4	10.1	13.0	12.7	16.1	23.0	20.6	40.5	37.3	9.6	11.4	14.1	16.1	21.4	24.1	60
D Regress	10.1	10.9	12.5	11.4	13.6	12.7	14.7	19.9	14.9	34.0	32.7	12.0	12.4	13.0	12.8	13.8	14.6	60
WINTERS	8.2	8.9	9.8	10.5	13.6	11.5	14.7	19.0	14.9	34.0	32.7	10.7	11.5	13.5	14.2	16.0	18.1	60
Autom. AEP	11.1	10.3	11.6	13.6	15.0	15.3	20.2	26.0	14.6	27.7	29.0	12.6	14.3	16.4	16.4	17.4	18.6	60
Bayesian F	11.3	8.1	10.4	9.5	11.3	10.8	18.1	18.1	14.2	31.1	22.9	9.9	12.9	11.9	12.4	13.5	14.4	60
Combining A	7.3	6.9	9.4	10.0	12.4	10.8	14.5	18.6	12.7	30.9	28.7	9.7	10.7	12.6	13.0	14.7	16.2	60
Combining B	6.9	7.1	9.6	9.0	12.2	11.0	13.2	18.6	14.7	30.9	28.0	9.6	10.4	12.3	13.2	15.1	16.5	60
Box-Jenkins	8.4	10.5	10.1	10.0	13.5	13.7	17.3	20.6	15.1	23.3	30.6	11.0	12.3	14.3	13.2	15.8	17.2	60
Lewandowski	10.1	10.8	12.9	12.6	14.3	12.9	17.3	19.8	16.6	33.7	23.8	12.7	13.5	14.7	14.9	16.9	17.6	60
Parzen	8.5	11.0	10.0	10.1	12.8	12.2	13.7	16.1	13.9	19.2	24.8	11.0	11.6	12.7	12.9	13.8	14.9	60
Average	10.8	10.7	13.1	13.6	15.3	15.3	18.6	23.3	16.9	33.9	32.8	13.2	14.4	16.3	16.6	18.5	20.0	60

TABLE 18 Average MAPE: Non-Seasonal Data (51)

METHODS	MODEL FITTING	Forecasting Horizons										Average of Forecasting Horizons						n(max)
		1	2	3	4	5	6	8	12	15	18	1-4	1-6	1-8	1-12	1-15	1-18	
NAIVE 1	10.5	9.8	13.1	18.3	17.7	22.8	22.1	17.0	18.8	35.3	45.8	14.7	17.3	17.9	17.6	18.8	21.1	51
Mov.Averag	9.9	13.5	16.2	21.2	20.9	25.7	25.1	21.4	18.3	33.9	44.4	17.9	20.4	21.3	20.3	21.1	23.0	51
Single EXP	10.3	9.9	12.1	16.9	17.4	21.8	21.9	15.7	17.6	33.7	43.8	14.1	16.7	17.2	16.8	18.0	20.2	51
ARR EXP	12.1	11.0	14.8	18.9	21.0	23.6	24.2	17.5	18.3	33.5	42.3	16.4	18.9	19.2	18.5	19.4	21.4	51
Holt EXP	10.7	9.5	11.3	16.7	17.6	24.5	23.5	29.8	19.0	31.6	40.1	13.8	17.2	19.9	19.8	20.7	22.2	51
Brown EXP	9.9	10.8	12.3	17.2	17.2	25.3	24.5	32.0	29.6	56.2	74.6	14.4	17.9	20.9	21.8	24.6	28.7	51
Quad.EXP	10.1	12.0	14.6	20.7	21.5	32.6	33.6	58.6	57.0	103.0	142.8	17.2	22.5	25.2	31.6	37.8	46.3	51
Regression	14.3	14.4	17.7	24.0	24.9	28.6	29.1	32.4	56.1	115.6	161.4	20.3	23.1	28.4	28.4	34.4	45.9	50
NAIVE2	10.5	9.8	13.1	18.1	17.7	22.8	22.1	17.0	18.8	35.3	45.8	14.7	17.3	17.9	17.6	18.8	21.1	51
D Mov.Avrg	9.9	13.5	16.2	21.2	20.9	25.7	25.1	21.4	18.3	33.9	44.4	17.9	20.4	21.3	20.3	21.1	23.0	51
D Sing EXP	10.3	9.9	12.1	16.9	17.4	21.8	21.9	15.7	17.6	33.7	43.8	14.1	16.7	17.2	16.8	18.0	20.2	51
D ARR EXP	12.1	11.0	14.8	18.9	21.0	23.6	24.2	17.5	18.3	33.5	42.3	16.4	18.9	19.2	18.5	19.4	21.4	51
D Holt EXP	10.7	9.5	11.3	16.7	17.6	24.5	23.5	29.8	19.0	31.6	40.1	13.8	17.2	19.9	19.8	20.7	22.2	51
D brownEXP	9.9	10.8	12.3	17.2	17.2	25.3	24.5	32.0	29.6	56.2	74.6	14.4	17.9	20.9	21.8	24.6	28.7	51
D Quad.EXP	10.1	12.0	14.6	20.7	21.5	32.6	33.6	58.6	57.0	103.0	142.8	17.2	22.5	25.2	31.6	37.8	46.3	51
D Regress	14.3	14.4	17.7	24.0	24.9	28.6	29.1	32.4	56.1	115.6	161.4	20.3	23.1	28.4	28.4	34.4	45.9	50
WINTERS	10.7	9.5	11.3	16.7	17.6	24.5	23.5	29.8	19.0	31.6	40.1	13.8	17.2	19.9	19.8	20.7	22.2	51
Auton. AEP	10.4	9.3	10.8	13.9	15.2	18.8	17.2	16.0	20.9	38.0	48.7	12.3	14.2	15.4	15.9	17.4	19.9	51
Bavesian F	15.7	12.8	15.0	18.4	18.1	23.5	21.8	21.3	22.1	42.0	53.6	16.1	18.3	19.5	19.1	20.6	23.5	51
Combining A	9.0	9.1	10.3	14.1	14.9	20.8	19.6	21.2	18.6	36.3	47.0	12.1	14.8	16.5	16.6	17.9	20.3	51
Combining B	9.6	9.4	10.8	14.4	17.5	20.5	20.1	23.0	17.9	32.4	41.5	13.1	15.5	17.1	16.9	18.1	20.0	51
Box-Jenkins	N.A	10.0	11.3	13.0	15.7	19.5	18.4	15.5	20.4	35.1	45.0	12.5	14.5	15.3	16.8	17.4	20.5	51
Lewandowski	14.8	12.6	12.7	16.8	16.4	21.0	18.0	17.2	18.1	31.0	42.9	14.6	16.2	16.6	16.8	17.8	20.5	51
Parzen	9.3	10.1	11.5	11.4	14.3	16.9	15.9	15.9	12.8	32.4	31.6	11.8	13.3	14.1	14.1	15.0	16.3	51
Average	10.6	11.0	13.2	17.8	18.6	24.0	23.4	25.4	25.8	48.5	64.2	15.2	18.0	19.8	20.2	22.3	25.8	

TABLE 19 Average Ranking: Yearly Data (20)

METHODS	MODEL FITTING	Forecasting Horizons																		AVERAGE OF ALL FORECASTS	n(max)
		1	2	3	4	5	6	7	8	9	10	11	12	13	14	15	16	17	18		
NAIVE 1	16.7	13.0	14.9	15.6	15.9	15.6	15.3	0.0	0.0	0.0	0.0	0.0	0.0	0.0	0.0	0.0	0.0	0.0	0.0	15.06	20
Mov.Averag	15.5	14.8	15.4	16.8	16.6	16.9	15.6	0.0	0.0	0.0	0.0	0.0	0.0	0.0	0.0	0.0	0.0	0.0	0.0	16.01	20
Single Exp	14.9	14.3	15.2	16.3	16.8	16.5	17.0	0.0	0.0	0.0	0.0	0.0	0.0	0.0	0.0	0.0	0.0	0.0	0.0	16.01	20
ARR Exp	20.7	17.2	19.1	19.1	18.9	17.6	20.0	0.0	0.0	0.0	0.0	0.0	0.0	0.0	0.0	0.0	0.0	0.0	0.0	18.66	20
Holt Exp	6.0	8.5	8.6	9.5	9.7	10.0	8.6	0.0	0.0	0.0	0.0	0.0	0.0	0.0	0.0	0.0	0.0	0.0	0.0	9.15	20
Brown Exp	7.5	11.3	10.7	8.4	9.5	9.8	8.8	0.0	0.0	0.0	0.0	0.0	0.0	0.0	0.0	0.0	0.0	0.0	0.0	9.71	20
Quad.EXP	7.7	10.9	10.2	7.7	8.3	9.5	8.8	0.0	0.0	0.0	0.0	0.0	0.0	0.0	0.0	0.0	0.0	0.0	0.0	9.23	20
Regression	8.5	13.4	11.1	13.5	12.3	11.5	13.1	0.0	0.0	0.0	0.0	0.0	0.0	0.0	0.0	0.0	0.0	0.0	0.0	12.55	20
NAIVE2	16.7	12.9	15.0	16.2	15.9	15.6	15.3	0.0	0.0	0.0	0.0	0.0	0.0	0.0	0.0	0.0	0.0	0.0	0.0	15.16	20
D Mov.Avrj	15.5	14.0	15.5	16.7	16.6	16.5	15.6	0.0	0.0	0.0	0.0	0.0	0.0	0.0	0.0	0.0	0.0	0.0	0.0	15.84	20
D Sing Exp	14.9	14.1	15.6	15.7	17.2	16.3	17.3	0.0	0.0	0.0	0.0	0.0	0.0	0.0	0.0	0.0	0.0	0.0	0.0	16.07	20
D ARR Exp	20.7	16.8	19.1	18.8	18.9	17.3	20.0	0.0	0.0	0.0	0.0	0.0	0.0	0.0	0.0	0.0	0.0	0.0	0.0	18.51	20
D Holt Exp	6.0	8.5	8.6	9.5	9.7	10.0	8.6	0.0	0.0	0.0	0.0	0.0	0.0	0.0	0.0	0.0	0.0	0.0	0.0	9.15	20
D brownEXP	7.5	11.3	10.2	8.4	9.3	9.8	8.6	0.0	0.0	0.0	0.0	0.0	0.0	0.0	0.0	0.0	0.0	0.0	0.0	9.71	20
D Quad.EXP	7.7	10.9	10.2	7.7	8.3	9.5	8.8	0.0	0.0	0.0	0.0	0.0	0.0	0.0	0.0	0.0	0.0	0.0	0.0	9.23	20
D Regress	8.5	13.4	11.1	13.5	12.3	10.5	13.1	0.0	0.0	0.0	0.0	0.0	0.0	0.0	0.0	0.0	0.0	0.0	0.0	12.55	20
WINTERS	6.0	8.5	11.1	8.6	9.7	10.6	10.0	0.0	0.0	0.0	0.0	0.0	0.0	0.0	0.0	0.0	0.0	0.0	0.0	9.15	20
Autom. AEP	10.3	11.8	11.1	11.7	10.4	10.6	10.0	0.0	0.0	0.0	0.0	0.0	0.0	0.0	0.0	0.0	0.0	0.0	0.0	10.92	20
Bayesian F	19.0	14.0	12.1	9.8	8.9	9.4	10.5	0.0	0.0	0.0	0.0	0.0	0.0	0.0	0.0	0.0	0.0	0.0	0.0	10.77	20
Combining A	10.3	11.0	11.6	10.6	12.3	12.3	13.0	0.0	0.0	0.0	0.0	0.0	0.0	0.0	0.0	0.0	0.0	0.0	0.0	11.32	20
Combining B	11.5	13.0	13.6	12.5	12.9	12.7	13.3	0.0	0.0	0.0	0.0	0.0	0.0	0.0	0.0	0.0	0.0	0.0	0.0	12.99	20
Box-Jenkins	N.A	12.0	13.1	10.1	10.3	11.5	12.0	0.0	0.0	0.0	0.0	0.0	0.0	0.0	0.0	0.0	0.0	0.0	0.0	11.48	20
Lewandowski	17.5	12.0	9.9	9.0	10.2	8.3	10.3	0.0	0.0	0.0	0.0	0.0	0.0	0.0	0.0	0.0	0.0	0.0	0.0	10.28	20
Parzen	6.6	11.6	10.0	11.2	8.6	11.5	10.1	0.0	0.0	0.0	0.0	0.0	0.0	0.0	0.0	0.0	0.0	0.0	0.0	10.52	20
Average	11.5	12.5	12.5	12.5	12.5	12.5	12.5	0.0	0.0	0.0	0.0	0.0	0.0	0.0	0.0	0.0	0.0	0.0	0.0	12.50	

TABLE 20 Average Ranking: Quarterly Data (23)

METHODS	MODEL FITING	Forecasting Horizons 1	2	3	4	5	6	7	8	9	10	11	12	13	14	15	16	17	18	AVERAGE OF ALL FORECASTS	n(max)
NAIVE 1	17.1	9.9	11.5	14.7	12.2	12.3	12.5	13.2	11.5	0.0	0.0	0.0	0.0	0.0	0.0	0.0	0.0	0.0	0.0	12.22	23
Mov.Averag	14.6	12.6	14.1	13.7	14.5	14.2	15.0	14.0	14.0	0.0	0.0	0.0	0.0	0.0	0.0	0.0	0.0	0.0	0.0	14.09	23
Single EXP	13.2	12.1	13.1	12.2	13.3	12.4	13.0	11.7	11.7	0.0	0.0	0.0	0.0	0.0	0.0	0.0	0.0	0.0	0.0	12.45	23
ARR EXP	19.8	13.0	14.9	12.8	13.3	11.8	12.3	11.8	12.4	0.0	0.0	0.0	0.0	0.0	0.0	0.0	0.0	0.0	0.0	12.79	23
Holt EXP	9.1	12.0	12.1	15.8	12.7	14.0	13.7	16.2	14.1	0.0	0.0	0.0	0.0	0.0	0.0	0.0	0.0	0.0	0.0	13.84	23
Brown EXP	11.1	12.5	12.2	13.5	13.4	14.4	13.2	14.1	13.1	0.0	0.0	0.0	0.0	0.0	0.0	0.0	0.0	0.0	0.0	13.30	23
Quad.EXP	12.7	15.4	14.5	16.8	15.5	16.7	15.5	14.9	16.0	0.0	0.0	0.0	0.0	0.0	0.0	0.0	0.0	0.0	0.0	15.65	23
Regression	17.4	17.7	15.4	14.3	14.5	14.1	11.7	13.3	11.6	0.0	0.0	0.0	0.0	0.0	0.0	0.0	0.0	0.0	0.0	14.07	23
NAIVE2	12.4	10.2	12.4	12.6	12.5	11.3	12.1	10.5	11.6	0.0	0.0	0.0	0.0	0.0	0.0	0.0	0.0	0.0	0.0	11.66	23
D Mov.Avrg	11.1	10.6	13.5	14.3	13.4	13.1	13.5	13.5	13.0	0.0	0.0	0.0	0.0	0.0	0.0	0.0	0.0	0.0	0.0	13.12	23
D Sing EXP	8.7	11.8	12.9	11.7	11.9	11.1	12.4	9.8	10.9	0.0	0.0	0.0	0.0	0.0	0.0	0.0	0.0	0.0	0.0	11.51	23
D ARR EXP	16.7	12.5	15.0	12.7	13.9	12.1	12.0	10.4	12.1	0.0	0.0	0.0	0.0	0.0	0.0	0.0	0.0	0.0	0.0	12.65	23
D Holt EXP	3.9	11.4	10.5	12.1	11.4	11.4	11.8	12.9	12.5	0.0	0.0	0.0	0.0	0.0	0.0	0.0	0.0	0.0	0.0	11.76	23
D brownEXP	5.6	12.3	11.3	11.5	11.0	11.6	12.6	12.7	12.6	0.0	0.0	0.0	0.0	0.0	0.0	0.0	0.0	0.0	0.0	11.87	23
D Quad.EXP	8.8	14.9	13.7	13.2	13.1	16.2	14.1	17.5	0.0	0.0	0.0	0.0	0.0	0.0	0.0	0.0	0.0	0.0	0.0	14.89	23
D Regress	15.9	16.4	15.4	14.5	13.1	12.4	11.7	11.5	11.8	0.0	0.0	0.0	0.0	0.0	0.0	0.0	0.0	0.0	0.0	13.36	23
WINTERS	5.0	11.4	9.8	10.9	11.7	12.4	10.7	12.7	12.1	0.0	0.0	0.0	0.0	0.0	0.0	0.0	0.0	0.0	0.0	11.48	23
Auton. AEP	10.7	9.1	10.3	10.5	10.7	12.3	11.0	13.3	11.5	0.0	0.0	0.0	0.0	0.0	0.0	0.0	0.0	0.0	0.0	11.10	23
Bavesian F	17.0	13.3	15.4	13.6	13.5	13.3	14.5	15.7	15.0	0.0	0.0	0.0	0.0	0.0	0.0	0.0	0.0	0.0	0.0	14.30	23
Combining A	10.1	9.7	9.3	8.5	10.0	10.8	10.5	11.8	11.6	0.0	0.0	0.0	0.0	0.0	0.0	0.0	0.0	0.0	0.0	10.29	23
Combining B	10.3	11.9	11.4	10.3	11.6	10.6	11.2	10.6	10.7	0.0	0.0	0.0	0.0	0.0	0.0	0.0	0.0	0.0	0.0	11.12	23
Box-Jenkins	7.4	11.9	9.7	9.3	11.9	10.8	11.2	10.3	11.6	0.0	0.0	0.0	0.0	0.0	0.0	0.0	0.0	0.0	0.0	10.84	23
Lewandowski	16.2	15.9	13.3	10.2	11.3	12.2	11.2	9.7	10.5	0.0	0.0	0.0	0.0	0.0	0.0	0.0	0.0	0.0	0.0	11.79	23
Parzen	8.5	11.5	8.3	8.8	9.2	10.2	10.1	10.1	10.7	0.0	0.0	0.0	0.0	0.0	0.0	0.0	0.0	0.0	0.0	9.86	23
Average	11.5	12.5	12.5	12.5	12.5	12.5	12.5	12.5	12.5	0.0	0.0	0.0	0.0	0.0	0.0	0.0	0.0	0.0	0.0	12.50	

TABLE 21 Average Ranking: Monthly Data (68)

METHODS	MODEL FITTING	1	2	3	4	5	6	7	8	9	10	11	12	13	14	15	16	17	18	AVERAGE OF ALL FORECASTS	n(max)
NAIVE 1	17.8	14.8	15.2	15.4	13.8	15.5	14.1	14.8	15.0	14.2	12.5	13.3	11.1	12.5	14.6	14.0	13.4	13.6	13.2	13.95	68
Mov.Averag	15.8	13.2	13.5	13.5	12.3	13.8	13.1	12.6	12.4	12.1	12.6	11.7	13.3	12.0	12.9	12.7	12.0	11.9	12.1	12.66	68
Single EXP	15.3	13.6	14.1	13.6	14.1	14.2	13.7	13.0	13.0	12.7	12.5	12.4	12.7	12.4	12.4	13.1	13.7	12.8	12.8	13.04	68
ARR EXP	20.3	13.6	14.1	14.4	14.1	14.6	14.6	13.0	13.0	12.7	13.3	12.8	13.7	12.9	13.3	13.4	13.7	13.4	13.4	13.66	68
Holt EXP	14.6	13.1	14.2	13.0	13.6	14.5	13.4	13.2	13.1	13.6	13.7	14.2	13.9	12.8	14.4	14.3	13.1	13.4	12.6	13.56	68
Brown EXP	16.5	13.8	14.3	13.9	13.1	15.0	13.7	13.3	13.7	13.3	13.1	14.2	13.4	12.8	13.7	14.8	13.7	13.6	13.2	13.65	68
Quad.EXP	18.2	14.0	14.9	15.8	14.2	14.7	15.4	15.7	13.7	15.7	14.1	15.8	13.4	17.0	13.7	14.8	16.9	16.1	16.4	15.81	68
Regression	19.3	17.2	16.6	16.5	16.3	15.7	16.7	14.7	15.2	14.2	14.6	14.8	14.6	13.7	14.7	14.0	12.9	12.9	12.7	14.89	68
NAIVE2	10.1	11.5	11.1	11.7	12.5	11.7	12.0	13.0	12.3	12.7	12.3	12.3	11.2	13.4	12.2	12.0	12.3	12.0	12.6	12.14	68
D Mov.Avrg	6.8	13.1	13.1	14.2	14.9	14.2	14.3	13.6	14.2	14.2	13.8	13.3	12.9	13.3	12.2	12.9	13.8	14.4	14.2	13.78	68
D Sing EXP	6.4	9.6	10.9	10.6	11.3	11.5	11.5	11.3	10.9	11.0	11.2	11.4	11.2	11.8	10.5	11.0	11.3	11.5	11.9	11.13	68
D ARR EXP	12.0	11.5	11.9	12.1	13.0	12.5	12.0	12.0	11.5	11.6	12.7	11.9	10.9	10.9	11.4	12.6	11.8	12.4	12.9	12.17	68
D Holt EXP	4.5	9.9	10.2	9.9	10.5	10.6	12.3	11.3	11.5	12.0	11.5	12.0	12.0	10.9	11.6	12.1	11.6	11.8	11.8	11.26	68
D brownEXP	6.2	10.2	9.9	10.4	9.8	10.1	11.3	11.6	11.5	11.9	12.0	12.6	12.2	12.5	12.1	12.2	12.9	12.4	12.8	11.58	68
D Quad.EXP	7.8	10.4	10.9	10.6	10.6	11.6	13.7	13.1	14.7	14.0	14.0	15.1	15.2	15.4	12.0	15.2	15.8	15.3	14.9	13.37	68
D Regress	12.5	14.4	14.6	14.3	14.3	11.6	12.0	10.8	11.5	11.7	12.4	12.7	11.6	12.0	12.4	12.5	11.9	12.7	11.9	12.92	68
WINTERS	9.0	13.1	11.2	10.6	11.3	10.3	10.7	10.8	11.5	11.7	12.1	12.6	12.0	11.6	11.1	10.5	12.5	11.3	10.8	11.44	68
Autom. AEP	10.1	12.2	12.9	12.9	11.8	11.4	11.1	11.7	11.3	12.8	12.3	11.7	12.2	12.1	12.5	11.0	11.8	11.6	12.7	11.95	68
Bavesian F	16.6	12.2	12.1	11.3	10.9	11.4	11.4	12.6	12.6	12.3	12.3	12.3	12.1	12.4	12.4	11.7	11.8	11.6	11.8	11.65	68
Combining A	5.5	9.9	10.6	10.1	10.2	10.9	10.4	11.1	10.5	10.0	10.4	10.1	10.3	9.5	10.8	10.5	10.2	11.0		10.33	68
Combining B	5.4	9.8	9.7	11.6	11.2	11.1	11.4	11.1	11.3	11.6	10.9	11.7	11.6	12.1	10.7	10.7	11.1	11.2	11.6	11.16	68
Box-Jenkins	N.A	12.7	11.3	10.9	12.1	10.5	10.7	10.6	10.6	11.0	12.7	12.6	12.4	13.2	10.7	10.7	11.9	11.8	12.1	11.63	68
Lewandowski	14.9	13.1	12.9	11.8	12.1	10.5	10.1	10.7	10.7	11.5	11.5	9.9	10.9	10.0	10.0	10.3	9.5	9.5	8.8	10.79	68
Parzen	10.6	13.1	11.8	11.4	11.6	11.5	11.3	10.7	10.8	11.9	12.0	10.8	11.0	12.7	11.5	10.1	10.9	11.6	12.1	11.49	68
Average	11.5	12.5	12.5	12.5	12.5	12.5	12.5	12.5	12.5	12.5	12.5	12.5	12.5	12.5	12.5	12.5	12.5	12.5	12.5	12.50	

TABLE 22 Average Ranking: Seasonal Data (60)

METHODS	MODEL FITTING	1	2	3	4	5	6	7	8	9	10	11	12	13	14	15	16	17	18	AVERAGE OF ALL FORCASTS	n(max)
NAIVE 1	10.8	14.7	15.1	15.6	13.8	15.8	14.2	15.1	15.2	15.0	13.1	13.9	11.2	13.1	15.4	14.8	13.8	14.1	13.7	14.36	60
Mov.Averag	17.2	17.3	13.6	13.7	12.7	14.4	13.6	13.0	12.9	12.4	13.1	11.8	13.9	12.5	13.5	13.3	12.1	12.2	12.5	13.07	60
Single Exp	17.2	17.5	13.7	14.6	14.3	14.7	14.1	13.5	13.6	13.4	13.4	13.0	13.7	12.9	13.1	14.0	12.8	13.6	13.6	13.63	60
ARR Exp	20.9	13.9	14.7	13.5	13.5	14.3	14.3	13.3	13.1	13.3	13.3	13.0	13.9	12.8	13.6	13.4	13.6	13.8	13.6	13.69	60
Holt Exp	16.5	13.7	15.1	14.7	14.4	15.7	14.7	14.5	13.9	13.2	14.5	14.0	13.9	13.4	14.9	13.5	13.5	13.8	13.2	14.27	60
Brown Exp	18.2	13.5	14.4	14.4	14.8	16.3	14.5	14.0	13.4	13.2	13.0	13.3	13.3	13.4	14.0	15.5	16.7	13.6	13.2	14.09	60
Quad.Exp	19.7	13.8	15.0	16.4	15.1	15.6	15.7	15.7	15.6	15.3	16.0	15.9	17.1	17.6	15.5	17.0	16.7	15.6	15.9	15.91	60
Regression	18.9	15.0	15.6	15.1	15.1	15.6	15.7	13.7	13.7	13.8	14.0	14.2	13.1	14.5	13.6	12.3	12.3	12.1	12.7	14.24	60
NAIVE2	9.3	11.2	10.3	9.8	11.1	11.6	12.1	12.0	13.1	12.8	12.7	11.2	12.9	12.2	13.1	12.1	12.1	12.1	12.7	11.97	60
D Mov.Avrg	5.7	12.5	13.0	14.1	15.2	15.1	14.8	14.7	14.9	14.7	15.1	15.5	13.4	14.3	12.9	14.6	15.7	15.4	15.4	14.40	60
D Sing Exp	5.3	9.0	10.8	10.3	11.8	11.0	11.0	11.2	11.7	11.6	11.3	11.5	12.3	10.6	11.9	11.5	11.9	12.5	12.5	11.14	60
D ARR Exp	10.3	11.4	12.2	12.3	11.8	12.0	11.8	11.2	11.7	12.4	12.3	11.4	12.0	12.3	13.0	11.1	11.1	11.5	12.5	11.83	60
D Holt Exp	3.3	9.3	11.7	10.8	10.2	11.5	12.2	11.2	11.6	11.2	10.7	11.7	10.6	11.1	12.1	12.1	11.4	11.7	11.5	11.04	60
D brownExp	1.5	9.5	3.5	9.8	9.4	11.3	11.6	11.6	11.5	11.8	12.7	13.6	14.7	12.0	13.8	14.5	15.3	12.0	12.6	11.27	60
D Quad.Exp	5.0	10.3	10.1	11.0	10.0	13.1	13.1	14.2	14.2	14.2	13.6	14.7	14.7	13.8	14.5	15.3	15.3	13.8	12.7	12.77	60
D Regress	10.6	13.5	12.8	13.0	11.7	12.4	11.5	11.0	10.8	10.7	10.9	11.7	10.4	10.9	11.6	10.4	12.0	12.0	11.0	11.67	60
WINTERS	8.8	13.5	10.8	11.8	10.3	10.5	10.5	11.3	11.1	11.7	12.0	11.8	11.9	10.6	9.9	11.6	11.7	11.0	10.3	11.21	60
Autom. AEP	12.6	13.0	11.9	12.6	11.9	11.0	11.0	10.8	12.6	11.9	11.1	11.5	11.9	12.3	10.4	11.7	11.2	11.2	12.5	11.72	60
Bavesian F	15.5	12.2	12.1	11.8	11.9	12.0	11.5	13.5	11.7	12.0	11.6	12.1	10.2	11.8	10.6	11.7	11.0	11.0	11.0	11.64	60
Combining A	5.5	10.1	9.9	10.3	9.7	10.7	10.4	10.4	9.5	10.1	10.2	9.5	12.4	9.0	10.6	10.6	9.6	10.8	10.8	10.13	60
Combining B	4.4	10.3	9.9	11.1	11.1	11.4	11.2	12.2	12.2	10.9	11.6	11.6	12.4	10.7	11.4	11.9	12.4	12.1	12.1	11.37	60
Box-Jenkins	N.A.	13.3	11.7	11.8	10.5	11.0	11.0	11.0	11.0	12.6	12.0	12.0	12.4	12.2	12.0	12.0	11.8	12.1	12.1	11.64	60
Lewandowski	14.4	13.3	13.5	12.1	12.7	12.0	11.6	11.4	12.0	10.5	10.5	10.8	12.2	11.4	10.5	10.3	10.7	10.5	9.5	11.61	60
Parzen	11.7	13.8	11.1	12.2	11.0	11.5	10.4	10.2	11.2	11.9	10.8	11.5	12.9	11.7	9.2	10.6	10.6	11.7	12.1	11.35	60
Average	11.5	17.5	12.5	12.5	12.5	12.5	12.5	17.5	12.5	12.5	12.5	12.5	12.5	12.5	12.5	12.5	12.5	12.5	12.5	12.50	

TABLE 23 Average Ranking: Non-Seasonal Data (51)

METHODS	MODEL FITTING	Forecasting Horizons																		AVERAGE OF ALL FORECASTS	n(max)
		1	2	3	4	5	6	7	8	9	10	11	12	13	14	15	16	17	18		
NAIVE 1	14.6	12.0	13.8	14.6	13.8	13.8	13.7	13.0	12.0	11.8	10.9	11.4	10.8	10.7	12.1	11.7	12.3	12.0	11.9	12.85	51
Mov.Averag	13.5	13.5	14.1	14.7	14.5	14.6	14.0	13.5	12.6	11.2	11.2	11.4	11.5	10.6	11.0	10.9	11.7	11.0	10.8	13.13	51
Single Exp	12.0	13.3	13.6	13.5	14.5	13.7	14.2	11.5	10.9	10.2	9.9	10.5	9.8	10.9	10.4	10.4	11.1	10.2	10.3	12.42	51
ARR Exp	19.6	14.5	15.8	13.5	14.5	13.7	14.2	11.5	12.6	13.4	13.4	13.7	13.3	13.9	12.8	12.2	11.9	12.7	12.4	14.41	51
Holt Exp	6.3	10.1	10.0	10.7	10.2	11.1	10.2	12.9	12.3	13.5	13.4	13.7	12.6	11.8	12.8	12.2	11.9	12.1	12.4	11.36	51
Brown EXP	8.5	12.6	11.4	10.6	10.3	11.0	10.5	12.5	12.1	13.0	12.5	12.1	13.5	13.5	12.6	13.0	13.2	13.5	13.2	11.84	51
Quad.Exp	10.4	13.6	12.7	11.4	11.4	13.0	12.3	14.5	16.4	16.1	13.4	15.1	16.3	16.6	16.2	17.5	17.2	17.4	17.9	14.09	51
Regression	14.7	15.7	14.2	15.5	15.3	14.6	15.4	15.6	15.6	15.6	17.1	16.8	15.7	15.3	15.2	15.1	14.9	15.0	14.6	15.29	51
NAIVE2	14.6	11.8	14.1	14.9	14.1	13.8	13.9	12.7	12.2	11.5	10.9	11.2	11.1	10.8	12.3	12.1	12.6	12.0	12.3	12.98	51
D Mov.Avrg	13.5	13.1	13.7	14.6	14.6	13.7	13.8	12.7	12.2	10.6	10.1	11.2	11.1	10.4	10.6	10.5	11.1	10.6	10.8	12.87	51
D Sing Exp	12.0	13.1	13.7	13.3	14.4	13.7	14.2	11.5	10.8	10.2	10.1	10.7	10.0	10.8	10.2	10.5	10.7	10.2	9.9	12.36	51
D ARR Exp	19.6	14.3	15.8	16.0	16.5	15.0	16.0	11.3	12.6	11.4	13.6	12.3	13.3	13.3	12.6	13.4	13.9	13.5	12.7	14.37	51
D Holt EXP	6.3	10.1	10.0	10.6	10.3	11.1	10.5	12.5	12.1	13.0	12.5	13.7	12.6	11.8	12.8	12.0	13.2	13.5	12.4	11.36	51
D brownEXP	8.5	12.6	11.4	12.3	11.4	11.1	12.3	12.5	12.1	13.0	13.4	12.2	13.5	13.5	12.8	13.0	13.2	13.5	13.2	11.84	51
D Quad.EXP	10.4	13.6	12.7	12.3	11.4	13.0	12.3	14.5	16.4	16.1	13.4	15.1	16.3	16.6	16.2	17.5	17.2	17.4	17.9	14.09	51
D Regress	14.7	15.7	14.2	15.5	15.3	14.6	15.4	15.6	15.6	15.6	17.1	16.8	15.7	15.3	15.2	15.1	14.9	15.0	14.6	15.29	51
WINTERS	6.3	10.1	10.1	11.1	11.5	11.0	10.2	12.9	12.3	13.4	13.4	16.8	12.6	12.8	13.1	15.1	14.9	15.0	14.6	11.36	51
Autom. AEP	7.5	9.7	11.1	11.7	11.5	11.0	10.8	12.9	12.5	13.4	13.4	13.7	12.6	12.8	13.1	12.8	12.1	12.8	13.2	11.86	51
Bayesian F	19.0	13.4	12.2	11.1	10.2	11.3	11.6	13.8	11.5	11.6	11.5	11.2	11.8	11.3	14.2	15.2	15.2	15.2	14.4	12.39	51
Combining A	9.7	10.2	10.5	9.8	11.1	10.1	10.1	12.0	11.5	11.6	11.4	11.2	11.4	10.4	11.1	11.4	11.6	11.6	11.6	10.90	51
Combining B	10.7	11.4	12.2	10.8	12.6	11.5	12.2	10.8	9.5	9.9	10.9	12.1	11.4	11.5	10.5	9.1	9.2	9.3	9.9	11.16	51
Box-Jenkins	N.A	11.3	10.8	9.7	10.5	10.5	11.2	9.6	10.5	10.9	10.1	14.3	13.6	15.8	13.8	11.9	11.6	12.0	12.1	11.31	51
Lewandowski	17.1	13.7	11.1	10.5	10.3	8.7	9.6	8.3	9.8	10.2	13.2	6.3	6.8	8.8	8.2	5.8	6.3	6.8	6.8	9.50	51
Parzen	6.8	10.9	10.2	9.3	10.1	10.9	11.5	10.8	11.8	14.1	12.1	10.6	9.6	12.2	11.0	12.6	11.9	11.4	12.1	10.98	51
Average	11.5	12.5	12.5	12.5	12.5	12.5	12.5	12.5	12.5	12.5	12.5	12.5	12.5	12.5	12.5	12.5	12.5	12.5	12.5	12.50	

TABLE 24 Average Ranking: Micro Data (33)

METHODS	MODEL FITTING	||	1	2	3	4	5	6	7	8	9	10	11	12	13	14	15	16	17	18	||	AVERAGE OF ALL FORECASTS	n(max)
									Forecasting Horizons														
NAIVE 1	16.9		13.8	14.4	15.1	15.8	15.6	15.2	14.4	14.6	14.8	14.2	14.4	11.6	13.3	16.3	14.7	15.3	15.3	14.6		14.68	33
Mov.Averag	14.9		12.7	11.3	12.3	12.6	15.0	12.9	11.6	10.6	11.2	12.2	12.4	13.6	10.0	12.2	11.7	12.4	12.5	11.6		12.11	33
Single EXP	14.2		12.4	12.5	12.7	14.3	14.0	13.7	11.1	9.9	10.4	13.4	13.3	12.0	10.9	12.1	11.4	13.2	12.5	12.3		12.43	33
ARR EXP	19.7		12.8	14.0	14.0	13.7	13.2	14.0	11.1	10.9	12.0	14.3	13.3	13.3	10.5	12.1	11.4	12.6	12.3	12.7		12.76	33
Holt EXP	12.1		12.4	12.1	13.8	13.0	14.6	13.0	12.9	13.0	13.5	14.1	14.6	13.8	11.7	12.3	15.6	14.9	14.5	12.3		13.40	33
Brown EXP	14.6		14.4	13.8	13.8	13.0	14.7	13.2	12.2	11.8	12.7	13.6	14.2	12.9	12.8	12.5	15.1	14.3	14.7	11.9		13.31	33
Quad.EXP	15.5		14.4	14.3	15.3	14.3	15.3	14.1	14.0	17.0	16.7	15.0	15.4	18.6	17.7	16.8	18.2	17.1	16.4	16.4		15.74	33
Regression	17.0		15.3	14.1	14.1	13.7	13.9	13.6	13.2	11.6	10.6	13.2	10.5	12.3	11.4	9.3	10.5	10.5	10.2	9.0		12.10	33
NAIVE2	11.0		10.7	12.6	12.6	13.8	12.9	13.2	13.3	14.3	13.7	11.8	10.5	11.6	13.3	11.8	11.3	11.8	12.7	9.0		12.73	33
D Mov.Avrg	7.8		11.7	11.5	14.5	14.9	15.0	14.5	15.5	15.2	15.7	13.9	14.3	13.8	14.0	14.4	12.7	13.6	14.4	15.2		14.11	33
D Sng EXP	6.9		9.8	11.5	10.6	11.9	11.2	12.1	11.6	10.6	12.9	10.9	11.9	11.6	13.0	10.8	10.8	10.4	11.4	11.9		11.18	33
D ARR EXP	13.8		10.8	13.8	11.5	11.3	13.1	13.1	11.6	13.1	12.5	11.8	13.0	13.0	13.0	10.1	11.3	12.0	12.5	12.9		12.06	33
D Holt EXP	4.6		9.9	11.3	11.5	11.3	11.4	11.0	13.7	12.9	12.5	10.8	11.6	9.8	11.5	12.8	13.0	12.0	11.5	11.5		11.63	33
D brownEXP	6.5		12.2	12.6	11.7	9.6	10.5	11.8	14.7	12.4	15.8	15.9	13.7	18.3	13.4	13.0	13.5	14.6	12.5	12.8		12.41	33
D Quad.EXP	7.5		12.5	14.0	12.5	10.7	12.5	12.7	14.7	17.2	15.8	15.9	16.4	18.2	17.5	17.0	18.4	17.2	16.7	15.8		14.96	33
D Regress	12.8		12.1	13.1	13.1	11.9	12.5	12.0	10.6	11.8	8.1	8.5	7.3	8.9	9.2	8.0	9.0	9.0	9.0	9.8		10.63	33
WINTERS	7.9		12.7	13.5	13.0	11.9	11.8	10.7	12.9	12.4	12.5	11.6	13.1	11.7	11.8	12.7	11.7	13.5	12.2	12.4		12.36	33
Autom. AEP	11.0		14.6	13.2	13.4	12.5	11.8	11.6	13.9	16.0	15.0	11.6	12.8	15.0	14.4	13.0	11.9	12.3	13.2	14.1		13.24	33
Bayesian F	17.7		12.9	12.5	13.5	11.5	10.7	12.5	15.1	16.0	13.2	12.1	14.2	13.9	12.1	14.5	14.8	14.1	14.1	14.1		13.29	33
Combining A	7.6		10.5	11.0	11.0	11.2	10.9	10.2	12.2	11.3	11.2	10.0	11.1	9.6	10.5	11.4	12.0	12.0	11.0	11.8		11.04	33
Combining B	8.0		11.4	11.5	9.8	11.5	10.7	11.3	11.6	10.7	10.5	11.3	11.9	10.5	11.2	10.9	10.5	10.1	11.1	11.5		10.84	33
Box-Jenkins	N.A		13.3	11.2	11.2	12.6	12.0	11.0	12.4	10.7	10.5	13.0	11.9	11.5	13.9	12.3	12.3	11.9	11.6	12.1		11.96	33
Lewandowski	16.2		11.2	10.7	9.5	9.1	8.3	11.6	10.6	10.9	11.7	9.8	9.3	9.3	9.5	9.0	8.5	7.4	6.2	7.5		9.56	33
Parzen	11.9		15.0	12.6	11.4	12.6	11.3	11.5	9.8	11.2	11.7	11.1	9.9	10.5	13.8	10.9	8.6	9.5	10.1	13.0		11.48	33
Average	11.5		12.5	12.5	12.5	12.5	12.5	12.5	12.5	12.5	12.5	12.5	12.5	12.5	12.5	12.5	12.5	12.5	12.5	12.5		12.50	

TABLE 25 Average Ranking: Macro Data (35)

METHODS	MODEL FITTING	Forecasting Horizons 1	2	3	4	5	6	7	8	9	10	11	12	13	14	15	16	17	18	AVERAGE OF ALL FORECASTS	n(max)
NAIVE 1	16.9	14.2	15.4	15.1	13.2	14.4	13.2	13.7	13.3	14.6	12.1	13.1	11.4	11.8	13.9	13.1	12.5	11.8	12.3	13.54	35
Mov.Averag	15.6	14.6	16.1	14.6	14.7	15.6	14.6	14.8	13.4	12.4	14.4	11.7	13.5	11.5	13.9	12.7	12.9	11.9	11.6	13.98	35
Single EXP	14.4	14.4	10.3	14.0	15.5	14.8	16.2	13.7	12.9	13.7	15.3	13.0	14.5	12.1	13.4	13.5	14.0	12.9	13.2	14.24	35
ARR EXP	20.6	14.4	17.3	16.0	17.1	14.9	16.2	14.4	14.6	15.2	17.1	13.9	16.4	13.9	13.3	15.3	15.8	14.9	14.8	15.53	35
Holt EXP	9.2	11.2	11.1	10.3	9.8	11.7	10.7	11.8	11.6	10.8	13.8	13.1	14.6	14.1	15.6	12.9	12.0	11.6	11.2	11.76	35
Brown EXP	11.5	12.3	11.6	11.5	11.6	13.0	11.5	13.1	12.4	13.1	13.2	11.8	13.7	14.2	15.8	13.2	12.2	11.9	13.0	12.52	35
Quad.EXP	13.2	11.8	11.3	12.0	12.0	13.9	12.4	14.2	14.6	13.8	11.8	14.1	12.9	14.5	16.8	14.5	15.0	14.9	16.4	13.42	35
Regression	16.8	17.7	16.4	16.3	16.6	15.3	15.1	15.9	15.4	16.5	15.5	15.9	15.4	16.5	18.1	15.1	13.6	12.8	13.6	15.79	35
NAIVE2	13.3	11.1	13.2	13.8	13.6	13.3	13.9	12.5	12.8	12.9	13.5	12.6	11.8	11.4	11.7	12.5	12.8	13.2	13.3	12.82	35
D Mov.Avrg	12.1	13.3	15.0	13.3	14.8	14.4	14.9	13.4	13.8	13.5	14.6	15.4	12.7	13.1	13.1	13.1	13.1	15.1	14.8	14.11	35
D Sing EXP	10.7	12.3	13.8	14.2	14.1	13.6	14.6	12.5	11.9	13.1	13.3	12.5	11.5	11.0	11.3	12.0	12.8	12.9	13.2	13.02	35
D ARR EXP	16.8	13.8	15.5	15.7	16.5	14.3	15.9	12.9	13.4	14.0	14.8	13.3	13.2	12.5	13.2	13.7	14.8	14.8	15.1	14.49	35
D Holt EXP	4.5	9.2	6.5	8.2	8.7	9.9	9.2	10.2	11.0	10.1	11.2	10.9	13.2	11.1	11.3	10.2	10.9	11.5	9.6	9.94	35
D brownEXP	6.9	10.1	9.0	8.9	9.9	10.4	11.0	11.1	11.5	9.1	10.6	10.3	13.2	10.5	9.8	10.2	10.2	11.5	12.0	10.39	35
D Quad.EXP	8.7	9.8	9.5	10.1	10.5	11.6	11.8	13.0	13.7	12.0	8.8	10.8	10.6	10.9	12.1	11.8	14.5	12.8	14.2	11.42	35
D Regres	14.7	16.8	14.3	15.7	15.3	13.8	13.3	13.4	14.1	14.2	14.2	15.1	14.4	15.3	13.7	14.2	12.8	13.5	14.2	14.54	35
WINTERS	5.7	11.0	8.2	9.1	9.8	10.0	9.5	10.7	10.8	10.3	10.8	11.5	12.0	11.8	11.5	11.2	10.5	11.8	10.7	10.36	35
Autom. AEP	9.2	10.2	10.2	10.5	9.6	10.6	9.7	11.5	12.1	12.0	11.7	13.4	10.9	11.8	11.5	10.5	10.5	10.6	11.9	10.78	35
Bavesian F	16.5	12.1	11.4	11.0	10.1	11.6	12.5	12.1	10.6	8.6	10.6	10.3	11.2	9.9	10.5	11.9	8.8	8.6	10.6	11.47	35
Combining A	7.9	10.1	10.1	10.0	10.3	9.7	10.9	10.9	10.0	8.6	10.6	9.7	9.7	9.9	9.4	9.4	8.8	8.6	9.6	9.89	35
Combining B	8.5	11.4	11.8	12.1	12.4	11.6	12.0	11.8	10.0	11.8	11.8	10.4	11.5	12.8	9.4	9.7	11.6	11.2	11.7	11.49	35
Box-Jenkins	N.A	12.0	10.2	10.5	10.6	10.3	10.5	11.3	12.9	13.4	12.2	15.5	13.3	16.2	14.2	14.8	13.8	15.0	13.9	12.26	35
Lewandowski	15.8	15.6	14.6	13.6	13.3	13.6	10.1	9.2	11.2	10.6	6.6	8.0	9.3	8.8	9.6	9.8	9.8	10.1	7.5	11.06	35
Parzen	6.6	10.7	8.7	9.7	9.3	10.7	9.7	11.7	12.2	12.3	11.1	13.8	11.2	14.0	12.5	12.9	11.6	14.2	12.8	11.19	35
Average	11.5	12.5	12.5	12.5	12.5	12.5	12.5	12.5	12.5	12.5	12.5	12.5	12.5	12.5	12.5	12.5	12.5	12.5	12.5	12.50	35

Table 26 Median APE: Yearly Data (20)

METHODS	MODEL FITTING	FORECASTING HORIZONS										CUMULATIVE FORECASTING HORIZONS						
		1	2	3	4	5	6	8	12	15	18	1-4	1-6	1-8	1-12	1-15	1-18	N(MAX)
NAIVE 1																		
MOV.AVERAG																		
SINGLE EXP																		
ARR EXP																		
HOLT EXP																		
BROWN EXP																		
QUAD.EXP																		
REGRESSION																		
NAIVE2																		
D MOV.AVRG																		
D SING EXP																		
D ARR EXP																		
D HOLT EXP																		
D BROWNEXP																		
D QUAD.EXP																		
D REGRESS																		
*INTERS AEP																		
AUTOM.AEP																		
BAYESIAN F																		
COMBINING A																		
COMBINING B																		
BOX-JENKINS																		
LEWANDOWSKI																		
PARZEN																		

TABLE 27   Median APE: Quarterly Data (23)

| METHODS | MODEL FITTING | FORECASTING HORIZONS | | | | | | | | | | CUMULATIVE FORECASTING HORIZONS | | | | | N(MAX) |
		1	2	3	4	5	6	8	12	15	18	1-4	1-6	1-8	1-12	1-18	
NAIVE 1																	
MOV.AVERAG																	
SINGLE EXP																	
ARR EXP																	
HOLT EXP																	
BROWN EXP																	
QUAD.EXP																	
REGRESSION																	
NAIVE2																	
D MOV.AVRG																	
D SING EXP																	
D ARR EXP																	
D HOLT EXP																	
D BRO*NEXP																	
D QUAD.EXP																	
D REGRESS																	
WINTERS																	
AUTOM. AEP																	
BAYESIAN F																	
COMBINING A																	
COMBINING B																	
BOX-JENKINS																	
LEWANDOWSKI																	
PARZEN																	

TABLE 28   Median APE: Monthly Data (68)

METHODS	MODEL FITTING	1	2	3	4	5	6	8	12	15	18	1-4	1-6	1-8	1-12	1-15	1-18	N(MAX)
						FORECASTING HORIZONS						CUMULATIVE FORECASTING HORIZONS						
NAIVE 1	6.7	10.2	12.5	9.6	11.0	11.1	11.6	16.3	7.3	18.1	18.3	10.7	10.9	12.5	11.3	12.7	13.3	68
MOV.AVERAG	6.4	5.8	8.9	9.0	8.6	8.6	9.0	10.1	11.9	18.0	17.6	8.3	8.6	9.3	-9.6	10.2	11.1	68
SINGLE EXP	6.5	6.3	9.4	9.4	10.0	9.5	10.1	11.3	11.7	17.1	19.5	8.7	9.4	10.0	10.2	10.8	11.5	68
ARR EXP	7.6	4.5	9.2	7.6	8.6	9.1	11.7	11.5	11.3	15.5	16.0	8.0	8.9	10.1	10.4	11.1	11.7	66
HOLT EXP	6.2	6.1	11.2	7.9	11.0	10.7	10.6	12.5	10.9	18.9	17.2	8.9	10.0	10.6	11.1	12.3	12.9	68
BROWN EXP	6.5	6.0	10.3	8.8	10.7	10.3	11.3	11.6	12.1	21.0	16.1	8.9	10.0	10.4	11.3	12.1	12.5	68
QUAD.EXP	6.5	7.6	10.0	11.1	11.4	10.8	11.0	16.0	13.8	24.3	25.3	10.4	10.6	11.9	12.3	13.5	14.9	68
REGRESSION	8.8	11.1	13.3	11.1	12.6	12.8	14.6	13.0	12.3	17.9	15.7	11.7	12.4	12.7	12.9	13.5	13.7	68
NAIVE2	3.7	4.4	4.8	6.2	7.9	5.8	8.1	9.9	7.3	12.9	15.6	6.2	6.5	7.7	-7.9	8.8	-9.3	68
D MOV.AVRG	3.3	6.5	8.0	9.1	11.6	13.0	15.0	16.2	10.3	14.7	23.0	8.4	9.7	11.6	11.6	12.3	13.4	68
D SING EXP	3.4	2.7	5.8	5.5	6.0	6.1	8.1	9.8	8.6	12.9	15.6	5.4	5.6	6.8	-7.4	8.3	9.0	68
D ARR EXP	3.9	4.4	5.5	6.1	6.6	6.7	6.8	9.6	10.0	14.3	17.0	5.8	6.1	7.1	7.9	8.9	9.6	68
D HOLT EXP	3.3	3.8	5.8	7.3	7.6	5.4	8.1	8.9	7.4	12.4	15.0	5.9	6.3	7.1	7.7	8.3	8.7	68
D BROWNEXP	3.4	3.3	4.4	6.9	5.6	4.9	7.6	9.9	9.7	12.7	17.7	5.2	5.7	6.4	7.8	8.9	9.8	68
D QUAD.EXP	3.4	3.7	5.1	7.9	6.2	6.2	6.3	9.4	10.6	16.2	20.1	4.9	5.2	6.4	7.8	9.0	9.8	68
D REGRESS	5.4	6.0	7.6	9.2	8.6	7.7	11.9	11.6	8.8	14.3	13.7	8.3	8.7	9.3	9.5	10.1	10.7	68
WINTERS	3.8	6.0	6.0	5.7	4.4	6.9	9.4		7.4	10.5	10.7	5.9	5.7	6.4	7.1	7.9	8.5	68
AUTOM. AEP	4.4	5.6	6.6	7.8	6.6	6.5	6.5	9.5	9.2	13.7	15.2	6.6	6.6	7.0	8.1	9.0	9.6	68
BAYESIAN F	5.3	5.0	4.8	5.2	5.0	4.4	7.8	11.0	8.8	12.9	14.0	5.2	5.3	6.0	7.5	8.2	8.7	68
COMBINING A	3.2	3.0	5.8	5.6	4.6	4.8	6.1	9.2	6.6	12.9	14.8	5.5	5.6	6.1	7.0	7.6	8.3	68
COMBINING B	3.3	3.4	5.4	5.2	6.6	5.2	6.7	9.8	8.9	12.6	15.7	5.5	5.8	6.0	7.4	8.3	9.1	68
BOX-JENKINS	0.0	6.9	7.0	6.6	6.4	5.7	7.8	9.1	6.6	12.4	16.4	6.6	6.5	7.0	7.8	8.5	9.0	68
LEWANDOWSKI	4.8	5.7	5.6	5.8	5.9	5.8	6.7	8.6	5.4	10.3	10.3	5.9	5.9	6.3	6.7	7.3	7.5	68
PARZEN	4.1	6.0	7.1	7.3	8.3	6.7	7.6	8.3	6.6	11.5	11.6	7.4	7.4	7.5	7.7	8.7	9.4	68

TABLE 29 Median APE: Micro Data (33)

METHODS	MODEL FITTING	FORECASTING HORIZONS										CUMULATIVE FORECASTING HORIZONS						N(MAX)
		1	2	3	4	5	6	8	12	15	18	1-4	1-6	1-8	1-12	1-15	1-18	
NAIVE 1																		
MOV-AVERAG																		
SINGLE EXP																		
ARR EXP																		
HOLT EXP																		
BROWN EXP																		
QUAD-EXP																		
REGRESSION																		
NAIVE2																		
MOV-AVRG																		
SING EXP																		
ARR EXP																		
HOLT EXP																		
BROWN-EXP																		
QUAD-EXP																		
REGRESS																		
*INTERS AEP																		
AUTOM-AEP																		
BAYSIAN F																		
COMBINING A																		
COMBINING B																		
BOX-JENKINS																		
LEWANDOWSKI																		
PARZEN																		

TABLE 30 Median APE: Macro Data (35)

METHODS	MODEL FITTING	1	2	3	4	FORECASTING HORIZONS 5	6	8	12	15	18	CUMULATIVE FORECASTING HORIZONS 1-4	1-6	1-8	1-12	1-15	1-18	N(MAX)
NAIVE 1	2.5	3.0	5.3	7.2	7.3	9.2	9.8	9.9	9.0	12.8	16.7	5.3	7.1	7.7	8.2	8.8	9.1	35
MOV.AVERAG	2.6	3.8	6.6	7.0	7.5	9.0	11.1	9.9	11.3	12.9	12.3	5.6	7.3	8.2	8.5	8.9	9.2	35
SINGLE EXP	2.7	3.0	5.8	5.5	7.3	9.0	10.1	9.9	11.7	14.1	14.1	5.5	7.1	7.5	8.0	8.8	9.3	35
ARR EXP	3.0	3.0	8.4	6.3	9.1	9.0	11.7	11.5	12.2	14.9	15.0	5.8	7.9	8.4	8.9	9.5	9.7	35
HOLT EXP	2.1	2.5	4.1	5.5	6.0	7.0	7.3	8.5	11.2	12.6	13.1	3.6	4.9	5.0	6.2	7.3	7.5	35
BROWN EXP	2.1	2.7	3.7	4.8	7.1	7.6	8.0	9.0	12.1	11.0	11.9	3.7	4.9	6.0	6.7	7.8	7.9	35
QUAD.EXP	2.1	2.7	3.8	5.7	5.2	7.9	7.6	9.3	9.9	14.1	18.0	4.1	4.9	5.8	6.8	7.5	8.1	35
REGRESSION	4.7	6.6	9.2	8.1	10.0	12.5	11.1	13.0	14.2	13.6	15.4	8.6	9.6	10.3	11.0	12.5	12.5	35
NAIVE2	1.8	2.0	3.6	5.5	7.5	8.8	8.8	8.2	9.0	12.8	14.7	3.7	5.5	6.2	6.9	7.7	8.3	35
D MOV.AVRG	1.7	2.8	5.3	8.3	10.6	13.0	14.2	11.5	10.3	19.4	17.6	5.8	7.7	8.4	9.3	9.8	10.1	35
D SING EXP	1.8	2.2	3.6	5.5	7.3	8.8	9.8	8.1	8.8	12.9	12.3	3.7	5.5	6.2	6.8	7.7	8.3	35
D ARR EXP	2.0	2.7	4.5	5.5	8.2	6.8	10.8	8.4	10.4	13.9	12.9	4.9	6.0	6.8	7.5	8.4	8.9	35
D HOLT EXP	1.3	1.2	1.8	3.0	3.1	5.1	6.7	7.2	9.4	10.1	8.0	2.1	3.4	3.8	4.4	4.9	5.3	35
D BROWNEXP	1.3	1.5	1.8	3.2	4.0	5.6	7.4	9.0	9.8	7.7	6.3	2.6	3.4	3.8	4.4	4.9	5.3	35
D QUAD.EXP	1.3	2.0	3.2	2.6	4.8	6.7	8.9	8.9	3.9	7.7	14.6	2.7	4.2	4.7	4.8	5.1	5.8	35
D REGRESS	3.8	6.0	6.6	8.7	8.6	10.7	12.5	13.0	10.4	13.6	12.7	6.7	7.9	8.7	8.9	9.4	9.4	35
WINTERS	1.3	2.0	1.8	3.2	5.5	4.7	6.7	8.4	9.1	10.0	7.1	2.5	3.6	4.3	4.7	5.3	5.5	35
AUTOM. AEP	1.5	1.5	3.1	2.5	4.0	5.8	6.6	9.2	10.2	10.0	10.8	2.7	4.5	5.2	6.2	6.6	7.0	35
BAYESIAN F	2.1	2.2	2.7	4.6	3.0	5.8	8.0	10.5	7.9	12.8	14.0	2.8	4.1	4.5	5.0	5.7	6.5	35
COMBINING A	1.4	1.6	3.0	2.9	6.1	4.4	8.7	5.7	9.3	8.2	10.2	2.7	3.6	4.2	4.7	5.6	5.9	35
COMBINING B	1.5	1.6	2.9	3.4	5.8	4.5	5.8	4.7	10.7	12.4	13.5	2.9	3.5	4.2	4.9	5.8	6.5	35
BOX-JENKINS	0.0	1.7	2.8	3.0	4.3	4.9	7.8	8.2	10.0	18.4	15.8	2.9	4.2	4.5	5.1	6.0	6.5	35
LEWANDOWSKI	2.1	3.5	3.6	4.2	5.2	5.0	4.8	5.3	5.3	10.1	3.8	4.2	4.4	4.6	4.8	5.0	5.0	35
PARZEN	1.3	1.8	2.0	2.4	5.1	5.7	5.7	6.7	6.3	18.1	12.0	2.2	3.4	4.2	5.0	5.6	6.3	35

TABLE 31   Median APE: Seasonal Data (60)

METHODS	MODEL FITTING	FORECASTING HORIZONS									CUMULATIVE FORECASTING HORIZONS							N(MAX)
		1	2	3	4	5	6	8	12	15	18	1-4	1-6	1-8	1-12	1-15	1-18	
NAIVE 1																		
MOV.AVERAG																		
SINGLE EXP																		
ARR EXP																		
HOLT EXP																		
BROWN EXP																		
QUAD.EXP																		
REGRESSION																		
NAIVE2																		
D MOV AVRG																		
D SIN EXP																		
D ARR EXP																		
D HOLT EXP																		
D BROWN EXP																		
D QUAD.EXP																		
D REGRESS																		
*INTERS																		
AUTOM. AEP																		
BAYESIAN F																		
COMBINING A																		
COMBINING B																		
BOX-JENKINS																		
LEWANDOWSKI																		
PARZEN																		

TABLE 32 Median APE: Non-Seasonal Data (51)

METHODS	MODEL FITTING	FORECASTING HORIZONS										CUMULATIVE FORECASTING HORIZONS						N(MAX)
		1	2	3	4	5	6	8	12	15	18	1-4	1-6	1-8	1-12	1-15	1-18	
NAIVE 1																		
MOV-AVERAG																		
SINGLE EXP																		
ARR EXP																		
HOLT EXP																		
BROWN EXP																		
QUAD-EXP																		
REGRESSION																		
NAIVE2																		
D MOV-AVRG																		
D SING EXP																		
D ARR EXP																		
D HOLT EXP																		
D BROWNEXP																		
D QUAD-EXP																		
D REGRESS																		
*INTERS AEP																		
*AUTOM-AEP																		
BAYESIAN F																		
COMBINING A																		
COMBINING B																		
BOX-JENKINS																		
LEWANDOWSKI																		
PARZEN																		

It seems that the factors affecting forecasting accuracy are trend, seasonality and randomness (noise) present in the data. It is believed that the greater the randomness in the data, the less important is the use of statistically sophisticated methods. Furthermore, it seems that deseasonalizing the data by a simple decomposition procedure is adequate, making the majority of methods (both simple and sophisticated) perform about the same. Finally, it is believed that some statistically sophisticated methods extrapolate too much trend which can cause overestimation. This is why Naive 2 and single exponential smoothing do relatively well in comparison to some statistically sophisticated methods.

Effects of forecasting horizons

For short forecasting horizons (1 and 2 periods ahead) deseasonalized simple, Holt, Brown and Holt–Winters exponential smoothing do well. For horizons 3, 4, 5 and 6 deseasonalized Holt, Brown, and Holt–Winters, and Parzen perform relatively well in most accuracy criteria. Finally, for longer time horizons (i.e. 7, 8, 9, ..., 18) Lewandowski does the best.

The combining of forecasts

Combining A, a simple average of six methods (see Appendix 2), performs very well overall and better than the individual methods included in the average.

Combining B (using the same methods as Combining A but taking a weighted average based on the sample covariance matrix of fitting errors – instead of the simple average of Combining A) also performs well, but not as well as Combining A.

Combining can be profitably used to reduce forecasting errors by simply averaging the predictions of a few forecasting methods.

Significant differences

Are the differences in the relative performance of the various methods, discussed in the previous section, statistically significant? It is not easy to test statistically each of the statements presented in the previous section for two reasons. First, the errors are non-symmetric, which excludes using parametric statistics. Second, not enough data are available to test differences in subcategories. This is particularly true when the 111 series are used. However, several of the statements made in the previous paragraphs can be substantiated by statistical tests.

Assuming normality in the errors (an assumption which does not hold true), an analysis of various methods can be performed to test for statistically significant differences of how well these methods forecast the different series used, the various horizons and, overall, both series and horizons. These three aspects will be called series, horizons and methods respectively for which tests have been

TABLE 33 Analysis of Variance for Different Groupings of Methods

Grouping of Methods	F-Tests and Degrees of Freedom	Type of Data								
		Yearly			Quarterly			Monthly		
		Methods	Hori-zons	Series	Methods	Hori-zons	Series	Methods	Hori-zons	Series
24 Methods [1]	F-Test	7.73	31.18	256	4.36	49.69	119	10.45	12.11	293
	D.F.	23 2879	5 2879	19 2879	23 4415	7 4415	22 4415	23 29375	17 29375	67 29375
21 Methods [2]	F-Test	12.84	34.7	263	13.51	8.16	41.39	4.81	13.37	418
	D.F	20 22679	5 22679	179 22679	20 34103	5 34103	262 34103	20 233225	17 233225	616 233225
8 Methods [3]	F-Test	3.33	10.74	91.48	3.87	11.46	114	1.37 **	4.83	101.9
	D.F.	7 959	5 959	19 959	7 1471	7 1471	22 1471	7 9791	17 9791	67 9791
5 Methods [4]	F-Test	12.6	7.76	44	7.75	2.29 **	17.86	5.75	4.64	131
	D.F.	4 5177	5 5177	178 5177	4 7901	7 7901	4 1901	4 54887	17 54887	616 54887

All differences except those with ** are significantly different than zero, at least at a 99% level.

[1] For a list of the 24 methods see Table 1 (a).

[2] For a list of the 21 methods see Table 1 (b)

[3] The 8 methods are: Deseasonalized single exponential smoothing, Holt, Winters, Automatic AEP, Bayesian Forecasting, Box-Jenkins, Lewandowski, and Parzen.
These methods are considered to be the group containing the best methods, varying the least among themselves, for the 111 series.

[4] The 5 methods are: Deseasonalized single exponential smoothing, Holt, Winters, Automatic AEP, and Bayesian forecasting.
These methods are considered to be the group containing the best methods, varying the least among themselves, for all the 1001 series.

conducted by using a straightforward analysis of variance approach. Table 33 shows the *F*-tests together with the corresponding degrees of freedom. The great majority of the *F*-tests are significant at the 1% level. In general, variations due to series are much more significant than those due to horizons which in turn are more significant than those due to methods.

In order to perform the analysis of variance, the various methods were subdivided into four different groupings. The first grouping included all 24 methods (111 series), the second grouping the 21 methods for which all 1001 series have been used. However, comparisons involving all methods may be meaningless because some methods (e.g. simple methods when the data have not been deseasonalized) were only used as a yardstick to judge the relative

TABLE 34(a) Differences in Overall (i.e. Periods 1–18) Average Rankings from Deseasonalized Single Exponential Smoothing and Corresponding Value of d-Statistic (1001 Series)

Methods	All Data	Yearly Data	Quarterly Data	Monthly Data	Micro Data	Macro Data	Industry Data	Demographic Data	Seasonal Data	Non-Seasonal Data
D. Holt Exp.	-.01	1.04*	.27*	-.15*	-.12*	.38*	-.13	-.34*	-.03	.01
Winters	.02	1.04*	.26*	-.11*	-.14*	.40*	-.10	-.20*	.03	.01
Automatic AEP	-.08	1.10*	.30*	-.25*	-.32*	.26*	-.23*	.08	-.24*	.15*
Bayesian Forecast.	-.09	.75*	-.13	-.17*	-.14*	.18*	-.33*	-.08	-.12*	.05
d-Statistic	.23	.27	.22	.09	.12	.12	.14	.18	.09	.09

* Denotes significant differences at a 1% level.

TABLE 34(b) Differences in Overall (i.e. Periods 1–18) Average Rankings from Deseasonalized Single Exponential Smoothing and Corresponding Value of d-Statistic (111 Series)

Methods	All Data	Yearly Data	Quarterly Data	Monthly Data	Micro Data	Macro Data	Industry Data	Demographic Data	Seasonal Data	Non-Seasonal Data
D. Holt Exp.	.23	2.22*	.18	.04	-.07	1.14*	-.03	-.50	.09	.50*
Winters	.17	2.22*	.19	-.04	-.42	1.01*	.07	-.06	-.01	.50*
Automatic AEP	-.04	1.59*	.29	-.25	-.63*	.73*	-.09	-.23	-.21	.27
Bayesian Forecast.	-.05	1.38*	-.97	-.05	-.61*	.56*	-.14	.16	-.12	.08
Box-Jenkins	-.09	1.13	.39	-.06	-.19	.37	.01	.28	-.07	.36
Lewandoski	-.31	1.62*	.08	.22	.70*	.78*	-.30	-.34	-.11	1.08*
Parzen	-.20	1.57*	.88	-.04	-.03	.82*	-.15	.14	.01	56*
d-Statistic	.36	1.27	1.02	.40	.57	.55	.63	.82	.42	.46

* Denotes significant differences at a 1% level.

performance of the remaining methods. Thus, a third grouping of the eight most accurate methods was used, and a last grouping of five of these eight methods for which all 1001 series were available was also made. Table 33 presents the F-tests and gives the degrees of freedom for each of the four groupings.

Table 34 is more appropriate for the accuracy data in this study. It is a non-parametric multiple comparisons procedure for the average rankings (Hollander and Wolfe, 1973). Those differences in average rankings, which are statistically significant at the 1 % level, need to be bigger than the corresponding value shown in the last row of Tables 34(a) and 34(b). The base method for comparison was the deseasonalized single exponential smoothing. None of the differences in the average rankings are statistically significant as far as *all* of the data are concerned. This is true for each of the forecasting horizons *and* the average of all forecasts. However, the differences become significant when subcategories of data are used, which shows that there is not one single method which can be used across the board indiscriminately. The forecasting user should be selective. It is interesting to note that in Table 34(a) *all* differences in yearly, quarterly, monthly, micro and macro data are significant.

Furthermore, note that the signs in yearly, quarterly and monthly data are positive (meaning that the corresponding methods perform statistically better than deseasonalized single exponential smoothing) whereas for monthly and micro all the signs are negative.

Finally, fewer significant differences exist in Table 34(b) because there are only 111 data for the comparisons. However, the signs (with a few exceptions – e.g. Lewandowski) and the statistically significant values follow a pattern similar to that of Table 34(a). The implications of the results shown in Tables 34(a) and 34(b) are highly important as far as the practical utilization of extrapolative methods is concerned.

Non-significant differences

A most interesting aspect of making the comparisons has been those differences which turned out to be statistically non-significant. These cases are listed below and it is hoped that future research will explain the reasons why this is happening and what are the implications for forecasting.

1.   It was expected that forecasts before 1974 would be more accurate than those after 1974. In fact, when the data were separated into two corresponding categories, no significant difference, in post-sample forecasting accuracy, was found between pre and post 1974 data. Similarly, when the data were separated into a category which ended during or just before a recession and another including all other series, the differences between the two categories were not found to be statistically significant.

2.   The parameters of the various models were found by minimizing the one-step-ahead Mean Square Error for each of the series involved. All forecasts are therefore one-step-ahead forecasts. When the method required more values in order to obtain additional forecasts, the forecasts already found were used for this purpose. In addition to this one-step-ahead forecast, multiple lead time forecasts were also obtained for the deseasonalized Holt method. That is, optimal parameters for $1, 2, 3, \ldots, 18$ periods ahead were obtained and a single forecast was found, using these optimal parameters. Thus, for monthly data, each series was re-run 18 times, each time obtaining optimal parameters and one L-period ahead forecasts. For the method used to obtain multiple lead time forecasts, no significant differences were observed between their accuracy and that of one-period forecasts.

3. Several variations of Winters' Exponential Smoothing were run but no significant differences from the specific Holt–Winters model used in this paper were observed.

4. Two variations of Adaptive Response Rate Exponential Smoothing (ARRES) were run. The one which used a delay in the adaptation of alpha did not produce significantly more accurate forecasts than the non-delayed version. Furthermore, ARRES did not perform better than non-adaptive exponential smoothing methods, a finding consistent with that of Gardner and Dannenbring (1980).

5. In addition to deseasonalizing the data by a simple ratio-to-moving average (centred) decomposition method, the same deseasonalization was also done

    (a) by using the seasonal indices obtained by the CENSUS II method;
    (b) by using the one-year-ahead forecast of the seasonal factors obtained by the CENSUS II method.

Neither of these deseasonalized procedures produced forecasts which were better than those of the ratio to centred moving average method reported in Naive 2.

6. It makes little difference as to what method to use for industry-wide series, when there are demographic series, or for data that exhibit seasonality.

7. Finally, some preliminary work concerning the effect of the number of data points on accuracy has not produced evidence that, as the number of data points increases, relative performance is improved. This finding is consistent with that found in Makridakis and Hibon (1979) and raises some interesting questions about the length of time series to be used in forecasting.

CONCLUSIONS

The major purpose of this paper has been to summarize the results of a forecasting competition of major extrapolation (time series) methods and look at the different factors affecting forecasting accuracy. If the forecasting user can discriminate in his choice of methods depending upon the type of data (yearly, quarterly, monthly), the type of series (macro, micro, etc.) and the time horizon of forecasting, then he or she could do considerably better than using a single method across all situations – assuming, of course, that the results of the present study can be generalized. Overall, there are considerable gains to be made in forecasting accuracy by being selective (e.g. see Tables 34(a) and 34(b)). Furthermore, combining the forecasts of a few methods improves overall forecasting accuracy over and above that of the individual forecasting methods used in the combining.

The question that deserves further consideration is obviously this: why do some methods do better than others under various conditions? This could not be attributed simply to chance, given the large number of series used. Even though further research will be necessary to provide us with more specific reasons as to why this is happening, a hypothesis may be advanced at this point, stating that statistically sophisticated methods do not do better than simple methods (such as deseasonalized exponential smoothing) when there is considerable randomness in the data. This is clear with monthly and micro data in which randomness is much more important than in quarterly or yearly macro data. Finally, it seems that seasonal patterns can be predicted equally well by both simple and statistically sophisticated methods. This is so, it is believed, because of the instability of seasonal variations that dominate the remaining of the patterns and which can be forecasted as accurately by averaging seasonality as in using any statistically sophisticated approach.

The authors of this paper hope that the information presented will help those interested in forecasting to understand better the factors affecting forecasting accuracy and realize the differences that exist among extrapolative (time series) forecasting methods.

APPENDIX 1

The accuracy measures

This appendix presents the various accuracy measures used in the competition.

Two sets of errors were calculated for each method. The first was arrived at by fitting a model to the first $n - m$ values (where $m = 6$ for yearly, 8 for quarterly, and 18 for monthly data) of each of the series and calculating the error e_t as follows:

$$e_t = X_t - \hat{X}_t \tag{1}$$

where X_t is the actual value, and $\hat{X}_t$ is one-period-ahead forecasted value.

Two so-called errors of 'model fitting' were also calculated as follows, where all summations go from 1 to $n - m$:

(a)   The mean percentage error (MAPE) $= (n - m)^{-1} \sum (|e_t|/X_t)(100)$ (2)
(b)   The mean square error (MSE) $= (n - m)^{-1} \sum e_t^2$.   (3)
(c)   The percentage of time the error for method i was smaller than that for method j was also recorded.
(d)   The ranking of each method in relation to all others. (The best method received the ranking of 1, the second of 2, the third of 3 and so forth.) The rankings were then averaged for all series.
(e)   The median absolute percentage error.

The second set of errors involves the last m values, which were utilized as *post-sample* measures to determine the magnitude of the errors. The two measurements shown in equations (2) and (3), as well as the percentage of time method i was better than method j, and the average rankings were also computed for up to m forecasting horizons, starting at period $n - m + 1$. In addition, the median of the absolute percentage error was computed.

In *no* instance have the last m values been used to develop a forecasting model or estimate its parameters. The model fitting *always* involved only the first $n - m$ values for each series.

APPENDIX 2

The methods

(1) Naive 1.

$$\text{Model fitting: } \hat{X}_{t+1} = X_t, \tag{4}$$

where $t = 1, 2, 3, \ldots, n - m$

$$\text{Forecasts: } \hat{X}_{n-m+k} = X_{n-m}, \tag{5}$$

where $k = 1, 2, 3, \ldots, m$.

(2) Simple moving average.

$$\text{Model fitting: } \hat{X}_{t+1} = \frac{X_t + X_{t-1} + X_{t-2} + \cdots + X_{t-N+1}}{N}, \tag{6}$$

where N is chosen so as to minimize $\sum e_t^2$, again summing over t from 1 to $n - m$

$$\text{Forecasts: } X_{n-m+k} = \frac{X_{n-m+k-1} + X_{n-m+k-2} + \cdots + X_{n-m+k-N}}{N}. \tag{7}$$

When the subscript of X on the right-hand side of (7) is larger than $n - m$, the corresponding forecasted value is substituted.

(3) Single exponential smoothing.

$$\text{Model fitting: } \hat{X}_{t+1} = \alpha X_t + (1 - \alpha)\hat{X}_t, \tag{8}$$

where α is chosen so as to minimize $\sum e_t^2$, the mean square error where again summing is over t from 1 to $n - m$.

$$\text{Forecasts: } \hat{X}_{n-m+k} = \alpha X_{n-m} + (1 - \alpha)\hat{X}_{n-m+k-1}. \tag{9}$$

(4) Adaptive response rate exponential smoothing.

The equations are exactly the same in (8) and (9), except α varies with t. The value of α_t is found by

$$\alpha_t = |E_t/M_t|, \tag{10}$$

where $E_t = \beta e_t + (1 - \beta)E_{t-1}$ and $M_t = \beta|e_t| + (1 - \beta)M_{t-1}$. ($\beta$ is set at 0.2).

(5) Holt's two-parameter linear exponential smoothing.

$$\text{Model fitting: } S_t = \alpha X_t + (1 - \alpha)(S_{t-1} + T_{t-1}), \tag{11}$$

$$T_t = \beta(S_t - S_{t-1}) + (1 - \beta)T_{t-1}, \tag{12}$$

$$\hat{X}_{t+1} = S_t + T_t. \tag{13}$$

The values of α and β are chosen so as to minimize the mean square error. This was achieved by a complete search of all possibilities.

$$\text{Forecasts: } \hat{X}_{n-m+k} = S_{n-m} + T_{n-m}(k). \tag{14}$$

(6) Brown's one-parameter linear exponential smoothing.

$$\text{Model fitting: } S_t' = \alpha X_t + (1 - a)S_{t-1}', S_t'' = \alpha S_t' + (1 - \alpha)S_{t-1}'', \hat{X}_{t+1} = a_t + b_t, \tag{15}$$

where $a_t = 2S_t' - S_t''$ and $b_t = (1 - a)^{-1}(S_t' - S_t'')$.

The value of α is chosen so as to minimize the mean square error.

$$\text{Forecasts: } \hat{X}_{n-m+k} = a_{n-m} + b_{n-m}(k). \tag{16}$$

(7) Brown's one-parameter quadratic exponential smoothing.

$$\text{Model fitting: } S_t' = \alpha X_t + (1 - \alpha)S_{t-1}', \tag{17}$$

$$S_t'' = \alpha S_t' + (1 - \alpha)S_{t-1}'', \tag{18}$$

$$S_t''' = \alpha S_t'' + (1 - \alpha)S_{t-1}''', \tag{19}$$

$$\hat{X}_{t+1} = \alpha_t + b_t + 1/2c_t, \tag{20}$$

where

$$a_t = 3S_t' - 3S_t'' + S_t''',$$

$$b_t = \alpha\{2(1 - \alpha)^2\}^{-1}\{(6 - 5\alpha)S_t' - (10 - 8\alpha)S_t'' + (4 - 3\alpha)S_t'''\}$$

and

$$c_t = \alpha(1 - \alpha)^{-2}(S_t' - 2S_t'' + S_t''')$$

The value of α is chosen so as to minimize the mean square error.

$$\text{Forecasts: } \hat{X}_{n-m+k} = a_{n-m+k} + b_{n-m+k}(k) + 1/2c_{n-m+k}(k)^2. \tag{21}$$

(8) Linear regression trend fitting.

$$\text{Model fitting: } \hat{X}_t = a + bt, \tag{22}$$

where $t = 1, 2, 3, \ldots, n - m$, and a and b are chosen so as to minimize the sum of the square errors by solving the normal equations:

$$a = \frac{\sum X}{n - m} - b\frac{\sum t}{n - m} \qquad b = \frac{(n - m)\sum tX - t\sum X}{(n - m)\sum t^2},$$

where all summations go from 1 to $n - m$

$$\text{Forecasts: } \hat{X}_{n-m+k} = a + b(n - m + k). \tag{23}$$

(9) Naive 2 as Naive 1 (see (1)) but the data are deseasonalized and then seasonalized.

The seasonal indices for deseasonalizing and seasonalizing the data were done by the decomposition method of the ratio-to-moving averages. The specifics of this method can be seen in Makridakis and Wheelwright (1978, pp. 94–100).

(10) Deseasonalized single moving average as in (2) except the data have been deseasonalized and then reseasonalized.

(11) Deseasonalized single exponential smoothing as in (3) except for deseasonalizing.

(12) Deseasonalized adaptive response rate exponential smoothing as in (4) except for deseasonalizing.

(13) Deseasonalized Holt's exponential smoothing as in (5) except for deseasonalizing.

(14) Deseasonalized Brown's linear exponential smoothing as in (6) except for deseasonalizing.

(15) Deseasonalized Brown's quadratic exponential smoothing as in (7) except for deseasonalizing.

(16) Deseasonalized linear regression as in (8) except for deseasonalizing.

The deseasonalizing of the various methods (9) to (16) was done by computing seasonal indices with a simple ratio-to-moving average (centred) decomposition method. The $n - m$ data of each series were first adjusted to seasonality, as

$$X'_t = X_t/S_j$$

where X'_t is the seasonally adjusted (deseasonalized) value and S_j is the corresponding seasonal index for period t.

The forecasts for $\hat{X}'_{n-m-1}, \hat{X}'_{n-m-2}, \ldots \hat{X}'_n$ were reseasonalized as:

$$\hat{X}_{n-m+k} = \hat{X}'_{n-m+k}(S^j)$$

(17) Holt–Winters' linear and seasonal exponential smoothing.

If the data have no seasonality (i.e. significantly different to zero autocorrelation coefficient at lag 4, for quarterly data, or at a lag 12, for yearly data) then Holt's exponential smoothing is used (see (5) above). Otherwise, Winters' three-parameter model is used:

Model fitting:

$$S_t = \alpha \frac{X_t}{I_{t-L}} + (1 - \alpha)(S_{t-1} + T_{t-1}),$$

$$T_t = \gamma(S_t - S_{t-1}) + (1 - \gamma)T_{t-1},$$

$$I_t = \beta \frac{X_t}{S_t} + (1 - \beta)I_{t-L}, \tag{24}$$

$$\hat{X}_{t+1} = (S_t + T_t)I_{t-L+1},$$

where L is the length of seasonality.

The values of α, β and γ were chosen so as to minimize the MSE. This was done by a complete search of all possibilities, using a grid search method.

$$\text{Forecasts: } \hat{X}_{n-m+k} = (S_{n-12} + kT_{n-12})I_{n-12+k}. \tag{25}$$

Initial values for all exponential smoothing methods were computed by backforecasting on the data. This was done in order to eliminate any possible disadvantage of the exponential smoothing methods.

(18) AEP (Automatic) Carbone–Longini.[1]

The Carbone–Longini filtered method (1977) was developed to provide a practical solution to the problem of adapting over time parameters of mixed additive and multiplicative models without *a priori* information. The general model formulation to which the method applies is written as:

$$y(t) = \left[\left(\sum_{i=1}^{n} a_i(t)^{z_i(t)} \right) \sum_{j=1}^{P} b_j(t)x_j(t) \right] + e(t)$$

where, for time, t, $y(t)$ is the value of a dependent variable; $z_i(t)$ denotes the value assigned to the qualitative dimension i (1 if observed, 0 if not); $x_j(t)$ denotes the measurement of the quantitative feature j: $a_i(t)$ and $b_j(t)$ are the corresponding parameters at time t; and $e(t)$ is an undefined error term. In time series analysis the $z_i(t)$ could represent, for example, seasons (months, quarters, etc.), and the $x_j(t)$, different lag values of a time series.

A negative damped feedback mechanism is used to adapt the parameters over time. It consists of the following two simple recursive formulae:

$$b_j(t) = b_j(t-1) + |b_j(t-1)| \left[\frac{y(t) - \hat{y}(t)}{|\hat{y}(t)|} \cdot \frac{x_j(t)}{\bar{x}_j(t)} \cdot \mu \right]$$

$$a_i(t) = a_i(t-1) + a_i(t-1) \left[\frac{y(t) - \hat{y}(t)}{|\hat{y}(t)|} \cdot z_i(t) \cdot \mu l \right]$$

where $\hat{y}(t)$ is a forecast of $y(t)$ computed on the basis of the parameters at time $t - 1$; $\bar{x}_j(t) = sx_j(t) + (1-s)x_j(t-1)$ with $0 < s < 1$; μ is a damping factor between 0 and 1; and $l < 1$ is a positive constant for all t.

In this study, the method was applied under the most naive of assumptions (see Bretschneider, Carbone and Longini (1979)) in an automatic execution mode with no user intervention. For the 1001 series, the model formulation was one in which all $z_i(t)$ were assumed to be 0 and $x(t)$ represented lag values (autoregressors) of a time series. In other words, the model reduced to an autoregressive equation with time dependent parameters. Of the information available (series names, country, seasonal indicator and type of data), only the type of data (yearly, quarterly or monthly) was used. The number of autoregressive variables was at least 3, 4 or 12 for yearly, quarterly or monthly data respectively. The exact number for a specific series was established

---

[1] Carbone expresses his thanks to Serge Nadeau for his help in designing the AEP package which was specifically used for this study.

automatically as well as the data transformation applied (difference transformation when necessary) by internal program decision rules (automatic analysis of sample autocorrelation functions). In all cases, an identical initialization procedure was applied. Initial values of the parameters were set to the inverse of the number of autoregressors in a model. Start up values for the exponential smoothing means were always 100 with smoothing constant equal to 0.01. A damping factor of 0.06 was applied in all cases. Finally, the necessary learning process (iterating several times forward/backward through the data) was stopped by an internal program decision rule. A discussion of the internal decision rules can be found in Carbone (1980).

The results were obtained in a single run (around three (3) hours of CPU time on a IBM 370/158). Most of the computer time was devoted to reading and writing and report generation. The work could have been efficiently performed on a 64K micro-processor. Again, no revisions of forecasts through personalized analysis were performed.

(19) Bayesian Forecasting.

At its simplest, Bayesian Forecasting is merely a particular method of model estimation in which a prior probability distribution is assigned to the model's parameters and these are subsequently updated as new data become available to produce forecasts. (In the U.K., in particular, the term has recently become synonymous with the approach developed by Harrison and Stevens (1971, 1976). A program developed by Stevens has been used in this study.)

The basic transformed Bayesian forecasting model is:

$$Z_t = \mu_t + S_{i,t} + \varepsilon_t \qquad \varepsilon_t \sim N(0, V_\varepsilon)$$

$$\mu_t = \mu_{t-1} + \beta_t + \delta\mu_t \qquad S\mu_t \sim N(0, V_\mu)$$

$$\beta_t = \beta_{t-1} + \delta\beta_t \qquad \delta\beta_t \sim N(0, V_\beta)$$

$$S_{i,t} = S_{i,t-1} + \delta S_{i,t} \qquad i = 1, 2, \ldots, \tau$$

where $Z_t = \log Y_t$, and μ_t, β_t and S_t are the log transforms of the 'level', 'trend' and 'seasonal' factors.

In matrix notation these equations may be written

$$Z_t = X_t \theta_t + v_t : v_t \sim N(0, V_t) - \text{the observation equation}$$

$$\theta_t = G\theta_{t-1} + w_t : w_t \sim N(0, W_t) - \text{the systems equation}$$

$$\theta_t = (\mu_t, \beta_t, S_{1t} \ldots S_{\tau t})$$

For non-seasonal data, $X_t = (1, 0, 0 \ldots 0, 0, 0 \ldots 0)$

$$G = \begin{bmatrix} 1 & 1 & \\ 0 & 1 & 0 \\ 0 & & I \end{bmatrix}$$

For seasonal data, $X_t = (1, 0, 0 \ldots, 0 \ldots 0)$.

The model (M) is characterized by the four matrices $M \equiv (X_t, G, V_t, W_t)$. With these matrices assumed known it is possible to estimate θ_{t+k} and Z_{t+k} using the Kalman Filter to produce the k-period-ahead forecast.

The Bayesian model employed in the forecasting competition is the so-called multi-state model. Here it is supposed that in each and every period the process is in one of a number of possible states $M^{(i)}: i = 1, \ldots, 4$; the stable or no change state, the step change, the slope change and the transient. It is assumed that these states occur randomly over time with constant probability of occurrence independent of the previous state of the process.

These four different states can be characterized by the following set of parameters:

Model type	Prior weight	RE(i)	RG(i)	RD(i)	RS(i)
$M^{(1)}$: no change	1000	0	0	0	$\sqrt{(12/\tau)}$
$M^{(2)}$: step change	10	0	30% of level	0	0
$M^{(3)}$: slope change	10	0	0	12.5% p.a.	0
$M^{(4)}$: transient	100	33.3% of level	0	0	0

where τ denotes periodicity.

The variance of the raw observations is assumed to be $C^2(EY_t)^2$. The parameter C is estimated by defining a range of values within which it could lie. These were selected by individual examination of each of the 1001 series. With each new datapoint the posterior probability of each C value being correct is calculated. The estimated value of C is merely the average of the eleven C values weighted by their respective probabilities. This average value is then used to calculate the posterior probability distribution of θ_{t+k} and Z_{t+k}.

Bayesian forecasting is iterative in the sense that starting with a subjectively specified prior for the mean level of the series, the growth and seasonal factors, a set of forecasts can be produced. A new observation is then used to update the priors and generate new forecasts.

The priors used to start off the process were

	Low	Mean	High
Prior level (units per period)	1	1000	10%
Prior growth (% p.a.)	-33.3%	0	50%
Prior seasonality (if appropriate)	50%	100%	200%

Note that, as no fitting takes place, the entries in the various tables under 'model fitting' have no meaning for the method of Bayesian forecasting.

(20) Combining Forecasts (Combining A).

This method uses the simple average of methods (11), (12), (13), (14), (17), and (18).

$$\text{Model fitting: } \hat{X}_t = \frac{\hat{X}_t^{(11)} + \hat{X}_t^{(12)} + \hat{X}_t^{(13)} + \hat{X}_t^{(14)} + \hat{X}_t^{(17)} + \hat{X}_t^{(18)}}{6},$$

where $t = 1, \ldots, n - m$ and $\hat{X}_t^{(i)}$ is $\hat{X}_t$ for method (i).

Forecasts: $\hat{X}_{n-m+k} = \dfrac{\hat{X}_{n-m+k}^{(11)} + \hat{X}_{n-m+k}^{(12)} + \hat{X}_{n-m+k}^{(13)} + \hat{X}_{n-m+k}^{(14)} + \hat{X}_{n-m+k}^{(17)} + \hat{X}_{n-m+k}^{(18)}}{6}$

where $k = 1, \ldots, m$.

(21) Combining Forecasts (Combining B).

Here a weighted average of the six methods used in (19) is used. The weights are based on the sample covariance matrix of percentage errors for these six methods for the model fitting for each series.

$$\text{Model Fitting: } \hat{X}_t = \sum_i w_i \hat{X}_t^{(i)},$$

with

$$w_i = \sum_j \alpha_{ij} \bigg/ \sum_h \sum_j \alpha_{hj},$$

where all summations are over the set $\{11, 12, 13, 14, 17, 18\}$ and the d_{ij} terms are elements of the inverse of the covariance matrix of percentage errors. That is, if $S = (\beta_{ij})$, where

$$\beta_{ij} = \sum_{t=1}^{n-m} [u_t^{(i)} - \bar{u}^{(i)}][u_t^{(j)} - \bar{u}^{(j)}]/(n - m)$$

and

$$u_t^{(i)} = e_t^{(i)}/X_t = [X_t - \hat{X}_t^{(i)}]/X_t \qquad \text{and} \qquad \bar{u}^{(i)} = \sum_{t=1}^{n-m} u_t^{(i)}/(n - m),$$

then d_{ij} is the element in row i and column j of S^{-1}.

$$\text{Forecasts: } \hat{X}_{n-m+k} = \sum_i w_i \hat{X}_{n-m+k}^{(i)},$$

where $k = 1, \ldots, m$ and the summation is over the set $\{11, 12, 13, 14, 17, 18\}$.

(22) The Box–Jenkins Methodology.

The Box–Jenkins technique has become very popular since the publication of their book in 1970. In general, the process consists of a cycle of four components: data transformation, model identification, parameter estimation and diagnostic checking. Only after the diagnostic checks indicate that an adequate model has been constructed, are the forecasts produced. The methodology is well documented (see for example Box and Jenkins (1970), Granger and Newbold (1977), Nelson (1973), Anderson (1976) or Chatfield and Prothero (1973)) so that here we give only the broad outlines of what was done. The only data transformation considered was the natural logarithm, which was applied when there appeared to be an exponential trend, or heteroskedasticity in the errors. To look at wider classes of transformation appeared to be too expensive. Model identification was via the autocorrelation function in particular with the partial autocorrelation function used for confirming evidence, combined with a rigorously imposed 'Principle of Parsimonious Parameterization'. Once a tentative model had been identified, the parameters, together with a mean or trend constant were estimated. Diagnostic checking consisted of an examination of the 'important' lag residual autocorrelations and the original Box–Pierce X^2 statistic, together with limited overfitting. To produce the forecasts, the model was extrapolated, together with a correction factor applied, if the logarithms had been analysed.

Finally, the projections were examined to see if they seemed reasonable in light of the historic data. This last check was used mainly to distinguish between competing adequate models.

(23) Lewandowski's FORSYS System.

X_t, the time series, is decomposed as follows:

$$X_t = M_t S_t + e_t \tag{26}$$

The mean, M_t, is defined by a moving average process which is basically of exponential smoothing type. For instance, for a linear model, M_t is defined as:

$$M_t = 2(M1_t) - M2_t \tag{27}$$

where

$$M1_t = \sum^{\theta} \frac{X_{t-\theta}}{S_{t-\theta}} \alpha_{t-\theta} \prod^{\theta} (1 - \alpha_{t-\theta})^{\theta} \tag{28}$$

$$M2_t = \sum^{\theta} M1_{t-\theta} \alpha_{t-\theta} \prod^{\theta} (1 - \alpha_{t-\theta})^{\theta} \tag{29}$$

The smoothing constant α_t is given by:

$$\alpha_t = \alpha_{0_t} + \Delta \alpha_t.$$

The values of α_t vary as follows:

$$\alpha_{0_t} = \alpha_0 \rho^{f_1[\sigma_t^{(1)}]}$$

$$\Delta \alpha_t = \kappa_0 \rho^{f_2[\sigma_t^{(2)}]} - \kappa_1 \rho^{f_3(\Sigma_t)}$$

where $\sigma_t^{(1)}$ is a measure of the stability of the series and is defined as:

$$\sigma_t^{(1)} = \left| \frac{\mathrm{MAD}_t}{M_t} \right|$$

and where

$$\mathrm{MAD}_t = |\varepsilon_t| \gamma + (1 - \gamma) \mathrm{MAD}_{t-1}$$

$\sigma_t^{(2)}$ is a normalized measure of the randomness of the series. It is defined as:

$$\sigma_t^{(2)} = \left| \frac{\varepsilon_t}{\mathrm{MAD}_t} \right|$$

and finally, Σ_t^* is a tracking signal defined as follows:

$$\Sigma_t^* = \frac{\Sigma_t}{\text{MAD}_t}$$

where

$$\Sigma_t = \Sigma_{t-1}(1 - \gamma_{S_t}) + \varepsilon_t$$

where γ_{S_t} can be thought of as the coefficient of decay, that is:

$$\gamma_{S_t} = \gamma_{S_0}[1 - \rho^{f_4(\sigma_t^{(2)})}]$$

The seasonal coefficients are found by an exponential smoothing process similar to that of (28) and (29) which is:

$$S_t = \sum^\tau \frac{X_{t-\tau}}{M_{t-\tau}} \beta_{t-\tau} \prod^\tau (1 - \beta_{t-\tau})^\tau,$$

where

$$\beta_t = \beta_0 \bar{\rho}^{f_s(\Sigma_t^*)}.$$

The forecasting of the series is given by combining the components of (1), that is M_t and S_t. This results in the following projections:

$$\hat{X}_{t+\kappa}^{(1)} = M(\alpha)_t + \kappa T(\alpha)_t + \kappa^2 Q(\alpha)_t$$

$$\hat{X}_{t+\kappa}^{(s)} = M(a^\delta)_t + \kappa T(a^\delta)_t + \kappa^2 Q(a^\delta)_t$$

$$\hat{X}_{t+\kappa}^{(2)} = M(\alpha^*)_t + \kappa T(\alpha^*)_t$$

Finally, the forecasts are found by

$$\hat{X}_{t+\kappa} = \{\hat{X}_{t+\kappa}^{(\delta)} \delta_{t+\kappa}\} S_{t+\kappa}$$

For more details, see Lewandowski (1979).

(24) ARARMA Methodology.

The models used are called **ARARMA** models (see Parzen (1979), (1980)) because the model computed adaptively for a time series is based on sophisticated time series analysis of ARMA schemes (a short memory model) fitted to residuals of simple extrapolation (a long memory model obtained by parsimonious 'best lag' non-stationary autoregression).

The model fitted to a time series $Y(.)$ is an iterated mode.

$$Y(t) - \boxed{} \rightarrow \tilde{Y}(t) - \boxed{} \rightarrow \varepsilon(t).$$

If needed to transform a long memory series Y to a short memory series $\tilde{Y}$, $\tilde{Y}(t)$ is chosen to satisfy one of the three forms

$$\tilde{Y}(t) = Y(t) - \hat{\phi}(\hat{\tau})Y(t - \hat{\tau}),$$

$$\tilde{Y}(t) = Y(t) - \phi_1 Y(t-1) - \phi_2 Y(t-2), \tag{30}$$

$$\tilde{Y}(t) = Y(t) - \phi_1 Y(t - \tau - 1) - \phi_2 Y(t - \tau). \tag{31}$$

Usually $\tilde{Y}(t)$ is short memory, then it is transformed to a white noise, or no memory, time series $\varepsilon(t)$ by an approximating autoregressive scheme $AR(\hat{m})$ whose order $\hat{m}$ is chosen by an order determining criterion called CAT).

To determine the best lag, $\hat{\tau}$, a non-stationary autoregression is used; either a maximum lag M is fixed and $\hat{\tau}$ is chosen as the lag minimizing over all τ

$$\sum_{t=M+1}^{T} \{Y(t) - \phi(\tau)Y(t-\tau)\}^2$$

or $\hat{\tau}$ is chosen as the lag minimizing over all τ

$$\sum_{t=\tau+1}^{T} \{Y(t) - \phi(\tau)Y(t-\tau)\}^2$$

For each τ, one determines $\phi(\tau)$, and then one determines $\hat{\tau}$ (the optimal value of τ) as the value minimizing

$$\text{Err}(\tau) = \sum_{t=M+1}^{T} \{Y(t) - \hat{\phi}(t)Y(t-\tau)\}^2$$

or

$$\text{Err}(\tau) = \sum_{t=\tau+1}^{T} \{Y(t) - \hat{\phi}(\tau)Y(t-\tau)\}^2$$

The decision as to whether the time series is long memory or not is based on the value of $\text{Err}(\hat{\tau})$. An *ad hoc* rule is used if $\text{Err}(\hat{\tau}) < 8/T$, the time series is considered long memory. When this criterion fails one often seeks transformations of the form of (30) or (31), using semi-automatic rules described elsewhere (see Parzen (1982)).

REFERENCES

Anderson, O. D. (1976). *Time Series Analysis and Forecasting—The Box–Jenkins Approach*, Butterworth.

Armstrong, J. C. (1978). 'Forecasting with Econometric Methods: Folklore Versus Fact', *Journal of Business*, **S1**, 549–600.

Bates, J. M. and Granger, C. W. J. (1969). 'Combination of Forecasts', *Operational Research Quarterly*, **20**, 451–468.

Box, G. E. P. and Jenkins, G. M. (1970). *Time Series Analysis, Forecasting and Control*, San Francisco: Holden Day.

Bretschneider, S., Carbone, R. and Longini, R. L. (1979). 'An Adaptive Approach to Time Series Analysis', *Decision Sciences*, **10**, 232–244.

Carbone, R. (1980). *UNIAEP Program Documentation*, Pittsburgh: EDA Inc.

Carbone, R. and Longini, R. L. (1977). 'A Feedback Model for Automated Real Estate Assessment', *Management Science*, **24**, 241–248.

Chatfield, C. and Prothero, D. L. (1973). 'Box–Jenkins Seasonal Forecasting: Problems in a Case Study', *Journal of the Royal Statistical Society*, A, **136**, 295–336.

Hollander, M. and Wolfe, D. A. (1973). *Nonparametric Statistical Methods*, New York: Wiley.

Lewandowski, R. (1979). *La Prévision à Court Terme*, Paris: Dunod, 1979.

Makridakis, S. and Hibon, M. (1979). 'Accuracy of Forecasting: An Empirical Investigation (with discussion)', *Journal of the Royal Statistical Society*, (A), **142**, Part 2, 97–145.

Makridakis, S. and Wheelwright, S. C. (1978). *Forecasting: Methods and Applications*, New York: Wiley/Hamilton.

Makridakis, S. *et al. The Accuracy of Major Extrapolation (Time Series) Methods*, Wiley, forthcoming.

Nelson, C. R. (1973). *Applied Time Series Analysis for Managerial Forecasting*, San Francisco: Holden Day.

Nelson, H. L. Jr. and Granger, C. W. J. (1979). 'Experience with using the Box–Cox Transformation When Forecasting Economic Time Series, *Journal of Econometrics*, **10**, 57–69.

Newbold, P. and Granger, C. W. J. (1974). 'Experience with Forecasting Univariate Time Series and the Combination of Forecasts', *Journal of the Royal Statistical Society*, (A), **137**, 131–165.

Parzen, E. (1979). 'Time Series and Whitening Filter Estimation', *TIMS Studies in Management Science*, **12**, 149–165.

Parzen, E. (1980). 'Time Series Modeling, Spectral Analysis, and Forecasting', *Directions in Time Series Analysis*, ed. D. R. Brillinger and G. C. Tiao, Institute of Mathematical Statistics.

Parzen, E. (1982). 'ARARMA Models for Time Series Analysis and Forecasting', *Journal of Forecasting*, **1**, 67–82.

Pike, D. H., Pack, D. J. and Downing, D. J. (1980). 'The Role of Linear Recursive Estimates in Time Series Forecasting', Computer Sciences Division, Union Carbide Corporation (Nuclear Division), Oak Ridge National Laboratory.

Reid, D. J. (1969). 'A Comparative Study of Time Series Prediction Techniques on Economic Data', *Ph.D. Thesis*, Department of Mathematics, University of Nottingham.

Slovic, P. (1972). 'Psychological Study of Human Judgement: Implications for Investment Decision Making', *Journal of Finance*, **27**, 779–799.

Winkler, R. L. (1981). 'Combining Probability Distributions From Dependent Information Sources', *Management Science*, **27**, 479–488.

CHAPTER 5

Forecasting: The Box–Jenkins Approach

Allan Andersen and Andrew Weiss
University of Sydney, Australia

INTRODUCTION

The so-called 'Box–Jenkins' technique has gained great popularity since the publication of their book in 1970, both as a vehicle for theoretical developments, and as a method suitable for 'real-world' situations. Applications may be found in such diverse fields as astronomy, economics, accounting, psychology and criminology, and many others. In fact, there is hardly an issue of an applied type journal without a reference to their work. It seems quite natural, therefore, that this method (the exact meanings of such terms as method, technique, methodology or in fact mythodology, escape us, and thus we use all of these interchangeably) should be represented in this book.

Of all the techniques used here, it has the greatest theoretical base. It is also probably the most user-dependent, in that the analyst is required to use his judgement quite frequently throughout the process. Hence, it is important that part of this chapter be given over to our particular interpretation of the technique. The first six sections are devoted to a simple textbook type development of the theory, while the remainder deals with several case-studies, and comments on the results.

1. SOME THEORY

In no way is this section intended fully to describe the Box–Jenkins technique. It has been written for two purposes: to give an overview of the topic to those who may not be familiar with it; and to highlight where we, as the practitioners, have imposed our own philosophies within the general framework.

Let us begin by assuming that we have been given a set of equally spaced (in time) observations, and asked to predict the next value which will occur. Each of the techniques in this book is applicable in this situation.

The method we are using differs from most of the others in that it considers this particular set of data as a sample from some population. Our task as forecasters,

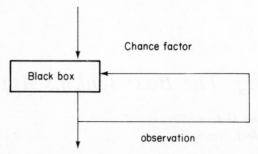

FIGURE 1   The generating mechanism

therefore, is to make inferences concerning the population on the basis of one sample and some assumptions, and then to use the inferred population characteristics to provide the required forecast. The first, and most basic, assumption is that there exists an underlying process, containing a chance factor, which produces the present observation (realization) in terms of past and present chance elements, and past realizations.

At the moment, there are no restrictions on the generating mechanisms. The first we impose is that of stationarity. Usually, we are given just one series, and asked to produce a forecast. Suppose instead, for a moment, that we are permitted to run the economy, factory, experiment or whatever, any number of times, thus obtaining that number of samples from our population, and hence the same number of 'observations' at a particular time point, t.

We may then define the random variable X_t, to be a variable whose possible values are all the realizations. Taken individually, each random variable has a probability distribution determined by that of the 'chance element'. Taken in pairs, there are many bivariate probability distributions, and, if considered n (say) at a time, there is a set of n-dimensional probability distributions. A generating mechanism is said to be stationary if the probability distribution function of $X_{t_1}, X_{t_2}, \ldots, X_{t_n}$ is the same as that of $X_{t_1+k}, X_{t_2+k}, \ldots, X_{t_n+k}$, for any integers $n, k, t_1, \ldots, t_n$. For the case $n = 1$, we see that this means that the average value and variance of any of the variables are time invariant. When $n = 2$, the covariances depend not upon the occurrence times of the variables but on the time 'distance' between them.

Returning to the one-realization case, suppose we consider that our data is non-seasonal, and that it was generated by a stationary process. From one realization, stationarity is impossible to 'prove', however, some indications are a lack of trend, or seasonal pattern, or that the variations in the data do not grow in time. Under these circumstances, we proceed by making the further assumption that the underlying process has the form

$$X_t = \mu + \phi_1 X_{t-1} + \cdots + \phi_p X_{t-p} + \varepsilon_t + \theta_1 \varepsilon_{t-1} + \cdots + \theta_q \varepsilon_{t-q}$$

and that the chance epsilons are identically and independently normally distributed with mean zero and constant variance σ^2. In the jargon, this process is referred to as a ARMA (p, q) process. To gain an insight into the theoretical properties of these processes, let us consider some simple cases first.

(1) White noise

$$X_t = \varepsilon_t + \mu.$$

This is the simplest of all generating mechanisms. The first question concerns stationarity. Quite obviously, if $\{\varepsilon_t\}$ are Normal, then so is $\{X_t\}$ so that if we can determine that all means, variances, and covariances are time invariant, then we have obtained stationarity.

$$E(X_t) = \mu$$
$$V(X_t) = \sigma^2$$
$$\text{Cov}(X_t X_{t-k}) = 0, \qquad k > 0.$$

The above calculations show that this is the case.

(2) The MA(1) process

$$X_t = \varepsilon_t + \theta_1 \varepsilon_{t-1} + \mu$$
$$E(X_t) = \mu$$
$$V(X_t) = (1 + \theta_1^2)\sigma^2$$
$$\text{Cov}(X_t X_{t-1}) = \theta_1 \sigma^2$$
$$\text{Cov}(X_t X_{t-k}) = 0, \qquad k > 1.$$

Once again, there is no problem with stationarity. The feature which allows us to distinguish between these two processes is the correlation between X_t and X_{t-1} (here referred to as the autocorrelation at lag 1, and given the symbol $\rho(1)$). In case (1) $\rho(k) = 0$ for all $k > 0$ while for the second case $\rho(k) = 0$, for all $k > 1$. The graph of $\rho(k)$ against k is called the theoretical autocorrelation function (ACF). Here, the two functions would look distinctly different.

(3) The AR(1) process

$$X_t = \phi_1 X_{t-1} + \mu + \varepsilon_t.$$

If we take expectations of both sides of this expression, we see

$$\mu_t = \phi_1 \mu_{t-1} + \mu$$
$$= \phi_1^{T+t} \mu_{-T} + \mu \sum_{j=0}^{t+T-1} \phi_1^j$$

which converges to

$$\mu_t = \frac{\mu}{1 - \phi}$$

for large T, fixed μ_{-T} (the starting up value), and $|\phi| < 1$, independent of either μ_{-T} or t.

Similarly, under the same type of conditions,

$$\text{Var}(X_t) = \frac{\sigma^2}{1 - \phi^2}, \qquad \text{also independent of } \sigma_{-T} \text{ or } t.$$

We also need to look at the covariances.

Multiplying through by $(X_{t-k} - \mu/(1 - \phi))$,

$$\left(X_t - \frac{\mu}{1 - \phi}\right)\left(X_{t-k} - \frac{\mu}{1 - \phi}\right) = \phi\left(X_{t-1} - \frac{\mu}{1 - \phi}\right)\left(X_{t-k} - \frac{\mu}{1 - \phi}\right)$$

$$+ \, \varepsilon_t\left(X_{t-k} - \frac{\mu}{1 - \phi}\right).$$

Thus,

$$\begin{aligned}
\text{Cov}(t, k) &= \phi_1 \text{Cov}(t - 1, k - 1) \\
&= \phi_1^k \text{Cov}(t - k, 0) \\
&= \phi_1^k \text{Var}(X_t)
\end{aligned}$$

giving

$$\rho(t, k) = \frac{\phi_1^k \text{Var}(X_t)}{\sqrt{\text{Var}(X_t)\,\text{Var}(X_{t-k})}} = \phi_1^k,$$

which is not a function of t. Figure 2 gives various theoretical autocorrelation functions for this process.

Since we have assumed that the epsilons have a Normal distribution, it may now be shown that the $\{X_t\}$ has a multivariate normal distribution uniquely specified by the means, variances and covariances obtained above. Thus, under the condition $|\phi| < 1$, a process of this type, beginning in the infinite past, is stationary.

(4) The AR(p) process

$$X_t = \phi_1 X_{t-1} + \cdots + \phi_p X_{t-p} + \mu + \varepsilon_t.$$

For notational convenience, we define the 'backwards' operator B, by the relation $B . X_t = X_{t-1}$. That is, instead of looking at the random variable at time t, by applying the operator B, we can concentrate on the previous one, X_{t-1}. We may define powers of B to indicate repetitive application, i.e.:

$$(B^k)X_t = B^{k-1}B . X_t = B^{k-1}X_{t-1} = \cdots = X_{t-k}.$$

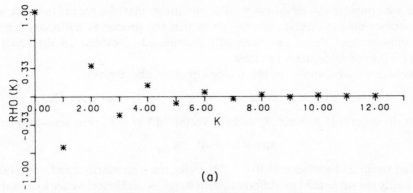

FIGURE 2(a) The ACF of an AR(1) process with $\phi = -0.6$

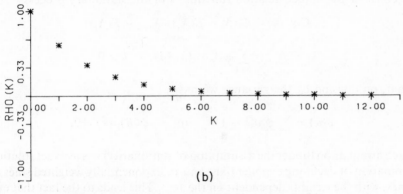

FIGURE 2(b) The ACF of an AR(1) process with $\phi = 0.6$

Thus, we may replace the above equation by

$$(1 - \phi_1 B - \phi_2 B^2 - \cdots - \phi_p B^p)X_t = \varepsilon_t + \mu$$

which is more succinctly written as

$$\phi(B)X_t = \varepsilon_t + \mu$$

where $\phi(y)$ is a polynomial of order p.

Another useful representation of an AR(p) processes is a vector AR(1) processes

$$\mathbf{X}_t = \begin{bmatrix} X_t \\ X_{t-1} \\ \vdots \\ X_{t-p+1} \end{bmatrix} = \begin{bmatrix} \phi_1, \phi_2, \ldots, \phi_p \\ 1, 0, \ldots, 0 \\ \vdots \\ 0, \ldots, 0, 1, 0 \end{bmatrix} \begin{bmatrix} X_{t-1} \\ X_{t-2} \\ \vdots \\ X_{t-p} \end{bmatrix} + \begin{bmatrix} \varepsilon_t \\ 0 \\ \vdots \\ 0 \end{bmatrix} + \begin{bmatrix} \mu \\ 0 \\ \vdots \\ 0 \end{bmatrix}$$

$$= \mathbf{A}\mathbf{X}_{t-1} + \varepsilon_t + \mu$$

If we consider the usual eigenvalue condition that the mean, variance and covariance do not diverge, we may show that the process is stationary (on the assumption that the ε^s are Normally distributed) provided all the roots of $\phi(y) = 0$ lie outside the unit circle.

Repetitive substitution in the vector equation above gives

$$\mathbf{X}_t = (\varepsilon_t + \mu) + \mathbf{A}(\varepsilon_{t-1} + \mu) + \cdots + \mathbf{A}^{t+T-1}(\varepsilon_{-T} + \mu) + \mathbf{A}^{t+T}\mathbf{X}_{-T-1}$$

with the difference between $\mathbf{X}_t$ and the vector MA $(t + T)$ representation being

$$\mathbf{R}(t + T) = \mathbf{A}^{t+T}\mathbf{X}_{-T-1}.$$

The mean and variance of $\mathbf{R}(t + T)$, under the stationarity condition above, are easily shown to tend to (different sized) $\mathbf{0}$ as $T \to \infty$. Hence, we see by equating corresponding matrix positions that X_t has an MA (∞) representation. Let us now consider the autocorrelation function. For the stationary process

$$\text{Cov}(k) = E\{(X_t - E(X_t))(X_{t-k} - E(X_t))\}$$

$$= \sum_{i=1}^{p} \phi_i \text{Cov}(k - i) \qquad k > 0$$

so that, on division by the variance we have

$$\rho(k) = \sum_{i=1}^{p} \phi_i \rho(k - 1) \qquad \text{or} \qquad \phi(B)\rho(k) = 0.$$

Such an equation (under the assumption of stationarity) has as exact solution a combination of declining exponentials, and/or exponentially weighted sines and cosines, with the weights dependent on the $\{\phi_i\}$. This leads to the fact that every different AR process has, in its autocorrelation function, a 'fingerprint'. However, it is also true that many of the 'fingerprints' are almost impossible to distinguish.

(5) *The MA(q) process*

$$X_t = \varepsilon_t + \theta_1 \varepsilon_{t-1} + \cdots + \theta_q \varepsilon_{t-q} + \mu = \theta(B)\varepsilon_t + \mu.$$

Under the assumption of Normal epsilons

$$X_t \sim N(\mu, (1 + \theta_1^2 + \cdots + \theta_q^2)\sigma^2)$$

and

$$\rho(k) = \sum_{i=0}^{q-k} (\theta_i \theta_{i+k}) \bigg/ \sum_{i=0}^{q} \theta_i^2, \qquad k < q \text{ and } \theta_0 = 1$$

$$= 0 \qquad k > q.$$

Now, since none of these formulae contain a reference to time, we see that all MA processes are stationary. We have also seen the fundamental difference between the two types of processes. For the AR case, the autocorrelation function decays exponentially, while for the MA processes, it cuts out suddenly.

(6) The ARMA (p, q) process

$$X_t - \phi_1 X_{t-1} - \phi_2 X_{t-2} - \cdots - \phi_p X_{t-p} = \varepsilon_t + \theta_1 \varepsilon_{t-1} + \cdots + \theta_q \varepsilon_{t-q} + \mu$$

or

$$\phi(B)X_t = \theta(B)\varepsilon_t + \mu.$$

Not surprisingly, the conditions for stationarity concern the AR parts only, and are precisely the same as those for the corresponding AR(p) process

$$\phi(B)X_t = \varepsilon_t + \mu$$

i.e. the solutions to $\phi(y) = 0$ all lie outside the unit circle. It is not difficult to show that the autocorrelation function for this process has the form

$$\phi(B)\rho(k) = 0 \qquad \text{for } k > q$$

and a considerably more complex formula for $\rho(k)$, when $k \le q$.

Now, if $q < p$, we require $\rho_0, \ldots, \rho_{p-1}$ as starting up values for the recursive relation above in exactly the same way as for the corresponding AR(p) process. However, unlike that case, these starting up values are functions of both the ϕ^s and θ^s, giving an 'overall' fingerprint different (at least theoretically) from that of any pure AR process.

Conversely, for $q \ge p$, there are $q - p + 1$ 'atypical' autocorrelations, before the recursive pattern commences. Two examples are given in Figure 3.

Hence, while the theoretical autocorrelation functions of different processes are different, it may be quite difficult to distinguish between them. The major differences are highlighted by the theoretical properties of a second statistic; the partial autocorrelation coefficient. This statistic, ϕ_{kk}, is related to the 'extra' proportion of the random variable X_t explained by approximating the true process by an AR(k), rather than an AR($k-1$). In line with previous terminology, we call the graph of ϕ_{kk} against k the partial autocorrelation function (PACF).

An AR(p) process, quite obviously, is exactly approximated by an AR(p), hence $\phi_{kk} = 0$, for $k > p$. On the other hand, since both the MA and ARMA processes have, in general, AR(∞) representations, any extra term will give a non-zero contribution, but its value will decrease as the number of terms increase.

	AR(p)	MA(q)	ARMA (p, q)
ACF	declines exponentially	cuts out after q lags	declines exponentially
PACF	cuts out after p lags	declines exponentially	declines exponentially

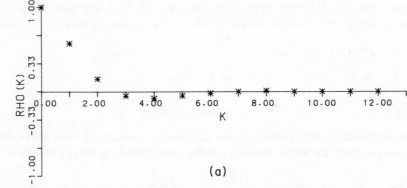

FIGURE 3(a)　The ACF of an ARMA (2,1) process with $\phi(1) = 0.8$, $\phi(2) = -0.3$ and $\theta = 0.1$

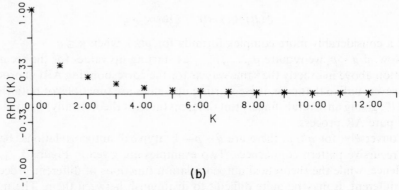

FIGURE 3(b)　The ACF of an ARMA (1,1) process with $\phi = 0.6$ and $\theta = 0.3$

Now, after all this, you might exclaim: 'we certainly know a lot of theory, and all I wanted was a forecast'. The next section ties in the practice with theory.

2.　THE APPLICATION

Of course, in any given situation, if we know the underlying process, then it is not difficult to produce the forecasts. However, all we have is a single sample from the (assumed stationary) process. It is our job to take the sample and make a 'best guess' concerning the process. This is done in a cycle of three parts.

(1)　IDENTIFICATION
(2)　ESTIMATION
(3)　DIAGNOSTIC CHECKING

The first stage involves guessing at p and q. In the second we estimate the $\{\phi_i\}$ $\{\theta_i\}$ and μ, while in the third, we check to see if our model (as our 'guess' is called) is adequate. If it is, we produce the forecasts; if not, we reidentify, re-estimate and apply the diagnostic techniques again, until the model becomes adequate. In our experience, this usually takes two or three examinations, but sometimes up to six. At that stage, it is possibly best to scrap the models already considered and start on some new tack!

Returning to the problem of tentatively identifying a model. As you may have guessed, the tools for doing this revolve around the functions we have already developed. By utilizing the stationarity assumptions we are able to estimate the autocorrelations as follows

$$r(k) = \frac{\sum\limits_{t=k+1}^{N} (x_t - \bar{x})(x_{t-k} - \bar{x})}{\sum\limits_{t=1}^{N} (x_t - \bar{x})^2},$$

where $\bar{x}$ is the mean and N is the number of observations.

This is not the only formula used; however, all the popular formulae are asymptotically equivalent. The estimates of the PACF, $\hat{\phi}_{kk}$, are derived from $\{r(k)\}$. Having computed these summary statistics, we need to choose the most appropriate model. The choice is very personal. There are few rules to follow; it is mainly a matter of 'eye-balling' the estimated correlation functions, picking up the patterns discussed in section 1, noticing the large coefficients, and by selecting the 'clues' felt to be important, putting together an appropriate model. There are two points to note here. Firstly, it should be remembered that this is a preliminary choice only; if it proves to be inadequate, we can easily discard it and try again. Secondly, as a matter of principle, we believe that it is better to be 'simply' wrong than 'complexly' wrong (parsimony).

There is some statistical knowledge to aid us in the model selection. The exact distributions of $\{r(k)\}$ and $(\hat{\phi}_{kk})$ are difficult to obtain, and are dependent upon the unknown parameters, except in the important case of white noise, in which case the analysis is finished, since then the optimal predictor is merely the mean. For large N, under this null-hypothesis, it may be shown that $r(k) \sim N(0, 1/N)$ for $k \neq 0$.

Thus the test (against the alternative $\rho(k) \neq 0$), involves comparing $r(k)$ against $1.96/\sqrt{N}$. Checking for lack of correlation at any lag is usually carried out by ignoring the problems of multiple hypothesis testing, and simply plotting $\pm 1.96/\sqrt{N}$ on a graph of the ACF, as shown in the example below.

In this example, the confidence intervals should have been modified for the fact that $r(1)$ is significantly different from zero. However, since the purpose of this exercise is to get a 'feeling' for the underlying process, this problem was ignored.

Our second statistic $\hat{\phi}_{kk}$, behaves very much like $r(k)$, in that, under the white noise null hypothesis, it too is distributed (in large samples) as $N(0, 1/N)$, allowing the same type of hypothesis tests.

Having decided upon preliminary values of p and q, we need to estimate

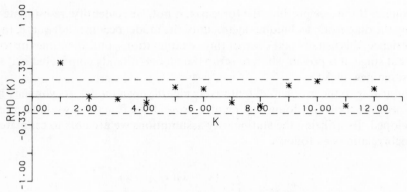

FIGURE 4   A possible residual ACF

parameters $\{\phi_i\}$ $\{\theta_j\}$ and μ. There are so many ways of doing the estimation that a whole chapter could easily be devoted to the problem. The criterion which we used in the experiment, and probably the one used by most practitioners, is least squares. That is, if

$$\hat{\phi}(B)X_t = \hat{\theta}(B)\hat{\varepsilon}_t + \hat{\mu}$$

is an estimated model (defining the residuals $\{\hat{\varepsilon}_t\}$), then we choose that set of $\{\hat{\phi}_i\}$ $\{\hat{\theta}_j\}$ and $\hat{\mu}$ which minimizes $\sum_{t=p+1}^{N} \hat{\varepsilon}_t^2$. The minimization is achieved by an iterative application of ordinary least squares to linear approximations of the chosen model.

For a purely AR model, there is no difficulty in obtaining the residuals, since

$$\hat{\varepsilon}_t = \hat{\phi}(B)X_t - \hat{\mu}.$$

On the other hand, when the model contains an MA part, the residuals are estimated recursively by

$$\hat{\phi}(B)X_t = \hat{\theta}(B)\hat{\varepsilon}_t + \hat{\mu}$$

using the q starting up values $\{\hat{\varepsilon}_p, \ldots, \hat{\varepsilon}_{p-q+1}\}$ which are most easily (and were for the 111 series) estimated by putting them equal to the unconditional expectation of $\{\varepsilon_p, \ldots, \varepsilon_{p-q+1}\}$, which of course is $\{0, \ldots, 0\}$. 'Better' estimates may be obtained by a technique known as 'back-forecasting'.

In general, to obtain forecasts for future observations, we will need to estimate the most recent epsilons. It is of interest to consider the effect of the 'starting up' approximations of those estimates. Let us for the moment pretend that we have the true process, and merely need to estimate ε_N, the last impulse. We know

$$\phi(B)X_N = \theta(B)\varepsilon_N + \mu$$

and require

$$\phi(B)X_N = \theta(B)\hat{\varepsilon}_N + \mu$$

where the latter incorporates the approximation. On subtraction, putting $e_N = \varepsilon_N - \hat{\varepsilon}_N$, we have

$$0 = \theta(B)e_N + 0$$

so that

$$\theta(B)e_N = 0,$$

a difference equation for e_N. The condition that $e_N \to 0$ as $N \to \infty$ is that all roots of $\theta(y)$ lie outside the unit circle. Such a condition is known as invertibility, and is essential for parameter estimation, model checking, and forecasting, and guarantees the existence of the $AR(\infty)$ representations mentioned previously. Naturally we do not have the true process, but impose the conditions of stationarity and invertibility on the parameter estimates. The next stage involves 'checking' the adequacy of the model.

The estimation procedure provides not only point estimates for the parameters, but asymptotic standard deviations as well, so that classical hypothesis tests may be used to gauge their significance. Because of their large sample nature, such tests were not always rigorously applied, especially when only a few observations were available.

Under the hypothesis that the model form is correct, and thus that the only error is the sampling variation in our parameter estimates, one may show (once again for large samples), that

(a) $r_{\hat{\varepsilon}}(k)$, the kth residual autocorrelation, has a $N(0, 1/N)$ distribution, and
(b) $R = N\sum_{k=1}^{M} r_{\hat{\varepsilon}}(k)^2$, the Box–Pierce portmanteau statistic, has a $\chi^2(M - p - q)$ distribution.

Note that the distribution of R is not dependent upon whether the mean was included in the estimation, an error which is present in the programme used for the experiment. Recently, there has been severe criticism of the Box–Pierce statistic, in that even for realistically 'large' samples its distribution is not close to χ^2.

The diagnostic checking stage is obviously very personalized. Our strategy was basically as follows. A model was considered inadequate if any of the low (or seasonal) order autocorrelations were significant. That is, lags 1, 2 and perhaps 3, were considered the most important, with little attention paid to other lags, provided the values were not 'huge'. The Box–Pierce statistic was then used for confirmatory evidence that the rest of the autocorrelations taken as a group were not wildly unacceptable. This rarely occurred (Prothero and Wallis, 1976).

If the model was adequate using (a) and (b) above, then the simplicity concept was considered. Insignificant parameters were dropped one at a time, and their effects were noted on the residual mean square and autocorrelation function. It is interesting to note that, in our experience, the MA parameters were important for removing autocorrelations but usually contributed little to the fit.

On the other hand, if the model was inadequate, then some appropriate action was taken; perhaps an extra parameter, or perhaps a whole new specification. In any case, the cycle proceeded until we were satisfied that the model provided a simple adequate representation of the process.

3.   FORECASTING

Having finally obtained an adequate model, we proceed to the forecasting stage. The theory of forecasting has been developed to a very general level. Here we consider only what is 'usually' done. Define the forecast of X_{N+h} made at time N (and thus lead h) to be $f_{N,h}$. We choose $f_{N,h}$ to minimize

$$E(X_{N+h} - f_{N,h})^2.$$

It is not difficult to show that if the model is the true process, then the forecasts are iteratively calculated by

$$\phi(B)f_{N,h} = \theta(B)\hat{\varepsilon}_{N+h} + \mu,$$

where B acts on h, and

$$\begin{aligned} f_{N,k} &= X_{N+k} \\ \hat{\varepsilon}_{N+k} &= \varepsilon_{N+k} \end{aligned} \qquad \text{for } k \leq 0 \text{ (recall invertibility)}$$

and

$$\hat{\varepsilon}_{N+k} = 0, \qquad \text{for } k > 0.$$

Simply put, producing the forecasts involves writing down the model for the appropriate lead, and substituting our best estimates. Quite obviously, the best estimates of things we have already seen are those observed values. For the epsilons and x^s yet to be seen we use the expected value, 0, and the previously calculated best forecast, f, respectively. In reality, when only the estimated model is available, the same procedure is followed, using the relevant estimated quantities.

As an example, consider the ARMA (1, 1) model

$$X_t = 0.5X_{t-1} + \varepsilon_t + 0.8\varepsilon_{t-1} + 2.$$

$$f_{T,1} = \hat{X}_{T+1} = 0.5\hat{X}_T + \hat{\varepsilon}_{T+1} + 0.8\hat{\varepsilon}_T + 2.$$

$$= 0.5X_T + 0 + 0.8\varepsilon_T + 2.$$

$$f_{T,2} = 0.5f_{T,1} + 2$$

and for $l \geq 2$,

$$f_{T,l} = 0.5f_{T,l-1} + 2.$$

Given stationarity, the long-lead forecasts revert to the mean, reflecting the fact that the information in the autocorrelation function is of little value for long-lead forecasting. This feature extends to the more general models.

4. SEASONALITY

The concept of seasonality is difficult to define. For ease in discussion, let us assume that we have quarterly data. We say that a process is seasonal if there is a correlation at the seasonal lags greater than that expected from the 'non-seasonal' components discussed above.

The processes we consider are of two types; with either multiplicative or additive seasonal components. The multiplicative type processes are probably the more popular, with the general form

$$\phi(B)\Phi(B^4)X_t = \theta(B)\Theta(B^4)\varepsilon_t + \mu,$$

where $\phi(B)$ and $\theta(B)$ are the usual AR and MA operators, and $\Phi(B^4)$ and $\Theta(B^4)$ are polynomials in the seasonal lags whose roots lie outside the unit circle (to maintain stationarity and invertibility respectively).

As one would expect, the three-part cycle of identification, estimation, and diagnostic checking is still used. The estimation stage presents no difficulties. On the other hand, the identification and diagnostic checking parts become much more difficult. Let us consider some simple cases:

(1) $X_t = \Phi_4 X_{t-4} + \varepsilon_t$

For this process, the theoretical autocorrelation function has the form

$$\rho(k) = \Phi_4 \rho(k-4), \qquad \text{for all } k \geq 0.$$

A little mathematics (or thought), will show that

$$\rho(k) = \Phi_4^{k/4}, \qquad \text{for } k = 0, 4, 8, \ldots,$$
$$= 0 \qquad \text{otherwise.}$$

We may complicate this a little more by considering the simplest 'mixed' multiplicative process:

(2) $(1 - \phi B)(1 - \Phi_4 B^4)X_t = \varepsilon_t + \mu$

so that

$$(1 - \phi B)(1 - \Phi_4 B^4)\rho(k) = 0, \qquad \text{for } k > 0.$$

Thus $\rho(k)$ is a declining exponential augmented by damped seasonal sine waves. For more complex processes the theoretical autocorrelation functions are made up from the correlation functions of the corresponding seasonal and non-seasonal processes. For the seasonal moving average process

(3) $X_t = (1 - \theta_1 B)(1 - \theta_4 B^4)\varepsilon_t + \mu,$
      $= (1 - \theta_1 B - \theta_4 B^4 + \theta_1\theta_4 B^5)\varepsilon_t + \mu,$

the only non-zero correlations are at lags 1, 3, 4, and 5, with the constraint that

$\rho(3) = \rho(5)$, distinguishing this process from the MA(s) process with some non-zero coefficients. For the seasonal ARMA processes, the autocorrelation function eventually has the same properties as the corresponding seasonal AR process, with the starting up values affected by the MA parameters.

The additive processes, as the name suggests, include the seasonal lags as additive powers of B in the AR or MA parts. For example

$$(1 - \phi_1 B - \phi_4 B^4)X_t = \varepsilon_t + \mu,$$

is an additive seasonal AR model. We may consider it as a more complex regular AR process, and develop the general class of theoretical autocorrelation and partial autocorrelation functions in the same way as the non-seasonal processes.

It has been our experience that the identification and diagnostic checking stages of the cycles for seasonal data involves a great deal of what may be called luck, or 'let's try this'. The analysis turns out to be not nearly as 'scientific' for seasonal data as it is for non-seasonal case. The principles of forecasting remain the same with seasonality present.

5. NON-STATIONARITY

Since most data series appear to be growing in time, or exhibit very stable seasonal patterns, it is necessary to widen our acceptable class of processes to include some that are non-stationary. Let us, once again, return to non-seasonal data, where the most general ARIMA (p, d, q) process has the form

$$\phi(B)(1 - B)^d X_t = \mu + \theta(B)\varepsilon_t,$$

with stationary and invertible operators $\phi(B)$ and $\theta(B)$ respectively; that is, we have generalized the AR part to allow roots on the unit circle, but not inside. It is not difficult to show that, for this type of process, the theoretical autocorrelations are dominated by the roots on the unit circle resulting in them remaining more or less constant over long lags; enabling us to differentiate them from the stationary processes whose autocorrelations eventually die out.

The identification procedure hence consists of two parts. Firstly, the differencing parameter, d, is determined. This is done by considering the estimated autocorrelations from successively differenced series. If the estimated autocorrelation function from the original data does not die out sufficiently quickly we conclude that the parameter d is at least one. If it has the value 1, then the variable $Y_t = (1 - B)X_t$ is stationary, so that the theoretical and (under most realizations) the estimated autocorrelations die out. If not, the 'next' differenced series $Z_t = (1 - B)Y_t = (1 - B)^2 X_t$ is examined and so on, until the estimated autocorrelations are indicative of a stationary process.

In the presence of seasonality, the procedure is substantially more complicated. Suppose s is the periodicity of the seasonality i.e. $s = 4$ for quarterly data, and 12 for monthly data. Then the (most) general seasonal process considered in this procedure has the form

$$\phi(B)\Phi(B^s)(1-B)^{d_1}(1-B^s)^{d_s}X_t = \mu + \theta(B)\Theta(B^s)\varepsilon_t,$$

so that a combination of d_s seasonal and d_1 regular differences is allowed. The presence of a non-zero d_s causes the autocorrelations at the seasonal lags to remain large, rather than die out. Hence, we can determine both d_1 and d_s by considering the autocorrelations of the applications of various combinations of the two differencing operators, until we come across one which has the features of a stationary process at both the ordinary and seasonal lags. In most cases, d, and d_s will equal 0 or 1.

Having determined these values, we proceed to analyse the appropriately differenced (and hence stationary) series. It is interesting to consider the forecasts from these types of models.

Let us firstly consider the very simple $(0, 1, 0)$ model

$$(1 - B)X_t = \mu + \varepsilon_t,$$

so that

$$\hat{X}_{T+l} = \hat{X}_{T+l-1} + \mu$$
$$= X_T + l\mu.$$

Hence we see that a trend is introduced into the forecasts, with a slope equal to the estimated mean, μ, but with an intercept at the last data point, X_T. Of course, if μ is zero, then a constant level is preserved.

Looking at second differences

$$(1 - B)^2 X_t = \varepsilon_t,$$
$$\hat{X}_{T+l} = A + Bl,$$

where A and B are obtained from

$$\hat{X}_{T+1} = A + B = 2X_T - X_{T-1}$$

and

$$\hat{X}_{T+2} = A + 2B = 2\hat{X}_{T+1} - X_T,$$

so we see that while we still have a trend in the forecasts, both parameters are now determined from the final two observations, with no 'deterministic' part present. Incorporating a mean into this relation provides a deterministic quadratic trend component, with the linear coefficients obtained from the data.

More generally, the presence of moving average parts disturbs the first few forecasts, and hence changes the parameters in the analytic form of the eventual forecast relationship. On the other hand, the stationary AR parts provide mixtures of diminishing sines and cosines around the trend introduced by the non-stationary components.

6.   TRANSFORMATIONS

In the previous sections we have described the procedure as if the reported data was analysed. There is no reason, of course, why the analyses cannot proceed after some transformation has taken place. Perhaps the most popular transformation (see Chatfield and Prothero, 1973) is that suggested by Box and Cox (1964) who propose the power transformation

$$Y_t = \frac{(X_t + M)^\lambda}{\lambda},$$

with M an additive constant. Usually, the parameter λ is constrained to be between zero and one, and is estimated at the same time as the other parameters, via a likelihood search routine. This is quite a desirable class of transformations; if λ is close to one, then the original data should be analysed, while for λ near zero, the natural logarithm is implied.

There is, however, some doubt as to the usefulness of this transformation – see Granger and Nelson (1978). In any case, the facilities to apply it to so many series were not available at the University of Sydney, so we proceeded with just the two extremes, the original data, and the logarithm. The data was transformed if the series appeared to be growing exponentially in time; or if the size of the residuals from an untransformed analysis appeared to be correlated with the general level of the series. This occurred in 31 series. If we put

$$Y_t = \ln X_t,$$

and if $f_{N,h,Y}$ and $f_{N,h,X}$ are respectively the unbiased lead h forecasts for Y and X, then it can be shown, e.g. Granger and Newbold (1977), or by using moment-generating functions in this particular case, that

$$f_{N,h,X} = \exp\left[f_{N,h,Y} + \tfrac{1}{2}\sigma^2(h)\right],$$

where $\sigma^2(h)$ is the lead h forecast error variance of the Y^s and may be estimated using the infinite moving average representation. In this experiment, the forecasts from the transformed series were calculated using this formula.

7.   SUMMARY

There is no doubt that the Box–Jenkins theory is elegant and complete. A major difficulty, however, is the need for substantial data. The method is not really applicable unless at least thirty observations are available (especially for the seasonal series), and is primarily designed for short-term forecasting. The questions which remain relate to the assumptions, and concern their relevance. The fundamental (unanswerable) question concerns the existence of stochastic processes. If these exist, we must question the representativeness of our models, especially the GAUSSIAN assumption, and so on. The next section deals with our application of these techniques via a set of case-studies.

8. SOME EXAMPLES

From the previous section, one may imagine that the derivation of an eventual forecasting model is a step-by-step scientific procedure. Unfortunately, this is not always the situation. In this section, we detail how we developed models for two of each of the yearly, quarterly, and monthly series. We have tried to include all the relevant details including wrong turns where they were retained. More details on our models for the other series may be obtained from the authors. The chosen series are:

  6: Production of Renault Tractors (yearly: 19 obs).
 11: National Product and National Expenditure (Yearly: 13 obs).
 26: Industrial Production: Textiles (quarterly: 32 obs).
 33: Norway Reserves (quarterly: 38 obs).
 48: Company data Switzerland (monthly: 105 obs).
 98: Industrial Production – Finished Investment Goods, Austria (monthly: 102 obs).

For each of the series, the following steps were undertaken:

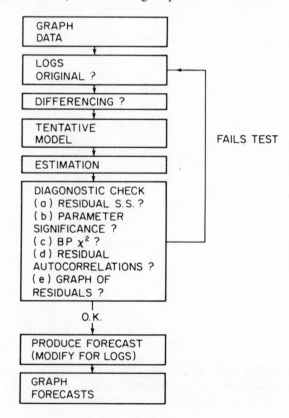

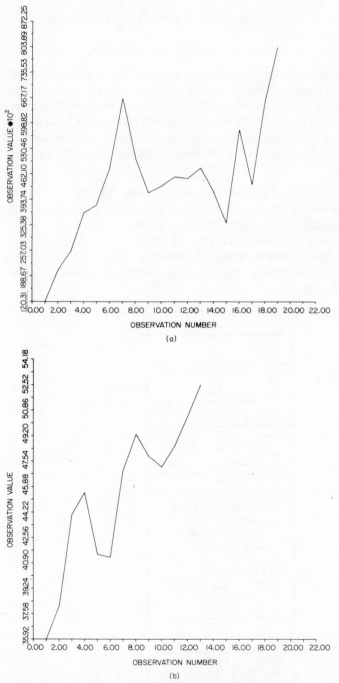

FIGURE 5   (a) Production Tractors VI—Renault; (b) National Product and National Expenditure; (c) Industrial Production—Textiles; (d) Norway—Reserves; (e) Company Data Swiss; (f) Industrial Production—Finished Investment Goods, Austria.

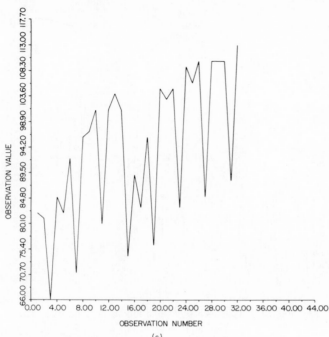

(c)

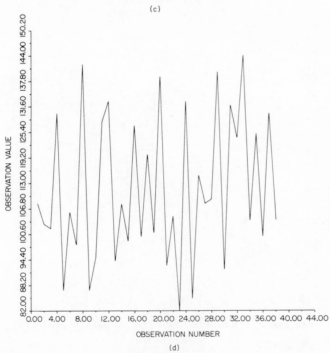

(d)

FIGURE 5—*contd.*

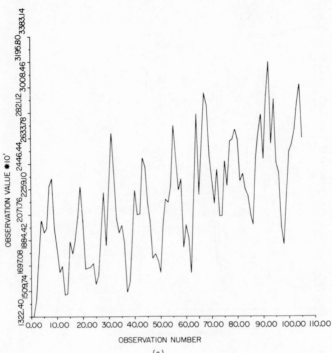

(e)

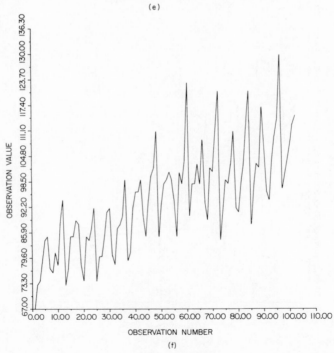

(f)

FIGURE 5—*contd.*

Let us now take a closer look at some of the series. A graph of each is given in Figure 5, while in Figure 6 the forecasts and out-of-sample actuals have been added. For each of the series considered, the first table gives the autocorrelations and partial autocorrelations at various lags for various differences. The means and variances are also supplied. The second table in each case gives the residual autocorrelations and the Box–Pierce statistic with the appropriate degrees of freedom.

Series 6: Production of Renault Tractors (19 obs)

As with the majority of the yearly series, this variable has a distinct trend, so that some differencing is necessary. From the autocorrelation function, it appears that the random walk model is satisfactory, giving

$$(1 - B)X_t - 3798.0 = \hat{\varepsilon}_t.$$

The resulting forecasts are given below.

TABLE 1   The Autocorrelations $(1/\sqrt{19} \approx 0.23)$

The original data	0.469	0.189	0.083	−0.165	−0.141	−0.122
M = 44285.	0.469	−0.040	0.013	−0.260	0.055	−0.063
V = 0.250 × 10⁹						
1st diff (Residuals)	−0.236	0.017	0.141	−0.144	−0.005	0.039
M = 3797.7	−0.236	0.041	0.144	−0.082	−0.063	0.008
V = 0.135 × 10⁹	BPχ^2(6) = 1.77					

Lead	Forecast	Actual
1	84187	75747
2	87984	72606
3	91782	71755
4	95580	70845
5	99377	72952
6	103175	97999

Series 11: National Product and National Expenditure (13 obs)

Here, we are faced with the problem of analysing a series with very few observations. Just from the graph of the series, it is apparent that the series is non-stationary, so that at least the first differenced series should be analysed. The autocorrelation function, however, is not so easy to interpret. Initially, we tried a simple AR(1) model, giving

$$(1 - 0.150B)(1 - B)X_t - 1.116 = \hat{\varepsilon}_t, \qquad (A)$$

with a residual mean square of 9.302.

As one would expect, the residual autocorrelation at lag 2 is not satisfactory.

TABLE 2(a)   The Autocorrelations ($1/\sqrt{13} \approx 0.3$)

	1	2	3	4	5	6
The original data	0.571	0.153	0.166	0.313	0.133	−0.235
M = 45.30	0.571	−0.258	0.323	0.125	−0.252	−0.285
V = 22.37						
1st diff.	0.149	−0.756	−0.276	0.417	0.253	−0.121
M = 1.38	0.149	−0.797	0.131	−0.369	0.074	0.105
V = 7.18						

TABLE 2(b)   The Residual Autocorrelations

	1	2	3	4	5	6
A	0.08	−0.78	−0.19	0.45	.18	−0.14
	$BP\chi^2(5) = 9.84$					
B	−0.22	−0.42	0.21	−0.14	−0.17	0.34
	$BP\chi^2(4) = 4.37$					

Thus, the model was extended to

$$(1 - 0.219B + 0.804B^2)(1 - B)X_t - 1.589 = \hat{\varepsilon}_t, \qquad (B)$$

and a resulting RMS of 9.569, slightly larger than that from model A. However, the a.c.f. of the residuals is now quite adequate, so that model B was retained. The forecasts are given below

Lead	Forecast	Actual
1	53.0	53.4
2	53.0	55.4
3	54.2	56.9
4	56.1	58.4
5	57.1	60.5
6	57.4	63.6

Series 26: Industrial Production: Textiles (30 obs)

This series exhibits a fairly strong seasonal pattern plus a trend. The autocorrelations indicate the need for seasonal differencing, and perhaps a regular difference as well.

To test this, we fitted an AR(1) on the seasonal difference, resulting in

$$(1 - 0.680B)(1 - B^4)X_t - 1.35 = \hat{\varepsilon}_t, \qquad (A)$$

TABLE 3(a)   The Autocorrelations $(1/\sqrt{30} \approx 0.18)$

	1	2	3	4	5	6	7	8
The original data	0.167	0.316	0.024	0.606	−0.127	0.051	−0.167	0.373
M = 93.56	0.167	0.296	−0.071	0.588	−0.504	−0.104	0.072	0.097
V = 171.68								
1st diff.	−0.572	0.205	−0.536	0.833	−0.513	0.164	−0.443	0.706
M = 1.00	−0.572	−0.181	−0.768	0.380	0.004	−0.291	−0.140	−0.127
V = 277.42								
1st seas diff.	0.679	0.336	−0.080	−0.395	−0.392	−0.416	−0.332	−0.285
M = 3.82	0.679	−0.234	−0.398	−0.232	0.270	−0.348	0.247	−0.176
V = 65.86								
1 diff. of each	0.074	0.092	−0.135	−0.493	0.027	−0.170	0.046	−0.080
M = 0.111	0.073	0.087	−0.149	−0.497	0.133	−0.104	−0.118	−0.381
V = 43.21								

TABLE 3(b)   The Residual Autocorrelations

A	0.20	0.14	−0.12	−0.46	−0.05	−0.21	−0.02	−0.12
		$BP\chi^2(8) = 9.38$						
B	0.11	0.12	−0.11	−0.06	−0.08	−0.16	−0.09	−0.06
		$BP\chi^2(7) = 2.36$						

with an RMS of 39.099, and an inadequate autocorrelation function. The inclusion of the MA at lag 4, removed the offending residual autocorrelation, thus the final model was

$$(1 - 0.685B)(1 - B^4)X_t - 1.239 = (1 - 0.693B^4\varepsilon)\hat{\varepsilon}_t, \qquad (B)$$

RMS = 26.24. The forecasts are given below

Lead	Forecast	Actual
1	111.90	113.00
2	115.07	124.00
3	93.21	93.00
4	117.61	127.00
5	116.31	128.00
6	119.33	130.00
7	97.37	89.00
8	121.71	116.00

Series 33: Norway – Reserves (38 obs)

Of all the series discussed here, this one appears to be the most stationary, with

TABLE 4(a) The Autocorrelations $(1/\sqrt{38} \approx 0.16)$

	1	2	3	4	5	6	7	8	9	10
The original data M = 111.37 V = 307.81	-0.411	0.235	-0.284	0.496	-0.197	-0.034	-0.177	0.212		
	-0.411	0.080	-0.196	0.392	0.178	-0.242	-0.204	0.060		
1st diff. M = -0.108 V = 890.64	-0.725	0.407	-0.452	0.514	-0.293	0.103	-0.183	0.203		
	-0.725	-0.251	-0.597	-0.262	0.163	0.092	-0.059	-0.212		
1st seas diff. M = 0.471 V = 310.31	-0.289	0.234	0.012	-0.285	0.028	-0.111	-0.014	-0.089		
	-0289	0.164	0.131	-0.331	-0.185	0.009	0.031	-0.220		
1 diff. of each M = 0.636 V = 810.29	-0.696	0.289	0.043	-0.251	0.169	-0.059	0.036	-0.053		
	-0.696	-0.380	0.128	-0.109	-0.300	-0.174	0.129	0.043		

TABLE 4(b) The Residual Autocorrelations

	1	2	3	4	5	6	7	8	9	10
A	0.04	0.19	0.02	-0.34	-0.13	-0.07	-0.09	-0.11	-0.02	-0.17
		$\text{BP}\chi^2(9) = 7.36$								
B	0.03	0.14	0.00	-0.11	-0.13	-0.05	-0.14	-0.03	0.00	-0.22
		$\text{BP}\chi^2(8) = 4.02$								

no lasting seasonality or trend. The autocorrelations indicate that the original data or perhaps one seasonal difference should be analysed. Thus, the two models considered were:

$$(1 + 0.290B)(1 - B^4)X_t - 1.262 = \hat{\varepsilon}_t \qquad (A)$$
$$\text{RMS} = 296.44$$

and

$$(1 + 0.326B)(1 - 0.486B^4)(X_t - 112.97) = \hat{\varepsilon}_t \qquad (B)$$
$$\text{RMS} = 224.22.$$

From the fit, and the residual autocorrelations, the latter model was superior, giving the forecasts below.

Lead	Forecast	Actual
1	120.32	108.00
2	106.17	108.00
3	121.41	128.00
4	108.56	80.00
5	116.59	92.00
6	109.66	91.00
7	117.07	102.00
8	110.82	83.00

Series 48: Company Data, Switzerland (105 obs)

The graph of this data shows a seasonal component and a trend which appears to die out towards the end of the observation interval. It would appear from the data that regular differencing is not required. This is confirmed by the autocorrelation function, which suggests seasonal differencing. However, since the auto-correlation at lag 12 of the seasonally differenced series is so negative, it appears that no differencing is at all appropriate. Thus, the first model considered was the multiplicative seasonal

$$(1 - 0.421B)(1 - 0.755B^{12})(X_t - 24940) = \hat{\varepsilon}_t, \qquad (A)$$

with the RMS at 0.398×10^7, and all the parameters significant.

The residual autocorrelation at the seasonal lag 12 is significant, suggesting that we should experiment with the following types of models

$$(1 - 0.298B)(1 - B^{12})X_t - 612.16 = (1 - 0.954B^{12})\hat{\varepsilon}_t \qquad (B)$$
$$\text{RMS} = 0.255 \times 10^7$$

and

$$(1 - B^{12})X_t - 893.33 = (1 + 0.277B)(1 - 0.938B^{12})\hat{\varepsilon}_t \qquad (C)$$
$$\text{RMS} = 0.259 \times 10^7$$

TABLE 5(a)   The Autocorrelations ($1/\sqrt{105} \approx 0.10$)

	1 /13	2 /14	3 /15	4 /16	5 /17	6 /18	7 /19	8 /20	9 /21	10 /22	11 /23	12 /24
The original data M = 21765. V = 0.164 × 10⁸	0.751	0.525	0.409	0.239	0.063	-0.055	0.026	0.231	0.373	0.457	0.608	0.706
	0.572	0.392	0.284	0.143	-0.017	-0.132	-0.053	0.103	0.199	0.278	0.412	0.515
	0.751	-0.087	-0.106	-0.203	-0.110	-0.066	0.366	0.413	0.235	0.005	0.157	0.127
	-0.166	-0.022	0.091	-0.100	-0.153	-0.195	0.064	-0.026	0.021	-0.043	0.010	0.057
1st diff. M = 126.17 V = 0.736 × 10⁷	-0.068	-0.201	0.159	0.007	-0.123	-0.398	-0.249	0.093	0.086	-0.173	0.094	0.515
	0.066	-0.105	0.088	0.034	-0.076	-0.401	-0.174	0.089	0.030	-0.126	0.063	0.439
	-0.068	-0.206	0.135	-0.015	-0.070	-0.462	-0.467	-0.287	0.035	-0.131	-0.131	0.209
	0.016	-0.087	-0.030	0.011	0.063	-0.167	-0.089	-0.134	0.009	-0.026	-0.075	0.049
1st seas. diff. M = 856.02 V = 0.473 × 10⁷	0.324	0.203	0.304	0.168	0.165	0.044	-0.218	-0.104	-0.128	-0.147	-0.156	-0.481
	-0.203	-0.033	-0.197	-0.229	0.019	-0.034	0.084	0.102	0.011	0.082	0.153	0.033
	0.324	0.109	0.237	0.004	0.072	-0.114	-0.314	-0.046	-0.072	0.038	-0.013	-0.406
	0.091	0.087	-0.008	-0.169	0.231	-0.033	-0.077	0.033	-0.027	-0.019	0.018	-0.265
1 diff. of each m = -12.96 V = 0.645 × 10⁷	-0.410	-0.165	0.182	-0.104	0.087	0.104	-0.274	0.103	0.004	-0.024	0.245	-0.449
	0.081	0.246	-0.106	-0.198	0.221	-0.131	0.074	0.080	-0.120	0.001	0.144	-0.107
	-0.410	-0.401	-0.112	-0.165	0.023	0.187	-0.110	-0.077	-0.167	-0.096	0.263	-0.275
	-0.217	-0.100	0.049	-0.322	-0.037	-0.002	-0.105	-0.045	-0.047	-0.084	0.206	-0.143

TABLE 5(b)   The Residual Autocorrelations

	1 /13	2 /14	3 /15	4 /16	5 /17	6 /18	7 /19	8 /20	9 /21	10 /22	11 /23	12 /24
A	-0.06	0.03	0.23	-0.00	0.09	0.04	-0.23	0.03	-0.01	-0.04	0.12	-0.30
	0.04	0.15	-0.08	-0.17	0.14	-0.12	0.04	0.06	-0.05	0.05	0.14	0.00

$BP\chi^2(22) = 32.2$

In both these cases, it was obvious that the seasonal MA parameter was too large, in effect cancelling the (unrequired) seasonal differencing. Thus, the next apparent model was

$$(1 - 0.303B)(1 - 0.994B^{12})(X_t - 0.157 \times 10^6) = (1 - 0.946B^{12})\hat{\varepsilon}_t$$

with $RMS = 0.259 \times 10^7$, with the same difficulties as before. This exhausted our possibilities, so we decided to stay with model A, accepting a slightly inferior residual autocorrelation function and deficient model fit. On reflection this is probably caused by a change in trend. The forecasts produced are given below.

Lead	Forecast	Actual
1	28847	22063
2	24740	24783
3	29125	25923
4	21195	20796
5	20139	21409
6	22957	23026
7	25259	26807
8	25683	25358
9	26447	27267
10	27889	32287
11	28990	31038
12	26007	27664
13	27596	26828
14	24793	22133
15	24327	26347
16	22111	22300
17	21313	20810
18	23444	26100

Series 98: Industrial Production: Finished Investment Goods – Austria

The graph of this series shows a distinct seasonal pattern, superimposed on a growth factor, with the seasonal fluctuations increasing as the overall level of the series increases. At this stage we considered logarithms but decided to produce models for both the transformed and untransformed data, and thus defer making a decision until a later stage.

Firstly, we need to obtain the differencing parameters. The ACF of the original data (transformed or untransformed), indicates that regular differencing and probably seasonal differencing is required. The first differenced series, on the other hand, indicate that a seasonal difference is necessary. If we consider the seasonally differenced series, then it is apparent that one seasonal differencing alone is appropriate, which is confirmed by the large and negative first and seasonal lags in the over differenced series. Hence, we decided to proceed with just the single seasonal difference.

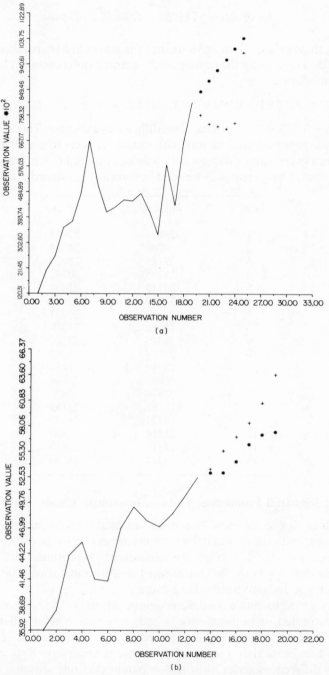

FIGURE 6   The data, the forecasts (*) and the actuals (+). (a) Production Tractors VI—Renault; (b) Production and National Expenditure; (c) Industrial Production—Textiles; (d) Norway—Reserves; (e) Company Data Swiss; (f) Industrial Production—Finished Investment Goods, Austria.

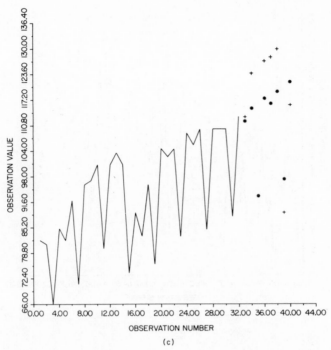

(c)

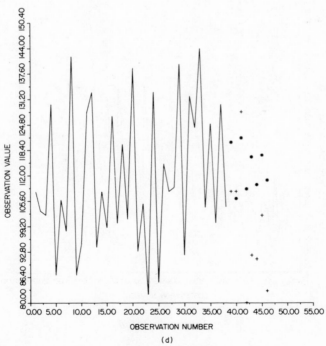

(d)

FIGURE 6—*contd.*

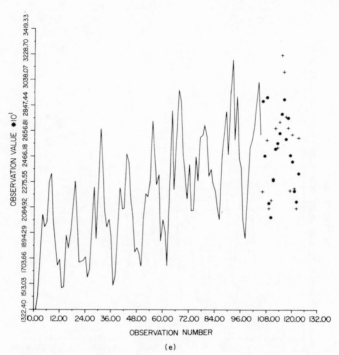

(e)

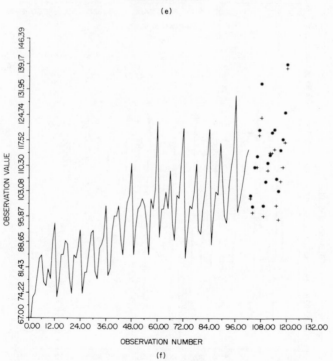

(f)

FIGURE 6—*contd.*

The first model chosen was an AR(2) on the original data, which resulted in the model

$$(1 - 0.149B - 0.35B^2)(1 - B^{12})X_t - 0.197 = \hat{\varepsilon}_t,$$

with the RMS at 18.72. The autocorrelations of the residuals given below indicate that this model is not adequate, and that some work is required at the seasonal lags. Another feature exposed at this point was that residuals appeared to be heteroscedastic, growing in relation to the size of the series. Hence, it was decided to proceed with the logarithms.

Building on the behaviour of the last model, it was decided to extend it to include a seasonal MA part, to give

$$(1 - 0.273B - 0.292B^2)(1 - B^{12}) \ln X_t - 0.017 = (1 - 0.682B^{12})\hat{\varepsilon}_t,$$

with the RMS of the logs now at 0.0016. The autocorrelations of these residuals, except for the outlier at lag 11, appear adequate.

Because of this large coefficient, the following model eventuated

$$(1 - 0.297B - 0.276B^2)(1 - B^{12}) \ln X_t - 0.016 = (1 + 0.239B^{11} - 0.548B^{12})\hat{\varepsilon}_t,$$

with the corresponding RMS at 0.0015, a very small reduction on the previous value. The residuals autocorrelations are now all insignificant, with the χ^2 statistic at 20.34.

At this stage a little overfitting was considered, giving

$$(1 - 0.214B - 0.391B^2)(1 - 0.370B^{12})(1 - B^{12}) \ln X_t = \hat{\varepsilon}_t,$$

with the inferior RMS of 0.0018, with a significant correlation at lag 11, and an almost significant one at lag 13, indicating that the MA approach was probably the better.

In view of the slight reduction in sum of squares, and the 'almost' insignificance of the 11th order MA parameter, we concluded that the model

$$(1 - 0.273B - 0.292B^2)(1 - B^{12}) \ln X_t - 0.017 = (1 - 0.682B^{12})\hat{\varepsilon}_t,$$

was adequate, with the following forecasts resulting

Lead	Forecast	Actual	Lead	Forecast	Actual
1	101.75	101	10	113.48	114
2	98.71	97	11	119.77	114
3	109.97	110	12	120.57	115
4	113.05	110	13	106.61	99
5	120.53	119	14	103.09	95
6	133.68	124	15	114.74	106
7	98.99	96	16	117.81	110
8	105.82	106	17	125.54	117
9	111.14	109	18	139.17	138

TABLE 6(a)　The Autocorrelations $(1/\sqrt{102} \approx 0.1)$

	1 /13	2 /14	3 /15	4 /16	5 /17	6 /18	7 /19	8 /20	9 /21	10 /22	11 /23	12 /24
The original data M = 94.108 V = 155.49	0.597	0.422	0.398	0.349	0.550	0.747	0.523	0.293	0.303	0.266	0.410	0.707
	0.388	0.230	0.214	0.170	0.355	0.533	0.311	0.138	0.126	0.105	0.223	0.473
	0.597	0.101	0.175	0.057	0.446	0.506	−0.112	−0.367	0.049	0.034	−0.051	0.436
	−0.330	−0.100	0.035	0.027	−0.006	−0.043	−0.100	0.110	−0.073	0.056	−0.104	0.072
1st diff. M = 0.475 V = 114.72	−0.296	−0.197	0.052	−0.324	0.014	0.495	0.032	−0.309	0.066	−0.208	−0.198	0.773
	−0.231	−0.152	0.038	−0.296	0.022	0.464	−0.064	−0.179	0.018	−0.171	−0.170	0.631
	−0.296	−0.312	−0.142	−0.515	−0.582	−0.100	0.288	−0.083	−0.067	0.031	−0.441	0.361
	0.054	0.058	0.025	0.005	0.034	0.094	−0.236	0.041	−0.115	0.068	−0.189	−0.190
1st seas. diff. M = 3.66 V = 20.45	0.221	0.326	0.278	0.243	0.173	0.146	0.211	0.050	0.188	−0.052	0.195	−0.226
	−0.167	0.008	−0.200	−0.117	−0.078	−0.002	−0.225	−0.091	−0.124	−0.184	−0.142	−0.268
	0.221	0.292	0.187	0.105	0.009	−0.011	0.104	−0.078	0.093	−0.162	0.153	−0.359
	−0.218	0.143	−0.103	0.003	0.112	0.064	−0.092	−0.103	0.091	−0.157	0.040	−0.189
1 diff. of each M = 0.034 V = 31.86	−0.548	0.076	0.005	0.020	−0.037	−0.059	0.166	−0.211	0.239	−0.291	0.426	−0.297
	−0.113	0.274	−0.193	0.025	−0.024	0.204	−0.238	0.111	0.027	−0.077	0.113	−0.257
	−0.548	−0.322	−0.197	−0.073	−0.068	−0.176	0.042	−0.140	0.113	−0.203	0.329	0.132
	−0.253	0.027	−0.087	−0.165	−0.096	0.036	0.064	−0.124	0.112	−0.079	0.143	−0.051

Autocorrelations and partial autocorrelations (each series: top line = autocorrelations, second line = partial autocorrelations; each pair of lines gives lags 1–12 and lags 13–24).

Series	lags	1	2	3	4	5	6	7	8	9	10	11	12
Logged data M = 4.50 V = 0.015	1–12	0.619	0.443	0.418	0.371	0.566	0.752	0.541	0.311	0.318	0.283	0.422	0.702
	13–24	0.399	0.239	0.225	0.185	0.367	0.536	0.329	0.145	0.135	0.117	0.228	0.468
(partials)	1–12	0.619	0.097	0.182	0.058	0.455	0.492	−0.122	−0.381	0.062	0.034	−0.045	0.418
	13–24	−0.353	−0.080	0.050	0.034	0.023	−0.047	−0.093	0.086	−0.085	0.053	−0.118	0.069
1st diff. M = 0.005 V = 0.012	1–12	−0.280	−0.215	0.057	−0.336	0.021	0.509	0.045	−0.330	0.061	−0.207	−0.197	0.777
	13–24	−0.218	−0.167	0.037	−0.306	0.031	0.466	−0.038	−0.207	0.017	−0.169	−0.177	0.642
(partials)	1–12	−0.279	−0.318	−0.133	−0.526	−0.563	−0.049	0.352	−0.057	−0.046	0.056	−0.399	0.381
	13–24	0.009	0.064	0.008	−0.036	0.006	0.049	−0.231	0.054	−0.102	0.098	−0.178	−0.047
1st seas. diff. M = 0.039 V = 0.002	1–12	0.288	0.348	0.321	0.279	0.184	0.161	0.227	0.077	0.176	−0.018	0.179	−0.233
	13–24	−0.166	0.001	−0.174	−0.114	−0.058	−0.004	−0.218	−0.095	−0.096	−0.152	−0.153	−0.229
(partials)	1–12	0.288	0.290	0.198	0.109	−0.020	−0.023	0.109	−0.069	0.078	−0.163	0.144	−0.393
	13–24	−0.175	0.168	−0.063	0.039	0.115	0.028	−0.117	−0.096	0.133	−0.133	0.027	−0.207
1 diff. of each M = 0.000 V = 0.003	1–12	−0.527	0.036	0.028	0.036	−0.056	−0.057	0.170	−0.198	0.206	−0.261	0.434	−0.336
	13–24	−0.100	0.272	−0.177	−0.003	−0.001	0.201	−0.247	0.092	0.044	−0.046	0.053	−0.221
(partials)	1–12	−0.528	−0.337	−0.195	−0.046	−0.049	−0.160	0.055	−0.114	0.114	−0.192	0.371	0.094
	13–24	−0.262	−0.002	−0.113	−0.169	−0.087	0.050	0.065	−0.157	0.091	−0.060	0.168	−0.066

TABLE 6(b) The Residual Autocorrelations

	1 /13	2 /14	3 /15	4 /16	5 /17	6 /18	7 /19	8 /20	9 /21	10 /22	11 /33	12 /24
AR(2) on seasonal diff.	−0.05	−0.07	0.14	0.10	0.01	0.03	0.13	−0.01	0.08	−0.03	0.29	−0.28
	−0.18	0.17	−0.14	−0.12	0.06	0.11	−0.20	−0.01	0.00	−0.07	−0.09	−0.19
AR(2), MA(12) on S.D. of logs	−0.06	−0.12	0.15	0.01	0.05	0.12	0.07	−0.01	0.15	−0.13	0.26	0.10
	−0.17	0.13	−0.08	−0.13	0.08	0.14	−0.20	0.02	0.03	−0.10	0.00	−0.11

$\mathrm{BP}\chi^2(21) = 30.00$

9.   SOME FINAL COMMENTS

This study is unique in several factors. It is the first piece of work of this nature in which the different methods were applied by different practitioners, so that a 'knowledge' bias evident in previous studies has been overcome. More importantly, though, the data was of sufficient diversity, both geographically and in subject matter, and from enough time intervals that the practitioners' knowledge of economic conditions was of little value. What we are saying here, is that it was the methods which were competing, not the methods confounded with the practitioners' outside knowledge. Thirdly, a further set of 'standard data' is now available. This means that such overworked series as the airlines data and the sunspot numbers are now not the only easily available descriptive examples.

The basic results presented confirm those of Makridakis and Hibon (1979), that is, that under the measures of performance used here, the extra effort involved in the Box–Jenkins approach, overall, is not reflected in improved forecasting accuracy. However, under different measures of performance, for example those used by Newbold and Granger (1974), the results may well be different. Perhaps this technique is more conducive to short-term forecasts since the parameters are estimated to give optimal one-step forecasts only. Research is proceeding on this matter.

REFERENCES

Box, G. E. P. and Cox, D. R. (1964). 'An Analysis of Transformations', *Journal of the Royal Statistical Society*, Series B, **26**, 211–243.

Box, G. E. P. and Jenkins, G. M. (1970). 'Time Series Analysis, Forecasting and Control', San Francisco: Holden Day.

Chatfield, C. and Prothero, D. L. (1973). 'Box–Jenkins Seasonal Forecasting: Problems in a Case Study', *Journal of the Royal Statistical Society*, Series A, **136**, 295–336.

Davies, N., Triggs, C. M. and Newbold, P. (1977). 'Significance levels of the Box–Pierce portmanteau statistic in finite samples', *Biometrika*, **64**, 517–522.

Ljung, G. M. and Box, G. E. P. (1978). 'On a Measure of Lack of Fit in Time Series Models', *Biometrika*, **65**, 297–303.

Granger, C. W. J. and Nelson, H. (1978). 'Experience Using the Box–Cox Transformation when Forecasting Economic Time Series', Working Paper 78 (1), Department of Economics, U.C.S.D.

Granger, C. W. J. and Newbold, P. (1977). 'Forecasting Economic Time Series', New York: Academic Press.

Makridakis, S. and Hibon, M. (1979). 'Accuracy of Forecasting; an Empirical Investigation', *Journal of the Royal Statistical Society*, Series A, **2**, 97–145.

Newbold, P. and Granger, C. W. J. (1974). 'Experience with forecasting univariate time series, and the combination of forecasts', *Journal of the Royal Statistical Society*, Series A, **137**, 131–146.

Prothero, D. L. and Wallis, K. F. (1976). 'Modelling Macroeconomic Time Series (With Discussion), *Journal of the Royal Statistical Society*, Series A, **139**, 468–500.

CHAPTER 6

AEP Filtering

Robert Carbone with Robert Bilongo*, Paul Piat-Corson* and Serge Nadeau**
Université Laval, Quebec, Canada

1. INTRODUCTION

In recent years, a number of forecasting techniques which are based on engineering principles in signal detection and noise cancelling have been proposed. Adaptive filtering, the State Space approach, Bayesian forecasting, and AEP filtering are among the most well-known.

Some of these techniques were developed as extensions or generalizations of previously published engineering methods such as Kalman filtering while others were derived with specific applications in mind.

The AEP filtering method owes its origin to a very specific problem. It was developed in the mid-1970s by Carbone and Longini (1977) for providing a practical solution to the problem of estimating mixed additive and multiplicative time-dependent parameter models to mass appraise real properties for property tax purposes (see also Carbone and Longini, 1976, and Carbone, Longini and Ivory, 1980). As a diagnostic tool of analysis of dynamic systems, the method was subsequently applied to such areas as environmental management (Carbone and Gorr, 1978) and marketing (Mahajan, Bretschneider, and Bradford, 1980). These works led to its natural extension to univariate (Bretschneider, Carbone, and Longini, 1979) and multivariate (Bretschneider, Carbone, and Longini, 1982) time series analysis and forecasting.

Our attention in this chapter will be restricted to univariate analysis and forecasting, or UNIAEP. The first part is devoted to a description of how UNIAEP was applied in the accuracy study. This is followed by a global evaluation of performance. From this analysis, a more general modelling framework is constructed in the second part of the chapter which embodies the identified desirable methodological characteristics. The application of the resulting generalized UNIAEP or GUNIAEP approach to the accuracy study data shows not only a significant improvement in the performance of AEP

Graduate students at Laval University* and Carnegie-Mellon University** who have contributed to the results reported in this chapter.

filtering but it also provides more accurate forecasts when compared to the most precise methods.

2.   ORIGINAL UNIAEP APPROACH

The extension of Carbone–Longini AEP filtering to the field of univariate forecasting as proposed by Bretschneider, Carbone, and Longini (1979) follows a similar simple philosophy as presented in Wheelwright and Makridakis (1973) adaptive filtering. It involves:

(a)   The adaptive (recursive) estimation of a univariate forecasting model which takes the form of an autoregressive equation with time-dependent parameters;

(b)   a parameter correction mechanism based on the concept of damped negative feedback;

(c)   a learning process;

(d)   a model identification stage relying on the analysis of sample autocorrelation functions to determine (i) whether the model should be formulated in terms of original data, change data (first difference), or rate of change data (second difference), and (ii), the number of autoregressors (order of the equation).

In mathematical terms, the forecasting model may be written as:

$$x(t) = \sum_{k=1}^{p} \beta_k(t)x(t-k) + e(t), \tag{1}$$

where $x(t)$ denotes the value of the time series (original or transformed) at time t; p is the number of autoregressors; $\beta_k(t)$ is the value of the time-varying parameter assigned to the kth autoregressor at time t; and $e(t)$ represents a random noise term.

What basically distinguishes UNIAEP from other filtering approaches is the parameter correction equation applied to (1), or what is technically known as the gain filter. Revised AEP parameter estimates of (1) are obtained with each new observation through the following equation:

$$\beta_k(t) = \beta_k(t-1) + |\beta_k(t-1)| \left[\frac{x(t) - \hat{x}(t)}{|\hat{x}(t)|} \cdot \frac{x(t-k)}{\bar{x}_k(t)} \cdot \mu \right] \quad \text{for all } k, \tag{2}$$

where μ is a damping factor to control stability $(0 < \mu < 1)$;

$$\hat{x}(t) = \sum_{k=1}^{p} \beta_k(t-1)x(t-k), \tag{3}$$

a one-step ahead forecast of $x(t)$; and

$$\bar{x}_k(t) = s|x(t-k)| + (1-s)\bar{x}_k(t-1), \tag{4}$$

is a revised smoothing average of the kth autoregressor with $s < 1$ (in cases of change or rate of change data, $\bar{x}_k(t)$ then becomes a smoothing average of absolute changes or absolute rates of change). In addition, to avoid overreaction to transient errors, a correction limit (CLIM) is imposed on the updating mechanism by setting

$$\left[\frac{x(t) - \hat{x}(t)}{|\hat{x}(t)|} \cdot \frac{x(t - k)}{\bar{x}_k(t)} \cdot \mu \right] \leq \text{CLIM} \qquad \text{for all } k, \tag{5}$$

where $0 < \text{CLIM} < 1$. For example, a correction limit (CLIM) of 0.1 would determine that the parameters could not be updated by more than 10 per cent from one observation to the next.

The implementation of UNIAEP presupposes a model identification stage involving the analysis of sample autocorrelation functions. Once the type of data and p are specified, it requires an initialization stage involving the determination of

(i) initial estimates of the parameters (β_ks);
(ii) initial estimates (positive) of the p smoothing averages ($\bar{x}_k$s); and
(iii) values for μ and s.

After the specification of the above values, a learning period then starts which consists in first iterating forward through the data from the first observation to the last followed by iterating backward beginning with the last observation to the first. This is denoted as one training cycle. The forward/backward mode of learning unique to AEP filtering has been shown by Longini *et al.* (1975) to eliminate potential phase shift problems. The learning period ends when one-step-ahead forecast accuracy between cycles does not improve significantly. Forecasts are then made on the basis of the parameters for the last observation of the time series obtained during the forward portion of the last training cycle.

The damping factor (μ) plays an important role in AEP filtering as with all other damped negative feedback algorithms. It determines the speed of adaptation, which in turn has a direct effect on fit and forecast performances. In general, faster adaptation (larger value of μ) causes some oscillations. Under static environments, best performance will result from slow adaptation (small value for μ) since overreactions to transient errors are reduced. In contrast, under dynamic environments, best performance will be achieved by a compromise between fast adaptation, necessary to track variations in the parameters, and slow adaptation for containing unwarranted oscillations.

The selection of an appropriate value for a given time series would then appear to be of paramount importance. Furthermore, the continual adjustment of this value as new observations are recorded would seem warranted since conditions may change as time elapses. In the absence of any mechanism to select a 'best' value for μ, it would appear more appropriate for forecasting purposes to arbitrarily fix μ to somewhat a high value whatever the type of series. As a result of faster adaptation, final parameter estimates used for forecasting should tend to be more reflective of prevalent conditions at that time since the method is then set to track variations over the time series. The setting of μ in fact draws attention to the distinction between fitting historical data and forecasting future values.

3. APPLICATION OF UNIAEP IN THE ACCURACY STUDY

UNIAEP was applied in an automatic execution mode with no user intervention. The automatic program was in fact developed in response to the study. The forecasts for the 1001 series were obtained in a single run (around (2) hours of CPU time on an IBM 370/158). No revisions of forecasts through personalized analyses were performed. Of the information available for each series (name or description, country, seasonal indicator, and type of data), only the type of data (yearly, quarterly, or monthly) was used.

Under automatic execution for the study, the order of differencing for each series was determined by an internal automatic examination of the estimated sample autocorrelation functions. It basically consisted of verifying the significance of a first-order lag model for the autocorrelations. The maximum order of differencing was set equal to 2. Having identified the order of differencing, the number of autoregressors (p) was then automatically set to the lag value corresponding to the most significant autocorrelation for the identified order of differencing data. However, p could not be set lower than 3, 4, or 12, for yearly, quarterly, or monthly data respectively.

For all series, an identical initialization procedure was applied. It consisted of setting:

(a) initial values of the parameters to the inverse of the order of the autoregressive equation selected (initial β's equal to $1/p$);
(b) start-up values for the exponential smoothing means ($\bar{x}$s) to 100 and the smoothing constant (s) to 0.01;
(c) the damping factor (μ) to 0.06 and CLIM to 0.1.

Finally, the learning process stopped whenever improvements in MSE (mean squared error) between cycles became negligible or the number of training cycles exceeded 60.

4. EVALUATION OF THE FORECASTING RESULTS

The value of an accuracy study of this magnitude is not to identify an overall winning or losing technique. It is rather to learn under which circumstances a particular type of approach appears more suitable. Also, one might learn from this kind of confrontation with real data how to improve a particular method or design a new approach which would embody the most desirable characteristics from perspectives of performance and facility of use.

Our analysis of the results will then take two directions. The first will be to examine the performance of AEP filtering in its automatic autoregressive execution mode in contrast to other methods. The second will focus on identifying some general conceptual conclusions which may be reached from all the measures reported in the various tables.

Given how the method was applied (type of automatic execution, uniform damping factor) to the 1001 series, the results of the accuracy study are quite encouraging. Looking at the tables which report the various accuracy measures for all 1001 series, Autom. AEP ranks among the leading methods. This is even more true if we take into account the fact that not all of the 1001 series were evaluated for certain methods. What is more important is to identify where the method failed. Clearly it is on seasonal data. Whereas it is revealed to have significantly generated the most accurate forecasts for non-seasonal data as compared to competitive methods, the performance on seasonal data is one of, if not the poorest. This certainly raises a question as to the effectiveness of a simple autoregressive structure (without seasonal differences) to capture seasonality.

One feature which appears to be universally shared by all methods is their relative poor performance on seasonal data. It is somewhat discouraging to find out that all the methods, even the more complex ones, could not provide significantly more accurate forecasts than deseasonalized single exponential smoothing.

The types of methods which obtained the most robust results are those methods which attempt to decompose a time series into various components. In addition to decomposition, adaptiveness is revealed to be a desirable characteristic along with trend saturation for long-run forecasts.

Based in part on the above subjective assessment of the results, a generalized modelling framework is developed and tested in the second part of the chapter.

5. GENERALIZED UNIAEP APPROACH

Several research activities were initiated as a result of the accuracy study. These activities have principally focused on two issues. The first consists in the formulation of a more general AEP modelling philosophy for univariate analysis which explicitly accounts for different components of a time series. The second involves the development of an automatic mechanism for adaptively adjusting the value of the damping factor depending upon prevalent conditions.

An adaptive decomposition model

Two components are common to all time series. The first focuses on the long-term pattern whereas the second centres on short-term variations. Under the assumption that (a) short-term variations take the form of an autoregressive scheme on the most recent observations, and (b) that the value of a time series at time t is a linear combination of the two components, the resulting forecasting model may then be written as

$$x(t) = (1 - \gamma)\bar{x}(t) + \gamma \sum_{k=1}^{p} \delta_k x(t - k) + e(t), \tag{6}$$

where $0 \leq \gamma \leq 1$ and $\bar{x}(t)$ denotes the long-term pattern.

It is interesting to note that several forecasting models are encompassed by the above formulation. For example $\gamma = 1$, $p = 1$, and $\delta_1 = 1$, results in the 'naive' model; $p = 1$, $\delta_1 = 1$, and $\bar{x}(t) = \hat{x}(t - 1)$ gives the single exponential smoothing model; finally, with $\gamma = 1$, (6) then becomes an autoregressive model of order p.

If it is further assumed that $\bar{x}(t)$ is a deterministic function which is approximated by a polynomial in t and that γ, the weight assigned to the short-term component, as well as all the other parameters, are time-varying, (6) becomes

$$x(t) = \sum_{j=0}^{q} \theta_j(t)t^j + \sum_{k=1}^{p} \beta_k(t)x(t - k) + e(t). \tag{7}$$

Our search for a general model must lead us to a formulation which also separately accounts for seasonal variations. In order to reflect these variations accurately, the final general model takes the form of a mixed additive and multiplicative model written as

$$x(t) = \left[\sum_{j=0}^{q} \theta_j(t)t^j + \sum_{k=1}^{p} \beta_k(t)x(t - k) \right] \prod_{i=1}^{L} \alpha_i(t)^{z_i(t)} + e(t), \tag{8}$$

where $z_i(t)$ equals 1 if the tth observation is of season i, otherwise 0; L is the length of the seasonality; and $\alpha_i(t)$ is the value at time t of the seasonal coefficient assigned to the ith season (the values of the time-varying seasonal indices are relative to a reference season which has a fixed coefficient value of 1 at all t).

The estimation of the time-varying parameters of (8) (the θs, βs and αs) is performed by AEP filtering. It involves the application of (2) for the θs and βs. As for the αs, they are updated by the following equation which was developed by Carbone–Longini (1977):

$$\alpha_i(t) = \alpha_i(t - 1) + \alpha_i(t - 1) \left[\frac{x(t) - \hat{x}(t)}{|\hat{x}(t)|} \cdot z_i(t) \cdot \mu \right] \qquad \text{for all } i. \tag{9}$$

The implementation of the generalized UNIAEP modelling approach or GUNIAEP also requires certain prerequisites. A user must specify a value for q which should be strictly smaller than 2. L and p must also be set as well as initial values for all the time-varying parameters (the θs, βs, and αs), the smoothing constant and averages. Finally, the selection of a damping factor initiates the forward/backward learning process.

There are several advantages to GUNIAEP besides its adaptive character. First, the identification tasks are greatly simplified. There is no need for data transformation or an analysis of the sample autocorrelation functions. The presence or absence of seasonality is automatically detected by the estimation of the seasonal component. Whatever is the strength of the seasonality, seasonality is taken into account. Only for the case of annual data would the estimation of a seasonal component be unnecessary.

As for the identification of the model's parameters q, p, and L, L and p should always be assigned a value corresponding to the type of data: for example, 4 for quarterly data, 12 for monthly data. On the other hand, q could always be set to 1. It should assume the value of 2 only in cases of trend saturation or explosion.

From a user standpoint, the model provides useful explanatory information on the behaviour of a time series over time. This information can, in fact, be used for obtaining a deseasonalized series. Deseasonalization is a by-product of the approach.

A third advantage resides in the ease of incorporating judgemental expertise into the modelling framework. For instance, a user could alter coefficient values to reflect future considerations or arbitrarily impose a trend saturation or explosion term. One or two additional multiplicative groupings (group of αs) could also be incorporated into a model to take into account special events or other types of interventions. Finally, forecasting without historical data can easily be performed by setting relative seasonal parameters, a mean and a trend term.

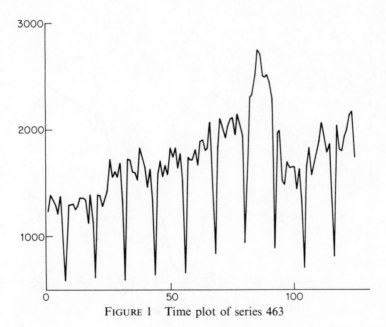

FIGURE 1   Time plot of series 463

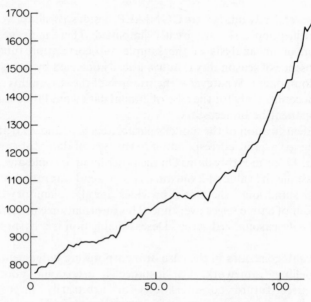

FIGURE 2   Time plot of series 805

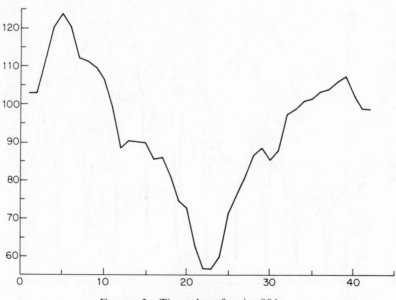

FIGURE 3   Time plot of series 886

To examine the behaviour of **GUNIAEP** closely, the method is applied to three monthly time series drawn from the 1001 sample. The selected series are series 463, 805, and 886. The first series is highly seasonal, the second shows no sign of seasonality whereas the third is too short for an examination of sample autocorrelation functions. Graphical representations of the series are displayed in Figures 1, 2, and 3.

Three models were generated for each time series. In all cases, the seasonal component and the autoregressive scheme were specified with $p = L = 12$. The three models for each series differed by setting q to 0, 1, and 2 respectively. To

TABLE 1   Comparison of Ending Value Parameter Estimates for Different AEP Model Formulations: Series 463

	Models			
Parameters	(0, 12, 12)	(1, 12, 12)	(2, 12, 12)	Autom. AEP
Polynomial				
Theta 0	4905.9687	4326.4375	3842.2036	
Theta 1	0.0	12.8379	5.0762	
Theta 2	0.0	0.0	0.0883	
Seasonal				
Alpha 1	1.000	1.000	1.000	
Alpha 2	1.059	1.053	1.045	
Alpha 3	0.995	0.993	0.989	
Alpha 4	0.946	0.943	0.932	
Alpha 5	0.901	0.901	0.896	
Alpha 6	0.990	0.987	0.985	
Alpha 7	0.825	0.823	0.825	
Alpha 8	0.417	0.414	0.411	
Alpha 9	1.187	1.179	1.185	
Alpha 10	1.108	1.105	1.101	
Alpha 11	0.971	0.967	0.959	
Alpha 12	0.983	0.979	0.968	
Autoregressive				
Beta 1	0.350	0.335	0.340	−0.016
Beta 2	0.178	0.170	0.168	0.014
Beta 3	0.119	0.113	0.133	0.074
Beta 4	0.042	0.038	0.036	0.033
Beta 5	0.040	0.039	0.041	0.025
Beta 6	0.027	0.025	0.025	0.023
Beta 7	0.009	0.008	0.007	−0.015
Beta 8	0.006	0.006	0.005	0.049
Beta 9	0.004	0.003	0.003	0.043
Beta 10	0.002	0.018	0.001	−0.006
Beta 11	0.003	0.002	0.002	0.039
Beta 12	0.008	0.006	0.005	0.671

TABLE 2   Comparison of Ending Value Parameter Estimates for Different AEP Model
Formulations: Series 805

Parameters	Models			
	(0, 12, 12)	(1, 12, 12)	(2, 12, 12)	Autom. AEP
Polynomial				
Theta 0	1654.773	3424.574	2993.565	
Theta 1	0.000	20.740	23.842	
Theta 2	0.000	0.000	0.014	
Seasonal				
Alpha 1	1.000	1.000	1.000	
Alpha 2	1.001	1.002	1.002	
Alpha 3	1.005	1.007	1.006	
Alpha 4	1.000	1.002	1.002	
Alpha 5	1.000	1.002	1.002	
Alpha 6	0.999	1.001	1.000	
Alpha 7	0.996	0.996	0.996	
Alpha 8	0.994	0.994	0.994	
Alpha 9	0.993	0.993	0.993	
Alpha 10	0.992	0.992	0.992	
Alpha 11	0.993	0.993	0.993	
Alpha 12	0.996	0.996	0.996	
Autoregressive				
Beta 1	0.121	0.067	0.074	0.070
Beta 2	0.108	0.065	0.070	0.079
Beta 3	0.098	0.062	0.066	0.104
Beta 4	0.089	0.060	0.063	0.080
Beta 5	0.081	0.058	0.060	0.068
Beta 6	0.075	0.056	0.057	0.062
Beta 7	0.071	0.055	0.055	0.119
Beta 8	0.067	0.053	0.053	0.101
Beta 9	0.064	0.052	0.052	0.109
Beta 10	0.062	0.051	0.051	0.077
Beta 11	0.060	0.050	0.050	0.137
Beta 12	0.057	0.049	0.048	0.090

adopt a familiar terminology, the (q, p, L) models for each series were $(0, 12, 12)$, $(1, 12, 12)$ and $(2, 12, 12)$.

As in the accuracy study, models were generated on the historical data only. Forecasts were then derived for the next 18 periods and compared with the post-sample values.

The initialization steps were identical in all cases. The seasonal indices (αs) were all initialized to 1.00. The initial value of each autoregressive parameter (βs) was 1/12. The initial values of the coefficients of the polynomial term were the mean of the series for $q = 0$, the mean of the first difference divided by T for $q = 1$,

TABLE 3   Comparison of Ending Value Parameter Estimates for Different AEP Model Formulations: Series 886

Parameters	Models			
	(0, 12, 12)	(1, 12, 12)	(2, 12, 12)	Autom. AEP
Polynomial				
Theta 0	17.520	18.421	32.003	
Theta 1	0.0	−0.009	−1.094	
Theta 2	0.0	0.0	0.029	
Seasonal				
Alpha 1	1.000	1.000	1.000	
Alpha 2	0.983	0.982	0.994	
Alpha 3	0.982	0.982	0.997	
Alpha 4	0.950	0.950	0.970	
Alpha 5	0.949	0.949	0.961	
Alpha 6	0.919	0.919	0.927	
Alpha 7	0.905	0.904	0.924	
Alpha 8	0.962	0.962	0.973	
Alpha 9	0.881	0.880	0.896	
Alpha 10	0.875	0.875	0.878	
Alpha 11	0.903	0.904	0.892	
Alpha 12	0.940	0.939	0.925	
Autoregressive				
Beta 1	0.698	0.713	0.461	0.39764
Beta 2	0.163	0.158	0.142	0.03915
Beta 3	0.041	0.038	0.049	0.07111
Beta 4	0.011	0.010	0.019	0.01729
Beta 5	0.004	0.003	0.009	−0.06167
Beta 6	0.002	0.001	0.006	−0.02145
Beta 7	0.000	−0.001	0.006	0.08344
Beta 8	−0.001	−0.002	0.004	−0.00299
Beta 9	−0.004	−0.005	0.003	−0.00763
Beta 10	−0.008	−0.010	0.001	0.05082
Beta 11	−0.016	−0.019	0.001	0.03365
Beta 12	−0.023	−0.028	−0.001	−0.00378

and the mean of the second difference divided by T^2 for $q = 2$. As for the damping factor (μ), it took the value of 0.1 with CLIM set at 10 per cent.

Tables 1 to 3 present the ending value AEP parameter estimates obtained for the different model formulations for each series respectively. For the purpose of comparison, the Autom. AEP ending parameter values taken from the competition results are also listed. Looking at Table 1, it can be observed that the estimated relative seasonal indices (αs) indicate strong seasonality in series 463. For example, values in season 8 are roughly estimated to be 58 per cent less than those for the reference season 1. What is also interesting to note is that no

TABLE 4   Comparison of Forecasting Performance of the Different AEP Model Formulations for the Three Series

	Mean absolute error	Mean absolute percentage error	Mean percentage error
Series 463			
(0, 12, 12)	1429.3	7.870	6.881
(1, 12, 12)	1006.8	5.862	3.376
(2, 12, 12)	1078.8	6.228	4.276
Autom. AEP	2909.6	14.818	11.789
Series 805			
(0, 12, 12)	974.197	5.676	−4.748
(1, 12, 12)	498.533	2.936	−1.166
(2, 12, 12)	559.165	3.286	−1.700
Autom. AEP	1791.93	10.375	−10.126
Series 886			
(0, 12, 12)	4.819	5.988	−5.216
(1, 12, 12)	4.320	5.354	−3.712
(2, 12, 12)	53.678	66.703	−66.703
Autom. AEP	16.408	20.204	−20.204

seasonality is present in the Beta 12 autoregressive coefficient. In contrast, the estimated Alphas for series 805 provide evidence of the method's ability to recognize no seasonality when none is present. Finally, the method detected only mild seasonality in series 886. Of particular interest is the robustness of the method in the estimation of the seasonal indices. They are very coherent in all cases from one formulation to another. Whatever GUNIAEP model was formulated ($q = 0, 1$, or 2), the near identical ending seasonal parameters resulted from the AEP estimation process.

A comparison of forecasting performance of the different AEP model formulations for the three series is presented in Table 4. Looking at this table, the improvement is at least of 50 per cent in contrast to the Autom. AEP results. In fact, for the three series, the $(1, 12, 12)$ formulation is revealed as the most accurate. What the results also show is the danger associated with a formulation incorporating a quadratic long-term component ($q = 2$). The performance of the $(2, 12, 12)$ formulation for series 886 bears evidence of that.

Automatic setting and adjustment of the damping factor

As we mentioned earlier, the damping factor plays an important role in the algorithm. Given that time series are products of different and time-varying environments, it would appear that performance could thus be improved with the specification and time adjustment of a different μ_t value for each series. How can one identify and time adjust the damping factor without *a priori* information on the environment present? What kind of tracking signal can be used to detect the magnitude and direction of the adjustment in μ_t?

Trigg and Leach (1967) have proposed, within the context of exponential smoothing models, a mechanism which provides partial answers to these questions. The tracking signal they have developed adjusts a smoothing constant ρ_t on the basis of the ratio of smoothed forecast errors over smoothed forecast absolute errors. Mathematically, the system is written as

$$\rho_t = |E_t|/M_t \tag{10}$$

where

$$E_t = \gamma e_t + (1 - \gamma)E_{t-1} \tag{11}$$

and

$$M_t = \gamma|e_t| + (1 - \gamma)M_{t-1}. \tag{12}$$

E_t represents a smoothed sum of forecast errors whereas M_t is a smoothed sum of forecast errors in absolute terms with $0 \leq \gamma \leq 1$.

One characteristic of the Trigg and Leach tracking signal is that it often results in very substantial changes in ρ_t from one observation to the next. The application of the same principle to μ_t, that is

$$\mu_t = \frac{|E_t|}{M_t},$$

could render the AEP correction process unstable. In order to alleviate drastic changes in the adjustment of μ_t from one observation to the next while still preserving the simplicity of the Trigg and Leach tracking signal, we have modified (10) such that

$$\mu_t = \lambda \frac{|E_t|}{M_t} + (1 - \lambda)\mu_{t-1} \tag{13}$$

where E_t and M_t are as previously defined, and $0 \leq \lambda \leq 1$.

A major advantage of our revised Trigg and Leach tracking signal lies in the fact that it can be applied to meet different objectives. On the one hand, by setting λ closer to 1, the system is geared to react more quickly to each signal received. On the other hand, if λ is set close to 0, the system is attempting more to identify a 'best' damping factor for a time series rather than reacting from one observation to the next.

6. APPLICATION OF GUNIAEP TO THE 1001 SERIES

The only test of a new forecasting methodology, however conceptually exact it may be, is its confrontation with real data and comparison with the performance of alternative approaches. In order to evaluate GUNIAEP, we have applied it without modifying the experimental design to the sample of 111 series for which forecasts were produced by all the methods involved in the accuracy study. Two runs of GUNIAEP on the 111 series were made in a completely automatic execution mode. For the first run, herein denoted GUNIAEP-1, a fixed damping factor was used to update all the parameters. In contrast, the second run, herein denoted GUNIAEP-A, used adaptive damping factors which were adjusted through the modified Trigg and Leach mechanism presented in the previous section.

The two runs were executed in the following way:

(a) Model specification parameters (q, p, L).
   (i) $q = 1$ for all series. The long-term component was then described by mean and trend terms.
   (ii) $p = 3$ for yearly series, 4 for quarterly series, and 12 for monthly series.
   (iii) $L = 0$ for yearly series, 4 for quarterly series, and 12 for monthly series.

(b) Initial values of the parameters.
   (i) Seasonal parameters.
   The reference season (r) was selected as the season with the most central mean value. Denoting S_r this value, the seasonal parameters were initialized as $\alpha_i(t) = S_i/S_r$ for all $i = 1, 2, \ldots, L$ at $t = 0$, where S_i represents the mean value for the observations of season i.
   (ii) Long-term parameters.
   For $q = 0$, the mean of the series: for $q = 1$, the mean of the first difference divided by T, the number of observations in a series. In the case of yearly data, the mean of the series divided by two was used for $q = 0$.
   (iii) Short-term parameters.
   The coefficients assigned to the autoregressors were all initialized at $1/p$.

(c) Correction limits (CLIM).
   (i) A correction limit of 0.3 was imposed on the seasonal parameters. The limit was 0.5 on all the other parameters.

(d) Smoothing constant (s).
   (i) The smoothing constant for the smoothing averages was set at 0.01. Initial values of these averages were the first value for the associated autoregressor or long-term component.

(e) Damping factors.
   (i) GUNIAEP-1: a constant of 0.1 was used in all cases.
   (ii) GUNIAEP-A: all seasonal parameters were updated with a fixed $\mu = 0.1$. The μ_1 for the long term was adjusted with $\lambda = 0.1$ and

TABLE 5 Comparison of MAPEs for Selected Methods (All, Seasonal, Non-seasonal Data)

	Forecasting horizon										Cumulative horizons					
	1	2	3	4	5	6	8	12	15	18	1–4	1–6	1–8	1–12	1–15	1–18
Seasonal data (60)																
Parzen	11.0	10.0	10.1	12.8	12.2	13.7	16.1	13.9	19.2	24.8	11.0	11.6	12.7	12.9	13.8	14.9
Bayesian F	8.1	10.8	9.5	11.3	10.0	13.1	18.1	14.2	22.7	22.9	9.9	10.5	11.9	12.4	13.5	14.4
Box–Jenkins	10.5	10.1	10.0	13.5	13.7	15.9	20.6	15.1	23.3	30.6	11.0	12.3	14.3	14.8	15.8	17.2
Winters	8.9	9.8	10.5	13.6	11.5	14.7	19.9	14.9	34.0	32.7	10.7	11.5	13.5	14.2	16.0	18.1
Lewandowski	10.8	12.9	12.6	14.3	12.9	17.3	19.8	16.6	33.7	23.8	12.7	13.5	14.7	14.9	16.9	17.6
Autom-AEP	10.3	11.6	13.6	15.0	15.3	20.2	26.0	14.6	27.7	29.0	12.6	14.3	17.0	16.4	17.4	18.6
GUNIAEP-1	9.1	10.3	10.6	13.6	11.6	14.0	17.5	12.2	28.2	26.5	10.9	11.6	13.0	13.1	14.5	15.8
GUNIAEP-A	8.6	9.6	10.4	12.5	12.1	14.9	18.9	12.7	16.4	19.8	10.3	11.3	13.0	13.0	13.2	13.7
Non-seasonal data (51)																
Parzen	10.1	11.5	11.4	14.3	16.9	15.9	12.8	12.8	32.4	31.6	11.8	13.3	14.1	14.1	15.0	16.3
Bayesian F	12.8	15.0	18.4	18.1	23.5	21.8	21.3	22.1	42.0	53.6	16.1	18.3	19.5	19.1	20.6	23.5
Box–Jenkins	10.0	11.3	13.0	15.7	19.5	18.4	15.5	20.4	35.1	45.0	12.5	14.7	15.3	15.6	17.1	19.5
Winters	9.5	11.3	16.7	17.6	24.5	23.5	29.8	19.0	31.6	40.1	13.8	17.2	19.9	19.8	20.7	22.2
Lewandowski	12.6	12.7	16.8	16.4	21.0	18.0	17.2	18.1	31.0	42.9	14.6	16.2	16.6	16.8	17.8	20.5
Autom-AEP	9.3	10.8	13.9	15.2	18.8	17.2	18.0	20.9	38.0	48.7	12.3	14.2	15.4	15.9	17.4	19.9
GUNIAEP-1	10.3	9.3	13.6	14.0	19.2	17.8	16.4	20.2	32.0	43.6	11.8	14.0	14.9	15.4	16.7	19.2
GUNIAEP-A	9.6	9.0	13.1	13.7	17.4	16.5	15.2	18.4	30.5	44.4	11.4	13.2	14.1	14.2	15.4	17.9
All Data (111)																
Parzen	10.6	10.7	10.7	13.5	14.3	14.7	16.0	13.7	22.5	26.5	11.4	12.4	13.3	13.4	14.3	15.4
Bayesian F	10.3	12.8	13.6	14.4	16.2	17.1	19.2	16.1	27.5	30.6	12.8	14.1	15.2	15.0	16.1	17.6
Box–Jenkins	10.3	10.7	11.4	14.5	16.4	17.1	18.9	16.4	26.2	34.2	11.7	13.4	14.8	15.1	16.3	18.0
Winters	9.2	10.5	13.4	15.5	17.5	18.7	23.3	15.9	33.4	34.5	12.1	14.1	16.3	16.4	17.8	19.5
Lewandowski	11.6	12.8	14.5	15.3	16.6	17.6	18.9	17.0	33.0	28.6	13.5	14.7	15.5	15.6	17.2	18.6
Autom-AEP	9.8	11.3	13.7	15.1	16.9	18.8	23.3	16.2	30.2	33.9	12.5	14.3	16.3	16.2	17.4	19.0
GUNIAEP-1	9.6	9.8	12.0	13.8	15.1	15.8	17.1	14.2	29.1	30.8	11.3	12.7	13.9	14.0	15.3	17.0
GUNIAEP-A	9.0	9.3	11.7	13.0	14.5	15.6	17.7	14.1	19.9	25.9	10.8	12.2	13.5	13.5	14.0	15.2

TABLE 6 Comparison of MAPEs for Selected Methods (Yearly, Quarterly, Monthly Data)

	Forecasting horizon										Cumulative horizons					
	1	2	3	4	5	6	8	12	15	18	1–4	1–6	1–8	1–12	1–15	1–18
Yearly data (20)																
Parzen	7.6	7.7	12.8	16.0	20.5	18.0					11.0	13.8				
Bayesian F.	12.2	12.6	14.9	18.0	20.6	20.6					14.4	16.5				
Box–Jenkins	7.2	10.8	13.7	18.6	23.2	22.3					12.6	16.0				
Winters	5.6	7.2	11.9	16.2	19.0	16.5					10.2	12.7				
Lewandowski	7.3	8.3	14.7	13.8	16.8	15.1					11.0	12.7				
Autom-AEP	7.1	8.8	14.1	17.8	21.8	19.1					11.9	14.8				
GUNIAEP-1	7.2	7.9	12.2	16.7	19.9	19.8					11.0	14.0				
GUNIAEP-A	7.1	8.3	12.2	15.9	18.2	18.0					10.8	13.3				
Quarterly data (23)																
Parzen	6.8	7.6	12.0	16.5	21.1	20.4	21.0				10.7	14.1	16.7			
Bayesian F.	12.7	18.6	20.4	24.7	27.8	26.8	28.8				19.1	21.8	24.6			
Box–Jenkins	7.6	8.2	13.9	21.3	26.1	26.1	25.4				12.7	17.2	20.1			
Winters	8.9	9.1	17.1	25.6	32.6	32.2	40.3				15.2	20.9	26.4			
Lewandowski	12.5	14.1	14.2	21.8	24.8	22.8	26.9				15.7	18.4	20.6			
Autom-AEP	8.3	8.8	15.4	22.4	29.2	34.7	40.2				13.7	19.8	25.9			
GUNIAEP-1	8.3	10.7	13.1	19.3	25.3	23.5	24.6				12.8	16.7	19.2			
GUNIAEP-A	8.5	10.8	16.5	21.8	27.4	28.7	32.4				14.4	19.0	22.7			
Monthly data (68)																
Parzen	12.7	12.6	9.6	11.7	10.2	11.8	14.3	13.7	22.5	26.5	11.7	11.4	12.1	12.6	13.9	15.4
Bayesian F.	8.9	10.8	10.9	9.9	10.9	12.8	16.0	16.1	27.5	30.6	10.1	10.7	11.8	12.6	14.5	16.6
Box–Jenkins	12.1	11.5	9.9	11.1	11.0	12.5	16.7	16.4	26.4	34.2	11.1	11.3	12.7	13.8	15.6	17.9
Winters	10.3	12.0	12.5	11.8	11.9	14.8	17.5	15.9	33.4	34.5	11.7	12.2	13.6	14.6	16.8	19.1
Lewandowski	12.6	13.6	14.6	13.5	13.8	16.6	16.2	17.0	33.0	28.6	13.6	14.1	14.4	14.9	17.1	18.9
Autom-AEP	11.2	12.8	13.0	11.9	11.2	13.4	17.6	16.2	30.2	33.9	12.2	12.2	13.4	14.2	16.1	18.4
GUNIAEP-1	10.8	10.1	11.6	11.1	10.3	11.9	14.0	14.2	29.1	30.8	10.9	11.0	12.0	12.8	14.8	17.0
GUNIAEP-A	9.8	9.1	9.9	9.3	9.1	10.5	12.7	14.1	19.9	25.9	9.5	9.6	10.4	11.4	12.5	14.2

TABLE 7 Comparison of Medians APEs for Selected Methods (All, Seasonal, Non-seasonal Data)

	Forecasting horizon										Cumulative horizons					
Seasonal data (60)	1	2	3	4	5	6	8	12	15	18	1–4	1–6	1–8	1–12	1–15	1–18
Parzen	6.0	6.5	6.3	7.6	7.2	7.6	6.7	6.2	11.4	11.8	7.0	7.2	7.2	7.2	8.0	8.5
Bayesian F.	4.9	6.6	5.6	6.8	4.6	8.6	11.2	7.8	12.6	13.1	6.1	6.3	7.5	7.7	8.3	8.5
Box–Jenkins	6.1	6.5	5.5	6.3	5.9	8.1	9.1	7.6	10.0	14.4	6.1	6.3	6.9	7.3	8.1	8.3
Winters	4.2	7.0	4.6	5.7	4.8	7.5	9.4	5.7	10.4	10.3	5.5	5.7	6.5	6.7	7.1	7.6
Lewandowski	5.7	7.4	5.8	7.6	7.9	7.9	9.5	6.3	12.3	10.6	6.3	6.9	7.3	7.3	7.9	8.1
Autom-AEP	5.3	6.5	6.8	8.2	6.9	7.4	9.5	6.5	12.5	11.8	6.4	6.6	7.1	7.5	8.4	8.8
GUNIAEP-1	3.7	5.7	5.3	6.7	4.4	7.8	9.1	5.2	9.9	12.7	5.4	5.4	6.5	6.7	7.2	7.5
GUNIAEP-A	3.9	4.7	6.6	6.4	6.1	7.7	9.2	6.4	10.4	14.1	5.5	6.1	6.9	7.1	7.5	8.1
Non-seasonal data (51)																
Parzen	2.6	4.7	5.3	7.2	9.2	11.6	13.9	7.8	18.1	11.5	5.0	6.6	8.1	8.4	9.2	9.8
Bayesian F.	5.6	4.4	5.9	7.9	12.2	10.8	16.3	18.5	17.6	21.5	5.3	7.0	7.9	8.3	9.6	11.0
Box–Jenkins	2.8	4.2	5.2	8.1	8.9	10.7	10.3	18.0	18.4	24.1	4.8	6.0	7.2	7.9	9.2	10.5
Winters	3.4	4.5	6.0	9.3	12.7	9.4	14.9	14.2	14.3	17.0	5.5	6.6	7.6	9.3	9.9	10.3
Lewandowski	5.0	3.4	4.8	7.2	7.9	7.6	8.2	5.1	7.2	7.8	4.4	5.2	5.6	5.6	5.7	5.9
AUTOM-AEP	2.4	4.5	7.3	9.2	10.9	8.4	10.9	14.5	17.8	28.5	5.4	6.6	7.5	8.9	9.4	10.3
GUNIAEP-1	5.8	6.1	8.2	8.4	12.2	10.9	13.1	13.0	19.2	16.6	6.9	7.8	8.3	8.5	9.5	10.4
GUNIAEP-A	4.2	4.1	8.0	7.1	9.6	10.8	12.4	9.9	10.9	13.3	5.9	7.1	7.3	7.4	7.7	7.6
All data (111)																
Parzen	4.8	6.1	6.3	7.6	8.5	9.0	9.1	6.6	11.5	11.6	6.0	7.0	7.0	7.6	8.4	9.0
Bayesian F.	5.0	5.3	5.9	7.0	7.5	9.3	12.1	8.8	12.9	14.0	5.7	6.7	7.6	7.9	8.5	8.9
Box-Jenkins	5.3	5.0	5.2	6.6	6.0	8.8	9.3	8.6	12.4	16.4	5.5	6.2	7.0	7.6	8.2	8.7
Winters	4.1	5.2	5.9	7.0	6.5	8.1	10.6	7.4	10.5	10.7	5.5	6.1	6.8	7.3	7.9	8.3
Lewandowski	5.4	4.8	5.5	7.5	7.9	7.9	9.3	5.4	10.3	10.3	5.6	5.9	6.6	6.8	7.3	7.4
Autom-AEP	4.6	6.2	7.0	8.7	8.3	7.6	10.5	9.2	13.7	15.2	6.1	6.6	7.4	8.1	8.8	9.2
GUNIAEP-1	5.1	5.9	7.0	7.9	7.4	9.2	10.2	5.6	11.6	12.8	6.5	6.7	7.3	7.3	7.8	8.2
GUNIAEP-A	4.0	4.6	6.8	7.0	7.3	9.3	9.6	7.1	10.4	13.6	5.7	6.6	7.0	7.3	7.6	8.0

TABLE 8   Comparison of Medians of AEPs for Selected Methods (Yearly, Quarterly, Monthly Data)

	Forecasting horizon										Cumulative horizons					
	1	2	3	4	5	6	8	12	15	18	1–4	1–6	1–8	1–12	1–15	1–18
Yearly data (20)																
Parzen	2.5	2.6	5.3	6.2	9.2	11.5					4.8	6.2				
Bayesian	3.3	4.4	5.8	7.0	7.5	10.9					5.7	7.0				
Box–Jenkins	2.8	4.2	5.2	6.7	13.0	10.7					5.2	7.8				
Winters	1.7	2.5	3.6	7.0	12.7	7.5					3.6	5.1				
Lewandowski	4.2	2.8	5.9	8.9	11.4	7.9					4.4	6.6				
Autom-AEP	2.3	3.9	3.6	6.4	11.8	9.3					4.5	6.4				
GUNIAEP-1	4.7	6.1	8.7	9.2	16.5	14.4					7.3	8.7				
GUNIAEP-A	3.1	4.3	8.7	7.6	9.7	10.8					6.6	7.6				
Quarterly data (23)																
Parzen	3.2	4.0	3.2	6.1	11.7	15.7	10.8				4.0	5.4	7.2			
Bayesian F.	4.5	8.1	7.6	10.0	14.7	16.3	20.2				7.6	9.4	11.7			
Box–Jenkins	2.6	2.8	4.4	9.6	9.1	12.2	10.3				4.3	5.2	6.5			
Winters	3.4	4.9	4.2	9.3	12.1	14.5	16.0				5.5	7.6	9.3			
Lewandowski	5.6	4.8	5.2	7.6	9.7	11.0	13.3				5.4	5.9	7.6			
Autom-AEP	2.1	4.2	5.4	9.2	14.0	11.9	11.7				5.3	7.1	8.4			
GUNIAEP-1	4.9	6.3	5.3	8.9	11.9	11.5	13.1				6.3	7.9	8.5			
GUNIAEP-A	3.4	5.9	6.7	9.6	9.2	11.2	14.0				6.3	7.3	8.9			
Monthly data (68)																
Parzen	6.0	7.1	7.3	8.3	6.7	7.6	8.3	6.6	11.5	11.6	7.4	7.4	7.5	7.7	8.7	9.4
Bayesian F.	5.0	4.8	5.2	5.0	4.4	7.8	11.0	8.8	12.9	14.0	5.2	5.3	6.8	7.5	8.2	8.7
Box–Jenkins	6.9	7.0	6.6	6.4	5.7	7.8	9.1	8.6	12.4	16.4	6.6	6.5	7.0	7.8	8.5	9.0
Winters	5.0	6.6	6.0	5.7	4.4	6.9	9.4	7.4	10.5	10.7	5.9	5.7	6.4	7.1	7.9	8.5
Lewandowski	5.7	5.6	5.8	5.9	5.8	6.7	8.6	5.4	10.3	10.3	5.9	5.9	6.3	6.7	7.3	7.5
Autom-AEP	5.6	6.6	7.8	6.6	6.5	6.5	9.5	9.2	13.7	15.2	6.6	6.6	7.0	8.1	9.0	9.6
GUNIAEP-1	5.8	5.0	7.0	5.0	4.6	6.5	9.1	5.6	11.6	12.8	5.8	5.4	6.3	6.7	7.4	8.0
GUNIAEP-A	4.1	3.9	6.7	5.2	6.0	6.0	8.8	7.1	10.4	13.6	5.0	5.6	6.3	6.9	7.4	8.0

$\gamma = 0.01$ whereas the μ_t for the short-term parameters was adjusted with $\lambda = 0.01$ and $\gamma = 0.01$. At $t = 0$, $\mu_t = 0.1$ in both cases.

In both runs, the training period for each series could not exceed 60 cycles. However, it is interesting to note that the GUNIAEP-1 run was terminated after 938 seconds of CPU time whereas the GUNIAEP-A was executed in 417 seconds (less than 7 minutes of CPU time).

The results of the two runs are displayed in Tables 5 to 8. Tables 5 and 6 look at the mean absolute percentage errors (MAPE) for the different forecasting horizons for all series, seasonal, non-seasonal, yearly, quarterly, and monthly series. Given the same data division, Tables 7 and 8 report medians of absolute percentage errors. To provide a basis of comparison, the tables display the measures reported for the techniques revealed as the most accurate. They are Parzen's ARARMA method, Bayesian Forecasting, Box–Jenkins, Winters Exponential Smoothing, Lewandowski's FORSYS system and Autom. AEP.

Looking at these tables, we first note a marked improvement in forecast accuracy realized with generalized UNIAEP as opposed to the Autom. AEP results. We also find the improvement is even more substantial when the damping is time-adjusted. For example, the MAPE over all data of GUNIAEP-A is over 20 per cent smaller than the MAPE reported for Autom. AEP, and 10 per cent smaller than the MAPE obtained for the run with a fixed damping factor. The type of data for which the improvement is the most significant is without any doubt, seasonal data.

What is also promising to observe is that the forecasts generated with the generalized UNIAEP modelling framework with adaptive damping factors are revealed as more accurate than those provided by the other methods. For example, over all series, the approach yields the smallest MAPE and ranks second on the basis of the median. This implies that in contrast to both Parzen's ARARMA method and Lewandowski's FORSYS system, GUNIAEP not only provides very accurate forecasts overall (low median) but also results in very few large errors (very low mean). This contrasts with Parzen, for instance, which, according to the measures reported, tended to provide less precise forecasts in general (high median) while limiting the number of large errors (low mean).

7. CONCLUDING REMARKS

This chapter is an example of conceptual development from empirical results rather than deductive reasoning. It presents a new methodological approach to univariate forecasting which is based on a decomposition model and relies on a feedback parameter estimation scheme. The results of the 1001 forecasting competition have, in fact, initiated and guided its development. It is apparent from the results reported in the previous section that this framework appears promising. These results demonstrate the ability of the method to obtain very, if not the most, accurate forecasts for any type of data and under different perspectives.

There are still some issues which remain unsolved. For example, how can one recognize automatically the need for including a quadratic parameter in the long-term component? How can the classical components of a time series be estimated from a GUNIAEP model? It is towards these refinements that our attention will be focused in the future.

REFERENCES

Bretschneider, S. I., Carbone, R., and Longini, R. L. (1979). 'An Adaptive Approach to Time Series Forecasting', *Decision Sciences*, **10**, 232–244.
Bretschneider, S. I., Carbone, R., and Longini, R. L. (1982). 'An Adaptive Multivariate Approach to Time Series Forecasting', *Decision Sciences*, **13**, 668–680.
Bretschneider, S. I. and Gorr, W. L. (1980). 'On the Relationship of Adaptive Filtering Forecasting Model to Simple Brown Smoothing', *Management Science*, **27**, 965–969.
Bretschneider, S. I. and Mahajan, V. (1980). 'Adaptive Technological Substitution Models', *Technological Forecasting and Social Change*, **18**, 129–139.
Carbone, R. (1980). 'AEP: A New Approach to Time Series Analysis and Forecasting', in *Analysing Time Series*, O. D. Anderson (ed.), Amsterdam: North-Holland, pp. 109–117.
Carbone, R. and Gorr, W. L. (1978). 'An Adaptive Diagnostic Model for Air Quality Management', *Atmospheric Environment*, **12**, 1785–1791.
Carbone, R., Longini, R. L., and Ivory, E. L. (1980). 'Competition used to Select a Computer-assisted Mass Appraisal System', *Assessors Journal*, **15**, 163–167.
Carbone, R. and Longini, R. L. (1976). 'Reform of Property Tax Administration for Achieving Intrajurisdictional Equity', *Assessors Journal*, **11**, 197–207.
Carbone, R. and Longini, R. L. (1977). 'A Feedback Model for Automated Real Estate Assessment', *Management Science*, **24**, 241–248.
Longini, R. L. *et al.* (1975). 'Filtering Without Phase Shift', *IEEE Transactions in Biomedical Engineering*, **41**, 432–433.
Mahajan, V., Bretschneider, S. I., and Bradford, J. W. (1980). 'Feedback Approaches to Modeling Structural Shifts in Market Response', *Journal of Marketing*, **44**, 71–80.
Makridakis, S. *et al.* (1982). 'The Accuracy of Extrapolative (Time Series) Methods: Results of a Forecasting Competition', *Journal of Forecasting*, **1**, 111–153.
Trigg, D. W. and Leach, D. H. (1967). 'Exponential Smoothing with an Adaptive Response Rate', *Operational Research Quarterly*, **18**, 53–59.
Wheelwright, S. L. and Makridakis, S. (1973). 'Forecasting with Adaptive Filtering', *R.A.I.R.O.*, **1**, 32–52.

CHAPTER 7

*Bayesian Forecasting**

Robert Fildes
Manchester Business School

1. INTRODUCTION

Bayesian methods have frequently been used in statistical applications, typically as a means of estimating the parameters of an econometric model (Zellner, 1971). However, their use in time series forecasting has been less frequent. This chapter considers the ideas behind Bayesian statistics very briefly. We then go on to consider their application in forecasting. The final section describes how Bayesian forecasting has been applied in the forecasting competition which is the primary focus of this book.

Adopting a time series framework, Bayesian methods can be thought of schematically by reference to Figure 1.

The essential features are a prior probability distribution representing the forecaster's initial view of the system, for example the forecaster may believe that sales growth is 5 percent p.a. ± 1 percent, and its distribution is normal. The second aspect of the Bayesian model is observation. If the current state of the system θ_t (possibly a vector of parameters) were known the probability distribution of the latest observation would also be known, i.e. $P(Y_t|\theta_t)$. This we call the observation distribution; (it is more usually called the sampling distribution).

A new observation is recorded and when combined with the forecaster's prior knowledge, a posterior probability distribution is obtained of the likely state of the system having observed Y_t, i.e. $P(\theta_t|Y_t)$. Bayes theorem is used to carry out the updating. If we drop the time suffix temporarily for ease of explanation this updating procedure can be represented by

$$P(\theta|Y) = \frac{P(Y|\theta)P(\theta)}{P(Y)}$$

(which is a statement of Bayes theorem). Now $P(Y|\theta)$, the observation

* This research was supported by the Science and Engineering Research Committee, UK and the Research Management Committee, Manchester Business School.

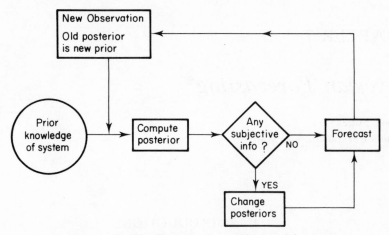

FIGURE 1   The Bayesian forecasting model

distribution, can be calculated from the assumptions made about how the observations are generated. $P(\theta)$ is of course the prior and $P(Y)$ can be calculated from $P(Y/\theta)$ and $P(\theta)$; (in fact it is the integral of the two). Thus we have derived the posterior distribution of θ once Y is observed. Leaving aside the question of subjective information for the moment, the next stage in the Bayesian procedure as shown in Figure 1 is to produce forecasts. The posterior contains all the information relevant for forecasting. To complete the loop, once a new observation is collected the old posterior, which encapsulated all known information related to the system, has now to be updated. The old posterior is therefore taken as the new prior and it in turn is updated by Bayes formula. The iteration has now been completed.

To make the above explanation more mathematically precise let us reintroduce the time suffix and see how $P[\theta_t|\mathbf{Y}_t]$ is calculated where $\mathbf{Y}_t$ represents the history of the series up to and including Y_t, the latest observation. Use of Bayes theorem shows that $P[\theta_t|\mathbf{Y}_t]$ depends on $P[Y_t|\theta_t, \mathbf{Y}_{t-1}]$ the observation distribution, $P[\theta_t|\mathbf{Y}_{t-1}]$ and $P[Y_t|\mathbf{Y}_{t-1}]$. But $P[Y_t|\mathbf{Y}_{t-1}]$, can be easily calculated from $P[Y_t|\theta_t, \mathbf{Y}_{t-1}]$ and $P[\theta_t|\mathbf{Y}_{t-1}]$. The key probability is therefore $P[\theta_t|\mathbf{Y}_{t-1}]$.

If the forecaster understands how the system evolves over time; that is, if the nature of the system's transition from the state at $(t-1)$, θ_{t-1} to the new state at t, θ_t is known, then all the right-hand side probabilities can be calculated, in particular $P[\theta_t|\mathbf{Y}_{t-1}]$. The probability $P[\theta_t|\mathbf{Y}_{t-1}]$ can be thought of as the forecast of θ_t based on the system's history to $(t-1)$. Bayes theorem has given us a means of updating our knowledge of the system once the new observation has come in.

Now before a further observation is recorded alternative information sources, perhaps sales force intelligence or market research, may alter the forecaster's posterior probability distribution of θ_t. Whether or not any subjective information is used this latest posterior can then be used to produce a forecast.

The framework described above is very general and can be used straightforwardly to estimate the parameters of a regression model or a Box–Jenkins ARIMA model. More interestingly it can be used to decide which of a number of distinct models is 'true' at a particular point in time.

Let us suppose the state of the system is represented by one of a number of distinct models $\{M(j)\}_{j=1}^n$. The forecaster usually has a set of prior beliefs representing his views concerning which of the models holds true, and these beliefs can be represented by a prior probability distribution. Let $\pi_j, j = 1, \ldots, n$ represent the probability that model j holds 'true' at any particular point in time. Once a set of data is observed, these beliefs can be updated to incorporate both the forecaster's priors and the observed evidence to derive the posterior distribution for the various models, i.e. $P[M(j)|\mathbf{Y}_t]$. One simple way of selecting which model could best be regarded as 'true' would be to choose the model with maximum posterior probability. Colloquially, this process can be described as choosing the model which is most compatible with the forecaster's initial beliefs *and* the observed data. In contrast with the first example where the states of the system were the possible parameter values of θ_t, the possible states are here the different models, $M(j), j = 1, \ldots, n$.

In summary, Bayesian methods offer an alternative procedure for estimating conventional statistical and econometric models. They also provide a formal statistical framework for discriminating between alternative models. But how can they be used most effectively for forecasting? The next section examines this question in detail.

2. THE BAYESIAN MODEL

A number of researchers in forecasting have characterized their own particular approach as Bayesian, e.g. Morrison and Pike (1977). In the UK during the 1970s the term became synonymous with the class of models introduced by Harrison and Stevens in 1971 and elaborated further in 1976. During the same period a working paper was written describing a number of applications. The papers elicited widespread interest when they appeared but relatively little research work was done evaluating the method's effectiveness, an exception being Fildes (1979). Extensions to the early work have been described by Fildes and Stevens (1978), Johnston and Harrison (1980) and Smith (1979).

The Bayesian forecasting models of Harrison and Stevens have their roots in the exponential smoothing models of Brown, Holt, and Winters. Let us suppose that the current observation Y_t can be represented in the non-seasonal case by a random fluctuation around its true current (but unknown) level A_t.

$$Y_t = A_t + \varepsilon_t \{\varepsilon_t \sim N(O, VE)\}. \tag{1}$$

However, instead of assuming the level is fixed A_t is updated from its previous level of A_{t-1} by a trend factor B_t, disturbed once again by noise:

$$A_t = A_{t-1} + B_t + \gamma_t \{\gamma_t \sim N(O, VG)\}. \tag{2}$$

To complete the description of how the parameters develop over time the trend is assumed to be a random walk:

$$B_t = B_{t-1} + \delta_t \{\delta_t \sim N(O, VD)\}. \tag{3}$$

The above equations can be written more generally in a matrix form familiar to control engineers:

$$Y_t = \mathbf{F}_t \boldsymbol{\theta}_t + v_t \qquad \text{– the observation equation } \{v_t \sim N(O, V_t)\}$$

$$\boldsymbol{\theta}_t = \mathbf{G}\boldsymbol{\theta}_{t-1} + \mathbf{w}_t \qquad \text{– the system equation } \{\mathbf{w}_t \sim N(O, \mathbf{W}_t)\}$$

where $\boldsymbol{\theta}_t$ is a $(p \times 1)$ vector of process parameters: $\mathbf{F}_t$ is a $(1 \times p)$ vector of independent variables, known at time t: $\mathbf{G}$ is a $(p \times p)$ known system matrix describing the transition of the process parameters and v_t and $\mathbf{w}_t$ are random normal vectors of dimension (1×1) and $(p \times 1)$ respectively with zero means and known variances at time t, V_t and $\mathbf{W}_t$.

In the previous example,

$$\boldsymbol{\theta}_t = \begin{pmatrix} A_t \\ B_t \end{pmatrix}: \qquad \mathbf{F}_t = (1 \quad 0): \qquad v_t = \varepsilon_t: \qquad \mathbf{G} = \begin{pmatrix} 1 & 1 \\ 0 & 1 \end{pmatrix}$$

and

$$\mathbf{w}_t = \begin{pmatrix} \gamma_t + \delta_t \\ \delta_t \end{pmatrix}. \qquad \text{F and G are constant for all } t.$$

3.   THE KALMAN FILTER

Paraphrasing Harrison and Stevens (1976), for the above model if we assume the distribution of $\boldsymbol{\theta}_0$ prior to the first observation is $N(\mathbf{m}_0, \mathbf{C}_0)$ the posterior distribution of $\boldsymbol{\theta}_t$ at time t is also normally distributed. This is easily represented. Let:

$$\mathbf{Y}_t = (Y_t, Y_{t-1}, \ldots, Y_1) = (Y_t, \mathbf{Y}_{t-1}),$$

then

$$(\boldsymbol{\theta}_t | \mathbf{Y}_t) \sim N(\mathbf{m}_t, \mathbf{C}_t)$$

and the values of $\mathbf{m}_t$ and $\mathbf{C}_t$ are obtained recursively from the Kalman filter as follows:

Let

$$\hat{Y}_t = \mathbf{F}_t \mathbf{G} \mathbf{m}_{t-1} \qquad \text{– The one-step ahead forecast}$$

$$e_t = Y_t - \hat{Y}_t \qquad \text{– The one-step ahead forecast error}$$

$$\mathbf{R}_t = \mathbf{G} \mathbf{C}_{t-1} \mathbf{G}^T + \mathbf{W}_t$$

$$\hat{\Omega}_t = \mathbf{F}_t \mathbf{R}_t \mathbf{F}_t^T + V_t \qquad \text{– The variance of the one-step ahead forecast}$$

$$\mathbf{D}_t = \mathbf{R}_t \mathbf{F}_t^T (\hat{\Omega}_t)^{-1} \qquad \text{– The smoothing constant}$$

then

$$\mathbf{m}_t = \mathbf{Gm}_{t-1} + \mathbf{D}_t e_t \qquad \text{– The updated mean}$$
$$\mathbf{C}_t = \mathbf{R}_t - \mathbf{D}_t \hat{\Omega}_t \mathbf{D}_t^T \qquad \text{– The updated variance–covariance matrix}$$

We have denoted the transpose of a matrix $\mathbf{F}$ by $\mathbf{F}^T$.

The posterior distribution of θ_t captures everything about the system that is known at time t and is relevant to forecasting its future values. The above equations show that the current posterior is calculated from the previous posterior $(\theta_{t-1}|\mathbf{Y}_{t-1})$, the current observation noise V_t and system noise $\mathbf{W}_t$.

4. THE MULTI-STATE MODEL

The above model is defined by the four matrices $\mathbf{F}$, $\mathbf{G}$, V, and $\mathbf{W}$ with the time suffix dropped for simplicity. This characterization of a time series can be extended by considering a number of alternative models $\{M(j)\}$ $j = 1, \ldots, N$ with

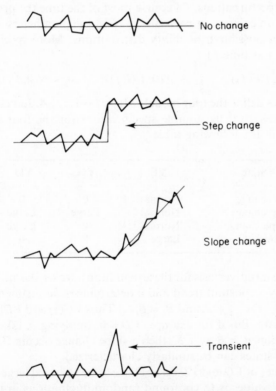

FIGURE 2   Possible states of the system. Reprinted with permission from *Operational Research Quarterly*, **22**, Harrison and Stevens, 'A Bayesian approach to short-term forecasting'. Copyright (1971) Pergamon Press Ltd.

associated probabilities of $P_t^{(j)}$ of the jth model being 'true' at time t. Two classes of problem are distinguished by Harrison and Stevens:

(1)   from the set of possible models $M(j)$, $(j = 1, \ldots, N)$ it is required to choose a single model to represent the time series;

(2)   it is required that forecasts and decisions be based on information relating to the whole set of models.

Zellner (1971), for example, has considered the first class of problem. However, it is a particular model in the second class to which the term 'Bayesian Forecasting' is usually applied. Let us suppose that any time series at a given moment can be characterized as being in one of four states: to put it another way, one of four models $M(j)$, $(j = 1, \ldots, 4)$ operates. The four states are 'no change', 'step change', 'slope change', and 'transient', as shown in Figure 2 (from Harrison and Stevens, 1971). In the 'no change' state the process described by equations (1)–(3) specifies probability distributions for A_t and B_t which are incrementally updated by new observations. However, in the other states these parameters are subject to violent perturbations. Of course most of the time the process is in the 'no change' state. In essence, at each observation the parameters are generated from a number of possible probability distributions. More specifically, if the system is in state j at time t let:

$$\varepsilon_t \sim N(0, VE(j)): \qquad \gamma_t \sim N(0, VG(j)): \qquad \delta_t \sim N(0, VD(j))$$

and these variances define the four models $M(j), j = 1, \ldots, 4$. Judicious choice of the above variances gives us suitable specifications for the four states named above, as shown in the following table.

State	VE	VG	VD
No change	Normal	0	0
Step change	Normal	Large	Large
Slope change	Normal	0	Large
Transient	Large	0	0

To illustrate the intuitive basis for these definitions, we see the 'no change' state is characterized by a constant trend and a deterministic adjustment to the level each period, i.e. $A_t = A_{t-1} + B_t$ and $B_t = B_{t-1}$. Thus $VG(1)$ and $VD(1)$ are zero in the 'no change' state. But if for example $VD \neq 0$, implying δ_t takes a non-zero value in the equation $B_t = B_{t-1} + \delta_t$, then a slope change occurs. The 'transient' and 'step change' states can be similarly characterized.

In principle values of VG and VD do not have to be assumed large but the effect of assuming small values is to confound random fluctuations in the data with fundamental changes in the parameters.

As before, in order to produce forecasts it is necessary to calculate the posterior distribution of θ_t conditional on current and past values of Y and F. Harrison and Stevens (1971) give the formulae in detail. We merely sketch out the argument here. Let us suppose that:

$$(\theta_{t-1}|M_{t-1}(j), \mathbf{Y}_{t-1}) \sim N(\mathbf{m}_{t-1}^{(j)}, \mathbf{C}_{t-1}^{(j)}),$$

i.e. the distribution of the parameter vector conditional on a specified model $M(j)$ holding 'true' at $t - 1$ is multivariate normal with mean $\mathbf{m}_{t-1}^{(j)}$ and variance–covariance matrix $\mathbf{C}_{t-1}^{(j)}$. Let us also suppose that $P_{t-1}^{(j)}$, the probability that model j holds at $(t - 1)$ is known. The posterior distribution of θ_t conditional on all the observations including the latest, Y_t, and also conditional on the state of system at $(t - 1)$, and its new state at t can be calculated using Bayes theorem, and the Kalman filter, i.e. the distribution of $(\theta_t|\mathbf{Y}_t, M_{t-1}(i), M_t(j))$ is calculated for all i and j. The final stage requires the use of the following approximation for the posterior:

$$(\theta_t|M_t(j), \mathbf{Y}_t) \sim N(\mathbf{m}_t^{(j)}, \mathbf{C}_t^{(j)}),$$

where

$$P_t^{(j)} = \sum_i P_t^{(i,j)} \quad \text{and} \quad P_t^{(i,j)} = \text{Prob. } (M_t(j), M_{t-1}(i)|\mathbf{Y}_t)$$

$$\mathbf{m}_t^{(j)} = \sum_i P_t^{(i,j)} \mathbf{m}_t^{(i,j)} / P_t^{(j)}$$

$$\mathbf{C}_t^{(j)} = \sum_i P_t^{(i,j)} \{ \mathbf{C}_t^{(i,j)} + (\mathbf{m}_t^{(i,j)} - \mathbf{m}_t^{(j)})(\mathbf{m}_t^{(i,j)} - \mathbf{m}_t^{(j)t}) \} / P_t^{(j)}.$$

To summarize the mathematics verbally, starting at $(t - 1)$ with a normal distribution for θ_{t-1} conditional on a particular model holding 'true' and the probability of that model holding, we can derive a conditional distribution for θ_t. Starting with a normal prior θ_0 an induction argument therefore shows that $(\theta_t|\mathbf{Y}_t)$ is approximately normal.

The above is only an approximation because in general a mixture* of multivariate normal distribution is not multivariate normal. Gathercole (1982) and Gathercole and Smith (1981) discuss its inadequacies and how best to resolve the problems posed.

5. FORECASTING

To produce forecasts of Y_{t+k} we first note that in the application of interest to us F_{t+k} is known, and constant and equal to F (say). Let

$$\hat{Y}_{t+k} = E(Y_{t+k}|\mathbf{Y}_t, \mathbf{F})$$

---

* A mixture of two probability distributions $F(x)$ and $G(x)$ is defined as $\alpha F(x) + (1 - \alpha)G(x)$ and is a probability distribution itself. This definition can be generalized to any number of distributions. In the above example of a mixture the weights are just $P(i, j)$ and the corresponding distributions are of $(\theta_t|\mathbf{Y}_t, M_{t-1}^t(i), M_t(j))$, the summation being over i.

where E is the expectations operator then

$$\hat{Y}_{t+k} = \mathbf{F}E(\boldsymbol{\theta}_{t+k}|\mathbf{Y}_t, \mathbf{F})$$
$$= \mathbf{F}\mathbf{G}^k\mathbf{m}_t.$$

The variance–covariance matrix is calculated iteratively as follows. Let:

$$\hat{\mathbf{C}}_{k,t} = \text{var}\,(\boldsymbol{\theta}_{t+k}|\mathbf{Y}_t, \mathbf{F})$$

$\hat{\mathbf{C}}_{0,t} = \mathbf{C}_t$ has already been estimated as described earlier. It can easily be shown that

$$\hat{\mathbf{C}}_{k,t} = \mathbf{G}\hat{\mathbf{C}}_{k-1,t}\mathbf{G}^T + \mathbf{W}_{t+k}$$

and

$$\text{var}\,(\hat{Y}_{t+k}|\mathbf{Y}_t, \mathbf{F}) = \mathbf{F}\hat{\mathbf{C}}_{k,t}\mathbf{F}^T + V_{t+k}.$$

Combining these equations gives an estimate of the variance–covariance matrix in terms of $\mathbf{F}$, $\mathbf{G}$, and the Vs and Ws.

6.   SPECIFYING THE MULTI-STATE MODEL

We have shown that starting with a prior probability distribution for $\boldsymbol{\theta}_0$ the parameter vector, as well as a set of prior probabilities $\{\pi_i\}$ of the four system states (or models) 'no change', 'step change', 'slope change', and 'transient', it is possible to derive the posterior probability distribution of Y_{t+k}. This forecast distribution is conditional on one making some assumptions about $\mathbf{F}$, $\mathbf{G}$, V_{t+k}, and $\mathbf{W}_{t+k}$.

The matrix $\mathbf{F}$ is known exactly in the applications we have considered here and for the non-seasonal growth model is $\mathbf{F} = (1 \quad 0)$ while $\mathbf{G} = \begin{pmatrix} 1 & 1 \\ 0 & 1 \end{pmatrix}$. Suppose for the moment that the variance in the 'no change' state is known and is V_o. We will specify the variances of the other states relative to this basic noise variance V_o.

$$VE(j) = RE(j)V_o: \qquad VG(j) = RG(j)V_o: \qquad VD(j) = RD(j)V_o$$

With the Rs and πs specified the only unknown parameter in the model is V_o. The values recommended in Harrison and Stevens' first paper for the Rs and the prior probabilities, the πs, are shown below.

State	Prior probability	RE	RG	RD
No change	0.900	1	0	0
Step change	0.003	1	100	0
Slope change	0.003	1	0	1
Transient	0.094	101	0	0

7. ESTIMATING THE NOISE VARIANCE

The final step in the process is to specify the noise variance V_o. The initial papers of Harrison and Stevens did not describe how this should be done. We describe the approach adopted by Stevens in a Bayesian forecasting computer package he has developed. Cantarelis (1979) describes alternative approaches.

First define a range of possible values V_{MIN} to V_{MAX} within which the variance is assumed to lie. This range is then divided into Q values $\{V_i\}$ (with Q typically specified as 11). The prior probability that V_i is the 'correct' variance is assumed: call it $q_i(o)$ $(i = 1, \ldots, Q)$. The posterior distribution $\{q_i(t)\}$ can now be calculated using Bayes theorem.

The final stage of the variance estimation routine is to estimate an 'average' V value which uses the V values around the median. This average V is used in the subsequent updating of the parameters in the forecasting. The calculation uses the value of $\{q_i(t)\}$: we call the value $V_o(t)$.

8. EXTENSIONS

The early work of Harrison and Stevens has been extended by Stevens although no published work is available describing the extensions. Stevens' programs are used in the empirical work reported in the next section. The most important change is that the model of the observation, $Y_t = A_t + \varepsilon_t$, is thought of as *multiplicative* in that the error variance is a power function of the mean level A_t of the series. More exactly, the noise variance is specified by: var $\varepsilon_t = (CA_t^P)^2$.

The parameter p is pre-set by the user according to whether the noise is viewed as absolute or dependent on the level of the series.

Now while the above model is used to develop an intuitive view of the parameters, the mean level, slope and error variance, it is *not* the model that forms the basis for using the Kalman Filter. The model used is:

$$\log Y_t = \mu_t + e_t \qquad \text{– the observation equation}$$
$$\mu_t = \mu_{t-1} + \beta_t + \gamma_t \qquad \text{– the system equations}$$
$$\beta_t = \beta_{t-1} + \delta_t$$

where $\mu_t = \log A_t$ and $\beta_t = \log B_t$.

The noise e_t is assumed to be normal with zero mean and variance given by $\log(1 + \tau^2)$ where:

$$\tau^2 = \text{var } \varepsilon_t / A_t^2.$$

The reader interested in the implications of these assumptions should refer to Fildes (1982).

Just as the error variance is specified in multiplicative form so the disturbance matrix **W** (equivalent to specifying the Rs) is defined somewhat differently. For precise details the reader is again referred to Fildes (1982). They are not given here as the packages publicly available through computer bureaux do not use the same specification.

A further extension is the inclusion of seasonality. In the additive model with seasonality, as Harrison and Stevens (1976, pp. 217–218) note, the observation equation becomes:

$$\log Y_t = \mu_t + \rho_{j,t} + e_t$$

where $\rho_{j,t}$ is the jth seasonal factor appropriate for period t.

$\rho_{j,t} = \rho_{j,t-1} + \delta\rho_{j,t}$ ($j = 1, \ldots, \text{NSF}$ – the number of seasonal factors) and these seasonal factors are updated as each new point becomes available. However, the seasonal factors are constrained to sum to zero, so

$$\sum \rho_{j,t} = \sum \delta\rho_{j,t} = 0.$$

If the joint distribution of the seasonal factors is assumed multivariate normal, each factor independent of the other, the distribution of each seasonal factor conditional on the sum of factors being zero is normal with the variance–covariances given below.

$$\text{cov}\left(\rho_i, \rho_j \middle| \sum \rho_k = 0\right) = \text{cov}\left(\rho_i, \rho_j\right) - \frac{\text{var}\left(\rho_i\right) \times \text{var}\left(\rho_j\right)}{\text{VSUM}}$$

where $\text{VSUM} = \sum_k \text{var}\left(\rho_k\right)$ and $\text{cov}\left(\rho_i, \rho_j\right) = 0, i \neq j$.

The error terms $(\delta\rho_{j,t})$ permit the seasonal factors to shift just as the level and slope parameters can shift. The inclusion of the seasonal parameters has the effect of increasing the dimension of the parameter vector and amending the disturbance matrix **W**. However, the Kalman Filter equations, the procedure for estimating the error variance etc., remain unaltered.

9.   SUMMARY

The above description has been a somewhat discursive introduction to the Bayesian multi-state growth model. The steps to be followed in developing such a model may be summarized in the form of an outline computer program, as follows:

(1)   Specify a prior probability distribution for the level and growth parameters (and seasonality, if any).

(2)   Select a set of prior probabilities for the four states, 'no change', 'step', 'slope change', and 'transient'.

(3)   Specify the variances of the step, slope, and transient states relative to the 'no change' variance (which implicitly defines the disturbance matrix **W**).

(4)   Specify the range of possible values for the 'no change' error variance V_{MIN} and V_{MAX}. A range of possible Vs are automatically calculated and the average value used in any further calculation $[V_o(0)]$.

(5)   When a new data point becomes available update the variance estimate $V_o(t)$.
(6)   Using the Kalman filter update the probability distribution for the level and growth (and seasonality).
(7)   Generate a set of forecasts and their estimated variances.
(8)   Repeat step 5 until all available data have been exhausted.

This generates forecasts and parameter estimates for every data point as it becomes available.

10.   SELECTING MODEL PARAMETERS

As noted above, before any Bayesian forecasting program can be run a number of parameters have to be fixed: (1) the prior probabilities of level, growth, and seasonality; (2) the long-run probabilities of the four states (the π_i); (3) the variance parameters (the Rs); and (4) the upper and lower limit on the error variance and the form of the variance law. Choosing subjective values for (1) above is easy for a forecaster familiar with the variables being forecast. The programs that the author is familiar with (which include the two publicly available in the UK), all have pre-set values for the πs and the Rs. However, before any empirical analysis of real data is carried out the forecaster must necessarily establish whether the default values specified in the program are adequate for the purpose to which the Bayesian model is to be put.

We will illustrate a suitable procedure with reference to the variance parameters. A wide range of parameters need to be considered. The reasons are that no earlier work had been carried out using the multiplicative model and there is difficulty relating the variance parameters when specified in the multiplicative framework to the work Harrison and Stevens (1971) and Cantarelis (1979) have done with the additive model.

The performance of these parameters was also evaluated under a wide range of simulated circumstances. Data were generated in a series of runs as follows:

(1)   with constant level and no trend, i.e. the 'no change' state: low and high error variance;
(2)   ARIMA $(0, 1, 1)$* data with low and high variance/low and high smoothing parameters;
(3)   step changes in the data as shown in Figure 2: various sizes of change, high and low variance;
(4)   slope changes as shown in Figure 2: various sizes of change, high and low variance;
(5)   transients in the data: various sizes of transient, high and low variance.

---

* ARIMA $(0, 1, 1)$ data is data generated in such a way as to make exponential smoothing the best forecasting method.

TABLE 1 Simulation Testing—Illustrative Ranking Derived from 16 Parameter Sets

Upper part of the table

Response measure	No change MSE			Step change T_μ			Slope change T_β			Transient Z		
	Fully specific			Size of change 5σ			Size of change σ			Size of change 10σ		
Simulation parameters	Low Var	High Var	Seasonality	Low Var	High Var	Seasonality	Low Var	High Var	Seasonality	Low Var	High Var	Seasonality
DEFAULT	9	9	11	7	8	8	3	7	7	5	10	6
SET1	1	1	1	*	*	*	*	*	*	1	4	1
SET2	16	14	5	10	12	7	12	9	12	2	2	4
SET3	3	3	2	14	14	15	*	*	*	9	1	10
SET4	5	5	11	4	2	1	8	1	1	14	8	14

Lower part of the table

Simulation parameters	No change MSE	Step change T_μ			Slope change T_β			Transient Z		
	Fully specific *	Size of change			Size of change			Size of change		
		2σ	5σ	10σ	0.5σ	σ	1.5σ	4σ	10σ	20σ
DEFAULT	NR	1	7	9	5	3	6	7	5	8
SET1	NR	*	*	*	*	*	*	1	1	1
SET2	NR	2	10	12	8	12	13	14	2	2
SET3	NR	14	14	14	*	*	*	12	9	4
SET4	NR	12	4	2	12	8	3	15	14	13

N.B: NR stands for not relevant. If the parameter set is completely unresponsive to a particular state a * is shown. In the upper part of the table each simulated state has three variants: data generated with a low variance, a high variance, and the Bayesian model run allowing the possibility of seasonality when no seasonality was present in the simulated data. The lower part of the table shows the rankings for two further specifications of the states. Note that the rankings in the middle column for each state repeat the low variance runs in the upper part of the table.

The error was multiplicative in all of the above cases. A larger scale test would have considered additive errors. Now we wish to evaluate the effects of choosing one particular parameter set over another. In the example described here 15 sets were considered plus the default set, a total of 16 in all. Their specific values are not given here and are referred to as DEFAULT, SET1, SET2, ..., SET15; (the particular values are given in Fildes, 1982).

In order to evaluate a parameter set's performance it was necessary to define suitable performance measures. Mean squared error was used only to a limited extent because different measures were thought appropriate depending on the way the data were generated. For the 'no change' state the error measure on which the rankings were based was the straightforward mean squared error. For the step change at time t_o, because the step size was determined by the experimenter when simulating the data, it was possible to calculate the number of periods between the step occurring (at t_o) and the parameter estimate ($\hat{\mu}_t$ say) first attaining the new level.

$$T_\mu = \{t_1 - t_o \text{ where } t_1 = \min_t \{t | \hat{\mu}_t \geq \mu_{t_o} : t \geq t_{o_o}\}\}.$$

Similarly T_β is the time taken for the slope estimate to attain the new, correct level.

The transient measure Z is based on the idea that a transient (at t_o) in the data should have no effect on the forecast made immediately after its occurrence. Let

$$Z = (\hat{Y}_{t_o + 1} - \hat{Y}_{t_o})/(\text{error standard deviation})$$

where $\hat{Y}_{t_o}$ is the one-step ahead forecast made at time $(t_o - 1)$. Then Z should be small.

Referring to Table 1 we can see exactly what the various entries mean. In row 1 the various response measures are listed, Mean Square Error (MSE), T_μ, T_β and Z as well as the state for which the measures are appropriate. Row two defines the data generation process in the simulation. In the top part of the table (Row 2; columns 2, 3, and 4) we see that for the no change state with data generated at random around a constant level three variants of the state were considered, low and high variance in the error, and seasonality. Taking a second example from the bottom half of the table, where the data series has a step change in it (columns 5, 6, and 7), three different sizes of step change were considered, 2σ, 5σ, and 10σ (row 8 of the table). Finally the various parameter sets, DEFAULT, SET1, etc., are listed in column 1.

While each of the four response measures, MSE, T_μ, T_β, and Z could have been used directly, instead it was decided to rank the measures. The entries in the table rank the various responses out of the 16 parameter sets under consideration. For example, looking at the bottom half of the table, for data simulated with a step change of size 10σ, the default parameter set gave a response (measured by T_μ)

which ranked 9 out of 16. Parameter SET1 did not respond at all (a ranking in effect of 16) while SET2 was 12th.

Use of these four error measures gives us the rankings shown in the table. From a study such as is briefly described above it is possible to find a subset of parameter sets which differ significantly in their performance characteristics. For example SET1 is best for the 'no change' and 'transient' states. But it is impervious to step and slope changes. SET4 has the opposite characteristic. The 'default' set has reasonable all round performance. Harrison and Stevens (1971) selected particular default values for the long-run probabilities (the πs) and the disturbance matrix (defined by the Rs). The argument they use in support is based on their experience. Cantarelis (1979) has in a detailed simulation experiment of the type described offered rigorous academic support for their recommended values. He first argues that the πs and the Rs are not independent and the forecasting system's responsiveness can be determined by altering either set. Intuitively this can be seen by observing that changing the π vector is equivalent to encouraging the system to identify the various states as more or less likely than before. Similarly with changes in the Rs. He concludes: 'equivalences between the πs and (R)s exist and further some variation in the πs can lead to large instability. Hence a good way of understanding the behaviour and controlling the responsiveness of the system is through the (R)s only'.

Accepting this conclusion as valid allows the researcher to limit consideration of possible parameter sets from which a short list is obtained of those with distinct characteristics (see Table 2).

It is also necessary to pre-set the upper (and lower levels) for the error variance. As stated in Section 2 of this chapter the model adapted for the variance was multiplicative with var $\varepsilon_t = (CA_t)^2$. An upper value of C was estimated by first calculating the variances of the raw data and the first differenced data.

TABLE 2   Short Listed Parameter Sets

Parameter sets	Response to state			
	No change	Step	Slope	Transient
DEFAULT	0	0	0	0
SET1	+ +	$-\infty$	$-\infty$	+ +
SET2	0	0	$-$ $-$	+
SET3	+ +	$-$	$-\infty$	$-$
SET4	+	+	0	$-$ $-$

NB: + or + + ($-$ or $-$ $-$) implies improved (diminished) response to a particular state changes compared with the default response, using the measures previously defined. 0 implies no substantial difference. $-\infty$ implies the state change could not be identified at all because the parameter set gave zero probability to that state, with a corresponding zero response to that state, when it occurred.

TABLE 3 Empirical Measures of Accuracy for Different Parameter Sets

Param- eter set	Lead time	RMSE					Median MAPE %					Response to state			
		1	4	8	12	18	1	4	8	12	18	No change	Step	Slope	Transient
DEFAULT		40.4	62.2	70.7	72.2	98.8	6.2	9.0	13.6	11.1	15.3	0	0	0	0
SET1		93.3	109.0	89.1	88.9	99.6	16.5	17.0	19.7	15.3	16.9	+	$-\infty$	$-\infty$	+
SET2		42.5	63.5	68.2	68.0	90.9	6.0	8.9	12.6	11.3	14.3	0	0	−	+
SET3		46.8	68.7	69.3	66.4	88.0	6.1	9.8	12.4	11.9	15.3	+	−	$-\infty$	−
SET4		40.1	68.6	83.8	95.5	136.0	6.1	11.4	16.1	15.7	22.2	+	+	0	−

The '+', '−', '0' and '$-\infty$' measures are explained fully under Table 2.

Now suppose the corresponding mean level of the series is approximately $\bar{A}$, then C_{MAX} was set $= \sqrt{\text{variance estimate}/\bar{A}^2}$.

Whatever the procedure used for estimating C_{MAX} it is important that the value used is sufficiently high so that the var $\varepsilon_t < (C_{MAX}A_t)^2$.

Through the procedures described the forecaster will have established plausible values for all the default parameters with the exception of the *R*s. For the *R*s a range of possible parameter sets with differing response characteristics will have been established instead. The final stage is to choose a *particular* set of *R*s.

In the forecasting competition because we had no prior knowledge of the nature of the series being forecast the default parameter set was accepted as a reasonable compromise. However, as Table 3 shows, the choice is important. In general it seems that parameter sets responsive to step and transient changes score well in short-term forecasting. However, parameter sets responsive to step and/or slope changes score very poorly in longer term forecasting.

Table 3 shows the (geometric) root mean squared error (RMSE) and median mean absolute percentage error calculated over 111 series for the last seven data points of each series. Fildes (1983) gives further details. A number of points emerge from the examination of the simulation results and the empirical results in this table:

(1)   that the empirical performance of Bayesian forecasting depends on the parameters selected;

(2)   the default parameters are a reasonable compromise;

(3)   different parameter sets are appropriate for different lead times and a simple modification of the program which takes this into account is likely to improve performance.

The figures presented above represent the performance of the particular variant of Bayesian forecasting. To test whether two other versions differed substantially their performance was also evaluated on the 111 series used here. A version available through SIA, London, a time-series computer bureau, performed better than the version used here. However the differences do not undermine the conclusions of the forecasting competition (Chapter 4).

11.   CONCLUSIONS

Bayesian forecasting offers a flexible and interesting extension of the conventional statistical approaches to time series analysis epitomized by Box and Jenkins (1976). Its major theoretical weakness lies in the need to specify *a priori* a wide range of parameters and most casual users (and even some experts) would find this difficult. Model performance depends critically on the choice of these parameters. Fortunately research is under way to estimate these parameters from the data (Harvey, 1982).

Two aspects of Bayesian forecasting highlighted by Harrison and Stevens (1971, 1976) make the method intuitively appealing: the notion that no single statistical model holds for all times and that the observed data can only be understood by reference to a number of distinct models (four in their formulation); and secondly, Harrison and Stevens argue about the importance of subjective intervention by the expert forecaster knowledgeable about the series being forecast. A subsidiary consideration is whether the method is fully automatic and self-monitoring; that is to say it requires no outside intervention to put it back on course if for example an outlier occurs in the data.

The proof of all these propositions can only be empirical. In carrying out this research for the 'forecasting competition' evidence was gathered which casts at least some light on them. While undoubtedly a cursory examination of just a few of the 1000 time series establishes the inability of a single statistical representation of the series, the Bayesian four-state model does not seem to represent the discontinuities any better than some of the alternative methods considered in this volume.

However, a careful examination of the data on mean response error, median absolute percentage error, and the rankings of the various methods supports the idea that the method is robust – it is seldom the best method, nor is it often the worst. This suggests that it offers a suitable self-monitoring forecasting method.

The importance of subjective intervention by the forecaster has not been assessed. Such a claim cannot be taken as self-evidently true. That said, the way the Bayesian model is specified in terms of the level and growth parameters (and their variances) is designed to make intervention easy.

In summary, my own verdict (biased of course in favour of a method with which I have become familiar) is that while its empirical performance is currently disappointing, modifications and improvements already being researched should prove fruitful. The very flexibility of Bayesian forecasting is its most attractive feature and offers the possibility of substantial improvements in on-line forecasting performance.

REFERENCES

Box, G. E. P. and Jenkins, G. (1976). *Time Series Analysis: Forecasting and Control*, revised ed., San Francisco: Holden Day.

Cantarelis, N. (1979). Bayesian Forecasting Models, Ph.D. thesis, University of Warwick.

Fildes, R. (1979). 'Quantitative Forecasting – the State of the Art: Extrapolative Models', *J. Operational Research Society*, **30**, 691–710.

Fildes, R. (1982). 'An Evaluation of Bayesian Forecasting', Manchester: Working Paper 77, Manchester Business School.

Fildes, R. (1983). 'An evaluation of Bayesian Forecasting', *J. Forecasting*, **2**, 137–150.

Fildes, R. and Stevens, C. F. (1978). 'Look – No Data: Bayesian Forecasting and the Effect of Prior Knowledge', in Fildes, R. and Wood, D. (eds), *Forecasting and Planning*, Farnborough, Hants: Teakfield.

Gathercole, R. B. (1982). 'Note for Bayesian Forecasters', in Anderson, O. D. (ed), *Time Series Analysis: Theory and Practice*, Amsterdam: North-Holland.

Gathercole, R. B. and Smith, J. Q. (1981). 'Approximations for Bayesian Forecasting', Research Report: London: Department of Statistical Science, University College.

Harrison, P. J. and Stevens, C. F. (1971). 'A Bayesian Approach to Short Term Forecasting', *Operational Research Quarterly*, **22**, 341–362.

Harrison, P. J. and Stevens, C. F. (1976). 'Bayesian Forecasting', *Royal Statistical Society (B)*, **38**, 205–247.

Harrison, P. J. and Stevens, C. F. (1975). 'Bayesian Forecasting in Action: Case Studies', Warwick: Statistics Research Report No. 14.

Harvey, A. C. (1982). 'Estimation Procedures for a Class of Univariate Time Series Models', Econometrics Programmes, Discussion paper A28, London School of Economics.

Johnston, F. R. and Harrison, P. J. (1980). 'An Application of Forecasting in the Alcoholic Drinks Industry', *J. Operational Research Society*, **31**, 699–709.

Morrison, D. G. and Pike, D. H. (1977). 'Kalman Filtering Applied to Statistical Forecasting', *Management Science*, **23**, 768–774.

Smith, J. Q. (1979). 'A Generalisation of the Bayesian Steady Forecasting Model', *Royal Statistical Society (B)*, **41**, 375–385.

Zellner, A. (1971). *An Introduction to Bayesian Estimation in Econometrics*, New York: Wiley.

CHAPTER 8

Naive, Moving Average, Exponential Smoothing, and Regression Methods

Michele Hibon
INSEAD

This chapter deals with the performance of nine simple time series methods. Eight of the nine methods were used in two versions. The ninth method – the Holt–Winters exponential smoothing – was run in a single version. In the first version the raw data was used by the methods to obtain the forecasts. In the second version the data was deseasonalized by first computing seasonal indices and then dividing the raw data by the seasonal indices. The deseasonalized data was then used by each method to obtain the forecasts which were consequently reseasonalized.

In total the following nine methods were employed:

(1) Naive 1;
(2) Moving average;
(3) Single exponential smoothing;
(4) Adaptive response rate exponential smoothing;
(5) Holts linear exponential smoothing;
(6) Brown's one-parameter linear exponential smoothing;
(7) Brown's one-parameter quadratic exponential smoothing;
(8) Linear regression trend fit;
(9) Holt–Winters linear and seasonal exponential smoothing.

The above nine methods were used with the raw data. However, some of the data was seasonal and some was not. For seasonal data a systematic error must occur since the first eight methods cannot deal with seasonality. In order to avoid systematic errors due to seasonality the data was deseasonalized. This was done by computing seasonal indices with a simple ratio-to-moving average (centred) decomposition method. If S_j is the appropriate seasonal index, the $n - m$ data of each series was adjusted for seasonality, (where $m = 6$ for yearly, 8 for quarterly and 18 for monthly data) as follows

$$X'_t = X'_t S_j$$

X'_t is the seasonally adjusted (deseasonalized) value on which a model and forecasts for the first eight methods were found. These methods are called 'deseasonalized', in addition to their name. Thus, 'moving average' is called 'deseasonalized moving average', 'single exponential smoothing', 'deseasonalized single exponential smoothing' and so on. The only exception is Naive 1 – its deseasonalized version has been called Naive 2.

In total seventeen methods were considered, that is the first eight methods in both versions and the Holt–Winter's linear and seasonal exponential smoothing which was run in a single version only – this method can also deal with seasonal data.

The technical details of the seventeen methods used are extremely simple and well known and there therefore is no reason to repeat them here. A summary of the methods can be found in Chapter 4. For more details the interested reader can consult Makridakis and Wheelwright (1978), and Makridakis, Wheelwright, and McGees (1983).

The remainder of this chapter will concentrate on some highlights of the methods 1 to 17 and will discuss the results of the competition in a comparative sense, first among themselves and secondly in relation to the remaining methods of the competition.

The major difference among the methods used is the way the trend of the data was accounted for. It is important to understand the principle of forecasting used by the methods examined in this chapter. Method 1 (Naive 1) cannot even be called a method: it simply states that the m future forecasts are equal to the latest available value X_t, where t denotes the most recent time period. This is the equivalent to what statisticians call the random walk model which assumes that trends and turning points cannot be predicted. It implies that the best available forecast(s) is the latest known data value. The forecast therefore becomes a horizontal line extrapolation.

The moving average model averages the data by giving equal weight to each one of the N values used in the average, where the value of N is found in such a way as to minimize the mean square error of model fitting. The forecasts of moving averages apply to the middle of the averaged data, they always trail the actual values when a trend is present in the data by $(N - 1)/2$ periods. Very rarely was the method of moving average superior to those of the Naive or exponential smoothing methods.

All the exponential smoothing methods also average the data. The averaging, however, is done in an exponential manner. That is to say, not all terms entering the averaging receive equal weights; rather, more recent observations receive more weight than less recent ones. That is, the weighting is exponentially decaying with the most recent data getting the highest weight and those further back receiving progressively (in an exponentially decreasing manner) less weight.

The main difference among the various exponential smoothing methods is the way they treat the trend and seasonality. Single and adaptive response rate exponential smoothing assume no trend. The forecast from these two methods is almost horizontal after an adjustment on the level of the weighted average is made. The difference between single and adaptive response smoothing is that the parameters of the former method are fixed while those of the latter are adaptive as new data becomes available. However, the way they both calculate the trend, or better, lack of trend, is the same: they extrapolate an almost horizontal trend. Between the two methods the single exponential in general did better than the adaptive response rate smoothing. This is a finding which is consistent with the forecasting literature (see Bunn, 1980; Gardner, 1983; Makridakis and Hibon, 1979).

Holt's and Brown's linear exponential smoothing extrapolates a linear trend which is computed by exponentially averaging the existing trend in the data. That is, the most recent trend receives more weight, in an exponential fashion, than less recent trends. The difference between Holt's and Brown's smoothing is that the former uses two parameters (one for the level of the series, the other for the trend) while Brown's uses one single parameter for both the level and the trend. From the results it seems that Holt's smoothing outperformed Brown's in most cases and is, therefore, to be preferred in practical applications.

Brown's quadratic smoothing extrapolates a quadratic trend. Depending on the value of the model's parameter this trend can be close to linear, or quadratical. The method uses a single parameter which adjusts the level of the series, and the linear and/or quadratic trends involved.

The Holt–Winters method is exactly the same as Holt's when there is no seasonality in the data. However, if there is seasonality a third parameter is used which computes L seasonal indices (where L is the length of seasonality) in an exponential smoothing fashion. That is, the weights given to seasonality decrease, in an exponential fashion, with recent seasonality receiving more weight and less recent ones progressively less. The difference between the deseasonalized Holt and the Holt–Winters is small. In the case of seasonal data in the Holt's method the data has been deseasonalized first and then the method run, while Holt–Winters computes its own seasonal indices. If the data is non-seasonal then Holt and Holt–Winters are exactly the same. For seasonal data the difference in forecasting accuracy is small but with Holt's is having a clear superiority over Holt–Winter.

Finally, in regression, the *average* trend in the data is found, by minimizing the mean square error, and extrapolated. The difference between regression and the linear exponential smoothing methods is that in the former the *average* trend of *all* data is being extrapolated while in the latter an exponentially weighted trend is used. From examining the results of the competition it is clear that the errors from regression are always greater than those of the exponential smoothing methods. This goes for almost all forecasting horizons and accuracy measures. It seems, therefore, that using the average linear trend is not a good way of forecasting. Forecasts depend more heavily upon the latest trend in the data which must, therefore, be given more weight in determining future forecasts.

Quadratic trend extrapolation does not produce as good results as linear ones. Even though the accuracy of the quadratic exponential trend is not as bad as that of the linear regression trend it is rarely better than the linear trend extrapolation of exponential smoothing. This is a surprising result since it was thought that quadratic exponential smoothing could adequately compete with the linear smoothing methods. This was not, however, the case with the 1001 series used in the competition. It seems that quadratic smoothing overreacts to fluctuations in the data and occasionally extrapolates too much or too little trend. The errors of this method, like those of linear regression, get very large as the time horizon of forecasting increases.

Brown's one-parameter linear exponential smoothing is on most measures and forecasting horizons inferior to that of Holt's. It seems that Holt's two parameters provide more flexibility and consequently more accurate results than Brown's one parameter. This applies to the present competition as well as the empirical study of the Makridakis and Hibon (1979) paper.

From the seventeen methods examined, those of Naive 1, Adaptive response rate exponential smoothing, Brown's one-parameter linear exponential smoothing, Brown's one-parameter quadratic exponential smoothing, and linear regression are, in the great majority of cases, inferior to the remaining methods. The author sees, therefore, little reason to be further concerned with these methods, unless some other procedure needs to utilize them (e.g. constraining the value of the smoothing parameters in quadratic exponential smoothing to less than 0.1 or 0.2).

The remaining methods are unique in some respects and provide advantages and disadvantages as far as their use is concerned. Naive 2 extrapolates a horizontal trend and it is amazing how well it does in comparison to the remaining methods. Since Naive 2 is a benchmark which other methods are to be compared with, it is interesting to see the improvement between a given forecasting method and Naive 2. By looking at the percentage better measure we can see that for the overall result the best method beats Naive 2 by not more than 7.2 per cent of the time (see Table 7(b) in Chapter 4). Moreover, for most of the methods the difference is less than 5 per cent. On the other hand, some of the differences in the various forecasting horizons and accuracy measures are very important from a practical point of view. Thus a difference in MAPE of a few percentage points can be worth millions of dollars, in reduced inventory costs or more effective production scheduling. This is why the benchmark of Naive 2 should be used to determine how much a given forecasting method improves over Naive 2 and whether such an improvement is worth the costs involved for utilizing the more sophisticated method.

Single exponential smoothing is superior (with very few exceptions) to Naive 1. The same is also true between Naive 2 and deseasonalized exponential smoothing, which is usually superior. As far as forecasting applications are concerned, therefore, exponential smoothing is superior to the Naive methods whose role should be limited to that of a benchmark.

The three remaining methods of single, Holt's, and Holt–Winters exponential smoothing are more accurate in some of the forecasting horizons and accuracy measures and less accurate in some others. In general, the little difference between the deseasonalized Holt's method and the Holt–Winters is clear in favour of Holt's but small. Single exponential smoothing does well when all data are used and better with MAPE than the other accuracy measures. Furthermore, it does very well for the one-period ahead forecasting horizon. Moreover, it does extremely well with monthly and micro data, while Holt and Holt–Winters do better with yearly data, average ranking, median MPE, and PB.

Exponential smoothing methods and, in particular, the single exponential smoothing, Holts, and Holt–Winters did extremely well in comparison with the more statistically sophisticated methods in the competition. For some forecasting horizons and accuracy measures they did the best and overall they did very well (see Tables 1, 2, 3, and 4 in Chapter 1). Two combining methods (combining A and B) which were constructed by mainly using exponentially smoothing methods also did extremely well and produced overall the best results of the competition with most forecasting horizons when the Percentage Better (PB) accuracy measure was used (see Table 4 in Chapter 1). This means that it might be possible to obtain accurate forecasts by averaging three or four exponential smoothing methods.

There is no reason why the more statistically sophisticated methods cannot modify the way they forecast by learning from the success of exponential smoothing methods. One thing which is obvious is that from a forecasting point of view it is desirable not to extrapolate 'too much' trend. This is clear from how well Naive 1 and 2, and single exponential smoothing do. These methods do not extrapolate any trend. In addition, making forecasts which are close to an exponential average of the data seems to work very well in comparison to the forecasts produced by alternative approaches. In this respect it is not that exponential smoothing methods capture superbly the pattern of the data, or accurately predict turning points. Rather, they do less badly than the other methods which overreact to random fluctuations by assuming that they are part of the pattern of the data. This author believes that the best advantage of exponential smoothing methods is their robustness to any and all types of data and forecasting situations. Finally, deseasonalizing the data with a simple ratio-to-moving average method captures the seasonality of the data and produces very similar results to those of any alternative method (see Tables 17, 22, and 31 in Chapter 4).

In conclusion, the methods of single and Holt's exponential smoothing can provide a powerful pair which can be used in almost any forecasting situation. In addition, the method of Holt–Winters can be used as an alternative to Holt's, if users do not want to first deseasonalize the data. These three exponential smoothing methods fare well in comparison with the statistically sophisticated methods and have shown that adequate post-sample forecasting accuracy can be achieved, easily and at a low cost, with exponential smoothing methods. The practical value and usefulness of this conclusion is considerable. It is important, therefore, that these methods are not dismissed as useless, just because they are simple. This and the Makridakis and Hibon (1979) competition have shown otherwise. Simple methods can provide accurate results at a low cost. Furthermore, they can be applied easily since their level of complexity is extremely low. The practising forecaster can, therefore, benefit a great deal from these methods and save a great deal in lower inventory costs and more effective planning and scheduling.

REFERENCES

Bunn, D. W. (1980). 'A Comparison of Several Adaptive Forecasting Procedures', *Omega*, **8**, No. 4, 485–491.

Gardner, E. S. Jr (1983). 'Automatic Monitoring of Forecast Errors', *Journal of Forecasting*, **2**, No. 1.

Makridakis, S. and Hibon, M. (1979). 'Accuracy of Forecasting: An Empirical Investigation (with discussion)', *Journal of the Royal Statistical Society*, Series A, **142**, Part 2, 79–145.

Makridakis, S. and Wheelwright, S. C. (1978). *Interactive Forecasting, Univariate and Multivariate Methods*, second Edition, San Francisco: Holden-Day.

Makridakis, S., Wheelwright, S. C. and McGees, V. (1983). *Forecasting Methods and Applications*, second Edition, New York: Wiley.

CHAPTER 9

Lewandowski's *FORSYS* Method

Rudolf Lewandowski
Marketing Systems

1. THE VALUE OF THE COMPETITION DESCRIBED IN THIS BOOK

An in-depth study of the value of major time series (extrapolative) forecasting methods, such as that carried out under the direction of Professor Spyros Makridakis, is not only a necessary piece of work, but also one of exceptional importance. From both a theoretical and, in particular, a practical point of view, the ability to judge objectively the forecasting accuracy of various methods is paramount in selecting an appropriate forecasting technique. Those familiar with forecasting will find in this competition extremely important information about the intrinsic value of existing methods. Forecasting practitioners will find therein a means of helping their company to resolve the forecasting problems which so often arise in actual business situations.

It should be noted that among forecasting practitioners and theorists alike, a certain 'positional war' has developed over the last several years. Confusion surrounding how accurately each forecasting method can predict future events results, inevitably, in great loss of intellectual energy, as insufficient knowledge concerning the selection of the most adequate method leads to apparent divergence of opinions. The present competition provides objective information to settle the issue of forecasting accuracy and method selection. Companies, still hesitating to adopt a forecasting system, should find in the present book an encouraging clarification of the issues of accuracy. It should perhaps be remembered that, before this study, the great majority of companies possessed no means of any objective comparison, other than the 'selling' talk of those advocating a given method. Now there is adequate information from this competition to make objective judgements.

At present no more than 5 per cent of large-size European organizations regularly employ statistical methods for the systematic analysis and forecasting of their sales. Furthermore, knowing the extremely difficult situation currently facing businesses which leads to a need to reduce management risks, we should realize that it is today more important than ever to be able to pinpoint the problems involved in forecasting and find satisfactory solutions. An improper selection of a forecasting method must be avoided at all costs. If a statistically sophisticated method is chosen which in fact fails to produce a substantial improvement over simple methods, this leads to a certain resentment, mistrust, and discredit as far as forecasting is concerned. This is a dangerous situation because management loses faith in forecasting and starts doubting the usefulness, not only of forecasting but also of all other management science methods in general.

Although this competition objectively assesses the forecasting accuracy of the different methods, the question of acceptance and ultimate use of a given method by management is also important. Furthermore, the ease with which a given method can be used, and the costs involved, cannot be ignored. These aspects of the forecasting methods will be discussed later on in this chapter. For the time being, let us avoid the discussion which could have arisen had the choice of forecasting method and type of series used not been sufficiently wide to cover every important forecasting situation. This has not been the case, however, since the wise selection of the number and diversity of series and methods amounts to the following:

(1) An adequate mix of different types of series, in particular micro-series coming from, say, sales of companies, and macro-series emanating from national economic statistics (Gross National Product, industry-wide data etc.).

(2) The choice of the time horizon of forecasting of up to 18 periods ahead has been chosen, thus covering not only short-term forecasting but also its use for the medium term, in preparing budget forecasts. Furthermore, for quarterly and yearly data the length was even longer, allowing comparisons on long-term (up to six years for yearly data) horizons.

(3) Many accuracy measures have been used thus allowing complete evaluation of the forecasting quality of the various methods used.

(4) All the principal techniques from the very naive and simple to the most complicated and statistically sophisticated forecasting methods have been employed in the competition.

It is thus, in the opinion of this author, the only comparison of forecasting methods which has from the start been able to cover, in a satisfactory manner, all the important issues involved in forecasting. Furthermore, the competition has been international, it has included recognized experts in each of the forecasting methods, and it was done in an extremely objective manner.

2. DISCUSSION OF THE RESULTS IN TERMS OF FORECASTING USERS

The principal results of the competition have been published in the *Journal of Forecasting* (see Chapter 4). It is considered necessary, therefore, to deal only

TABLE 1   Classification According to Average Ranking

Method	Monthly data	Quarterly data	Yearly data	Micro data	Macro data
Combine A	10.33	10.29	11.32	11.04	9.89
Combine B	11.16	11.12	12.99	10.84	11.49
Box–Jenkins	11.63	10.84	11.48	11.96	12.26
Lewandowski	10.79	11.79	10.28	9.56	11.06
Parzen	11.49	9.86	10.52	11.48	11.19
Mean of each method	12.50	12.50	12.50	12.50	12.50

Note: Underlined entries denote best method.

with certain important aspects that go beyond Chapter 4 as they are seen from the author's point of view.

Table 1 gives the results in terms of average ranking, for three elements of series periodicity – monthly, quarterly, and annually – and for micro- and macro-series. Furthermore, we have taken the two methods of combining simple approaches – that is, Combining A and Combining B – as well as the three more sophisticated methods of Box–Jenkins, Lewandowski and Parzen.

As Table 1 reveals, the method of Lewandowski provides the best results for monthly, annual, micro-, and macro-series. The method of Parzen is the most successful for quarterly series. The method of Box–Jenkins, however, fails to produce, apart from a very few exceptions, results superior to those of the more sophisticated techniques, or even of the simplest methods. In certain instances, as with micro- and macro-series, the results obtained are in fact considerably worse. The same bad performance of the Box–Jenkins method can be seen in Tables 1, 2, 3, and 4 in Chapter 1 and in the detailed tables of Chapter 4.

Given the indisputable evidence provided by this comparison, the results clearly, and most surprisingly, demonstrate that the highly complex ARIMA-type Box–Jenkins method is in no way superior to other methods as far as forecasting accuracy is concerned. From the point of view of practical forecasting such a method is, therefore, of no value, no matter what academic experts say in its defence.

From a practical point of view, as far as manufacturing, commercial, and service organizations are concerned the emphasis should, of course, be placed on micro-series. This is why Table 2 has been constructed, from which we can observe the following:

First, Lewandowski's method is clearly better than the academically popular ARMA models and a little more accurate than Parzen's ARARMA models. This is so for all horizons between 1 and 18 months. In fact, as shown by Table 2, with a few rare exceptions, the forecasting quality of Lewandowski's FORSYS method is better, particularly with regard to the Box–Jenkins method, for the

TABLE 2   Micro Data Results (MAPE)

Method	1 to 4 months	1 to 6 months	1 to 8 months	1 to 12 months	1 to 18 months
Combine A	9.9	10.1	10.2	10.2	11.5
Combine B	9.8	9.8	10.0	9.8	11.2
Box–Jenkins	11.1	10.7	10.0	9.9	12.2
Lewandowski	<u>8.3</u>	<u>8.7</u>	<u>8.7</u>	<u>8.7</u>	<u>9.7</u>
Parzen	11.3	11.3	10.9	10.4	11.3
Box–Jenkins compared with Lewandowski %	− 33.7	− 23.0	− 14.9	− 13.8	− 25.8
Box–Jenkins compared with Combine B %	− 13.3	− 9.2	− 0.0	− 1.0	− 8.9

Note: Underlined entries denote best method.

forecasting horizons of 1 to 4, as well as with all other forecasting horizons. This type of superiority is particularly valuable for short- and medium-term forecasting which involves inventory control, production planning, and yearly budgeting.

Second, at best, the Box–Jenkins method produces mediocre results. As a whole they are distinctly worse than the combination of simple methods, A and B, poorer than the Parzen approach, and about 20 per cent less accurate than the method of Lewandowski.

Figure 1 gives an idea of the comparison between the Lewandowski method and that of Box–Jenkins. These results are confirmed if other criteria of comparison are taken – for example, Average Rankings as shown in Table 1.

The comparison of the Box–Jenkins method with that of Lewandowski reveals some highly important, significant differences between the two. The forecasting accuracy of Box–Jenkins is about 34 per cent inferior for short-term horizons of

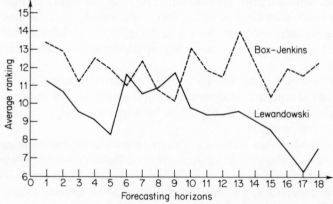

FIGURE 1   Comparison of micro-data results (average ranking) between Lewandowski and Box–Jenkins

up to 4 months ahead than that of Lewandowski, while for longer horizons the accuracy of Box–Jenkins is inferior by about 25 per cent.

The forecasting accuracy of the Box–Jenkins method is even lower than that of simple forecasting techniques such as exponential smoothing and the Combine A and B approaches. These results are extremely surprising. If the poor accuracy of the performance of the Box–Jenkins methods is coupled with the difficulty involved in identifying and testing an adequate Box–Jenkins model, then the current expectations about the usefulness and applicability of this method are wrong and highly misleading. The Box–Jenkins method is not only considerably more difficult and costly to use but produces results which are less accurate than most other methods.

3. COMPARATIVE ANALYSIS WITH RESPECT TO THE ARIMA MODELS

It is important to return to the question of the complexity of the Box–Jenkins (ARIMA) method and the ease of its use. Contrary to other methods, the Box–Jenkins approach cannot be automated. As it was stated in Makridakis *et al.* (1982, 111–154, see Chapter 4), the time required to deal with each series with the Box–Jenkins method was estimated at more than one hour per series. If account is taken of the fact that this analysis has been carried out by a university specialist, knowledgeable of the theory, and with considerable experience in using the Box–Jenkins approach, more time will be required if the Box–Jenkins method is used by companies. Moreover, the poor results obtained by the Box–Jenkins method are even more surprising in that it was practically the only approach which was not automated and required a considerable user intervention.

As has been indicated by the author (Lewandowski, 1982), it is more practical and efficient not to require judgemental inputs in the technical aspects of forecasting methods. However, methods should allow the user to incorporate exceptions such as marketing actions (publicity campaigns, promotions, price-changes, etc.). The automated treatment of special actions by FORSYS brings considerable improvements in forecasting accuracy. The authors believes that if his method was not run automatically and if special actions had been taken into consideration for the series of this comparison (i.e. effects readily identifiable from their specific profile), this would have improved the results of the analysis of FORSYS by around 20 per cent.

4. CRITERIA FOR ACCEPTING THE VARIOUS FORECASTING METHODS

While we cannot pretend to be able to define any indisputable criteria for the assessment of the degree of complexity or acceptance of a forecasting method, we will nevertheless attempt to take a forward step in this direction.

It should be remembered that the adaptation of any forecasting method or system can be made only if the user understands both the system employed and the results it provides. In other words, non-specialists such as production, marketing, or general managers must be able easily to understand and readily interpret the diversity of elements involved in forecasting.

The criteria for accepting given forecasting methods can be summarized under two main headings:

(1)   the degree of complexity of the method itself;
(2)   the degree of the difficulty in understanding the results.

(a) Measure of theoretical complexity of the method itself

In the experience of this author, drawn from working with more than 200 European companies, those in charge of forecasting usually have only a passing statistical or mathematical background. The use of a forecasting method therefore will require training to permit the potential forecasting users to learn about the method they will be using.

In order to assess the degree of complexity of the method itself, I have defined, in terms of hours, the training which would be required to teach a person, with an educational background not in forecasting or statistics, the principles of the method to be used.

As an example, teaching the Winters method to a non-specialist would require, in my opinion, the explanation of a dozen or so essential ideas. The time required for such explanations would be as follows:

	Minutes
Definition of the mean	5
Definition of the weighted mean	15
Definition of an exponential-type weighted mean	15
Definition of the smoothing coefficient α of the mean	5
Definition of the concept of short-term trend	15
Definition of the coefficient β for the trend	5
Definition of the concept of seasonality	15
Definition of the smoothing coefficient γ of seasonality	5
Problems involved in computing the optimal values of the smoothing coefficients α, β and γ	30
Interpretation of concepts such as initial parameter	60
Interpretation of forecasting results	20
Definition of certain statistical concepts such as MAD, etc.	30
Critical analysis and interpretation of forecasting errors	30
Thus the total time needed to give a non-specialist an understanding of a Winters-type method would be in the order of:	5 hours

(b) Degree of difficulty in interpreting the results

In order to make efficient use of forecasts obtained by quantitative methods it is necessary that the results be understood and consequently 'sold' to the various departments concerned, that is production, marketing, finance, and so forth. It is therefore proposed to introduce a second factor, measuring the degree of 'acceptance' of the method by the end user.

This degree of acceptance may be defined as the ability to explain, interpret, and understand the results of the forecasting method and its various characteristics. In order to evaluate the 'acceptance' factor a scale ranging from 0 to 10 has been defined. The greater the difficulty involved in understanding the results, the higher the value of this acceptance factor will be. Thus, while the naive method will score less than one on this value, the highly sophisticated methods of Box–Jenkins and Parzen will score around ten.

In order to define an overall coefficient of the degree of complexity and acceptance of each method the average of these two scales has been used, although the author believes that the degree of acceptance should have been given more weight.

5. THE RELATION BETWEEN COMPLEXITY AND ACCURACY OF THE METHODS

Taking into account the classification of the various methods according to their forecasting accuracy, and also their complexity and acceptance, it is possible to combine the accuracy and acceptance/complexity criteria. Table 3 and Figure 2 summarize the various elements of combining the two criteria. While the criterion

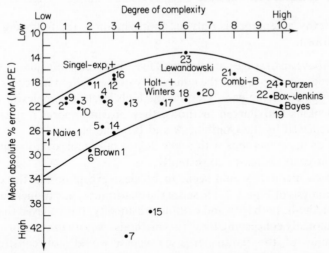

FIGURE 2   Classification of all methods according to MAPE and degree of complexity

TABLE 3   The Complexity Coefficients and Overall MAPE Value of Various Forecasting Methods

Method	Explanations coefficient	Acceptance coefficient	Overall complexity coefficient	MAPE value
1	0.2	0.2	0.2	26.8
2	0.5	1.0	0.75	21.4
3	1.0	1.0	1.00	21.6
4	2.0	2.0	2.00	20.4
5	2.0	2.0	2.0	25.5
6	1.0	2.0	1.5	29.5
7	3.0	3.0	3.0	43.1
8	2.0	2.0	2.5	20.2
9	0.5	1.0	0.75	20.4
10	1.0	1.0	1.0	22.2
11	1.5	1.0	1.25	18.0
12	2.5	2.0	2.25	17.3
13	2.5	3.0	2.75	21.7
14	1.5	3.0	2.25	26.3
15	3.5	4.0	3.75	39.0
16	2.5	2.0	2.25	16.7
17	4.0	5.0	4.50	21.7
18	5.0	6.0	5.50	20.9
19	10.0	10.0	10.0	22.7
20	5.0	7.0	6.0	19.8
21	7.0	8.0	7.5	16.6
22	10.0	10.0	10.0	20.2
23	5.0	6.0	5.5	13.7
24	10.0	10.0	10.0	18.4

of complexity/acceptance is only an approximation, the following conclusions can be drawn:

(1)   Above an average complexity of five, the forecasting quality of the model does not improve as the complexity of the method increases. In fact, the opposite is the case. Despite the high level of complexity of the Box–Jenkins and Parzen methods, they provide poorer results than those achieved by the Combine A and B methods.

(2)   For those methods with a low degree of complexity, the quality of the forecasts decreases consistently.

(3)   Two methods would seem to be clear exceptions in the comparative analysis of Figure 2. These are the quadratic-type exponential smoothing methods, both with and without seasonality. It is believed that there is an anomaly concerning these two methods. A probable reason could be the value of the parameter used which would not constrain. In the experience of this author, these two methods generally give results similar

to those obtained with the Holt–Winters or Single Exponential Smoothing methods if their parameter values are constrained to be less than 0.2.

CONCLUSIONS

It is important to reiterate certain essential points in this chapter. Firstly, an important and valuable forecasting competition was carried out, permitting for the first time the objective comparison of major time series forecasting methods. The result of the competition has been a considerable body of knowledge which will, inevitably, facilitate the choice of the most suitable forecasting method. Inevitably such a choice could save companies considerable sums of money and resources by providing them with objective information for selecting the best forecasting method themselves. Secondly, the results of other comparisons (Makridakis and Hibon, 1979), appear to be equally confirmed by the present competition.

The ARMA method, which because of its mathematical elegance has been adopted in almost every university over the last decade as an attractive method of statistical education, has, in fact, little practical value in the practice of forecasting. The Box–Jenkins method is extremely complex to use, its results are difficult to comprehend, and its forecasting accuracy is low. The forecasting accuracy of FORSYS, on the other hand, has provided overall the best results: that is improvements in forecasting accuracy of more than 20 per cent. Finally, it is important to stress that FORSYS is by far the best method of the competition where forecasting horizons of longer than five or six periods are concerned. This advantage of FORSYS has been achieved by the appropriate choice of the medium-term trend saturation which is done in the manner shown in Figure 3 (for more details about FORSYS see the next chapter which is reprinted from the *Journal of Forecasting* or Lewandowski, 1979).

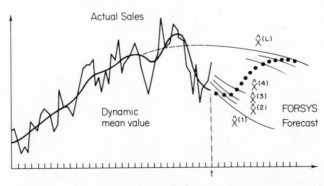

FIGURE 3 Trend extrapolation and forecast by FORSYS

REFERENCES

Lewandowski, R. (1979). *La Prevision a Court Terme*, Paris: Dunod.
Lewandowski, R. (1982). 'Sales Forecasting by FORSYS', *Journal of Forecasting*, 1, 205–214.
Makridakis, S. *et al.* (1982). 'The Accuracy of Extrapolative (Time Series) Methods: Results of a Forecasting Competition', *Journal of Forecasting*, 1, 111–153.
Makridakis, S. and Hibon, M. (1979). 'Accuracy of Forecasting: an Empirical Investigation', *Journal of the Royal Statistical Society*, Series A, 2, 97–145.

CHAPTER 10

Sales Forecasting by FORSYS

Rudolf Lewandowski
Marketing Systems

ABSTRACT

This paper describes a sales forecasting system widely used by European companies. The system, known as FORSYS, includes several unique characteristics which increase its use and applicability among practitioners. FORSYS is simple to use: its underlying rationale is clear to the user; it is adaptive, and it allows the incorporation of special events into the model in order to determine their influence on forecasting.

Many of the forecasting methods that are described in detail in academic textbooks and are very popular in university circles are rarely used in practice. This is particularly true with sophisticated forecasting methodologies which have failed to gain the confidence and acceptance of practitioners. As it has become obvious by the results of several surveys among practising forecasters (see Wheelwright and Clarke, 1976), most of the academically popular methods are used very little in practice. Why? In the opinion of this author, the answer is simple: these methods do not meet the needs of practitioners which are somewhat different from the academics' perception of them.

The purpose of this paper is to describe a sales forecasting method used by several hundred European companies over the past ten years. This forecasting system is known as FORSYS and has certain major attractions for practitioners. These include:

(1) It is simple to understand and easy to use, even though its mathematical formulation is rigorous and statistically correct.

(2) It is not a black box. Instead, the user is able to understand how the system works and can include explanatory variables which might influence the forecasts.

(3) The forecasting model is adaptive in the sense that it allows for automatic adjustments in the parameters of the model when there are non-random changes in the data.

Reproduced from the *Journal of Forecasting*, **1** (1982), 205–214.

(4)   Special events can be incorporated into the system and be subsequently integrated into the forecasts.

Each one of these major aspects of FORSYS will be discussed below. In addition, another major innovation of FORSYS will be described. This is its ability to forecast for varying time horizons by exploiting short-term trends without ignoring, however, longer term growth patterns. The result is a considerable improvement in the forecasting accuracy, in particular for forecasting horizons four or five periods ahead (see Makridakis *et al.*, 1982).

THE BASIC MODEL OF FORSYS

The objective of FORSYS is to decompose the time series X_t, usually sales, into each one of the several components that influence X_t. What remains after all the various factors have been extracted from X_t is random noise.

The basic equation of FORSYS is

$$X_t = f(M_t, S_t, S_t^{(c)}, \psi_t, E_t, \varepsilon_t) \tag{1}$$

where

M_t = trend (level)
S_t = seasonality
$S_t^{(c)}$ = influence of extra-ordinary calendar or climatic variations
ψ_t = influence of the special actions
E_t = influence of independent (exogenous) variables which might influence sales

and

ε_t = what remains after the decomposition, i.e. random noise.

Equation (1) provides a mixed time series explanatory forecasting model. The factors M_t and S_t are modelled as functions of time (see Appendix 1). However, such influences are usually intermixed with unusual weather conditions (e.g. consider the possible increase in consumption of beer during an unusually hot summer), special actions (e.g. a promotional campaign) or some independent (exogenous) factor that influences sales such as GNP, advertising or price of the product. The weather conditions can be quantified as 'temperature' and are included in $S_t^{(c)}$, ψ_t incorporates the special actions, while finally, E_t holds the independent variable. Figure 1 shows a general presentation of FORSYS and the various factors that have been mentioned.

In contrast to M_t and S_t, which are purely time dependent factors, ψ_t and E_t are included to 'explain' the behaviour of sales beyond fluctuations which are caused simply by historical momentum. Therefore, they can be called 'explanatory' since they allow the user to better understand fluctuations in sales. The user can then be involved in explaining and understanding sales by specifying the various special actions that have taken place in the past. Finally, there is the $S_t^{(c)}$ factor which

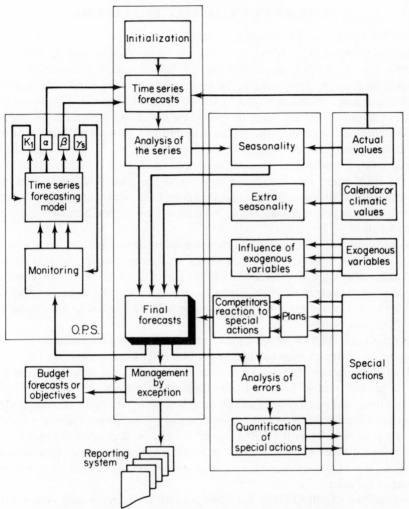

FIGURE 1   A general presentation of the FORSYS forecasting system

stands somewhere between time series and explanatory factors. It is time dependent but it can also explain the behaviour of sales for weather or calendar conditions that differ from the average. In my experience, this is a very important element in forecasting which is too often ignored and which, usually, considerably improves forecasting accuracy.

Thus, variations in sales can be caused by time factors (trend and seasonality), calendar or climatic fluctuations (working days, temperature), special events (promotions, advertising, price changes) or they could be caused by exogenous variables. Obviously, the influence of each factor needs to be separated so that the user can understand the influence of each factor on sales.

FORSYS AS A FORECASTING SYSTEM

FORSYS combines the advantages of time series forecasting with those of econometric modelling. If the user wishes, he or she can use FORSYS in a completely time series mode, by eliminating the influence of calendar/climatic or exogenous factors. Alternatively, he or she can use it as a multiple regression forecasting model where explanatory variables can become the only factors being used. In addition, the influence of special events allows the user to incorporate judgemental elements into the forecasting model. This is an aspect that most forecasting methods do not handle properly. No matter how accurate the forecasting method can be, the user may need to override the forecasts provided by the forecasting model. This occurs when the user possesses some inside information, knows about some special actions that will take place, is aware of structural changes in the economy or market, or simply because of political consideration.

In addition, forecasts generated by some quantitative method cannot deviate much from the budget or sales objectives set by top management. It becomes important, therefore, to be able to force the forecasts of the quantitative methods to become the same as those set in the budget or as the sales objective. Although such a thing is extremely simple to do, it can help tremendously if it can be done automatically.

In practice, FORSYS is used almost totally in its mixed time series explanatory mode since managers need to understand both the time component of their sales and the influence of exogenous factors such as promotions, advertising, reactions of competitors and so forth, into the forecasting system.

Equation (1) is not difficult to explain to non-technical users who can grasp its meaning, despite its mathematical complexity. After some training, managers can work personally with FORSYS, rather than be dependent on experts, and obtain forecasts themselves. Moreover, FORSYS is not only used for forecasting purposes but also for simulating various special actions and determining their influence on sales.

In summary, FORSYS has been designed to be a forecasting system rather than simply a method that generates forecasts and then leaves the user with no further help. I use the term 'forecasting system' to mean the ability to interact, understand, modify and simulate the model. These, in the opinion of the author, are critical aspects which must be part of every formalized forecasting effort.

ESTIMATING AND UPDATING THE PARAMETER OF FORSYS

The choice of parameters in a forecasting model is a well-defined statistical area. In practical terms, however, problems arise because most of the statistical methods assume that the data are stationary. This is rarely the case, however, with real life series which are non-stationary, exhibiting changes in trends, structural changes and discontinuities. FORSYS, as with all statistical methods, finds initial optimal least squares estimates for the parameters of the model. However, this is only a beginning. It is assumed that such parameters are a first approximation to a process which is inherently non-stationary. Subsequently, therefore, the values of the initial parameters are continuously updated when fluctuations occur that are thought to be non-random.

The method of optimization and the updating of the parameters used by FORSYS has been developed by the author (see Lewandowski, 1969) and is referred to as OPS.

A unique characteristic of OPS is that it allows for the adaptive optimization of seasonal parameters in addition to non-seasonal ones. Furthermore, it allows for the monitoring of the results so that, in cases of non-random errors, the system provides a warning and automatically adjusts the values of the parameters.

SALES FORECASTING AND SPECIAL EVENTS

Mechanistic forecasting approaches ignore a number of factors which might influence the sales (or whatever other variable is being forecast) of a company. For instance, when a company launches a sales promotion campaign for one of its products, the sales will inevitably be above average for several months during and after the campaign and then probably fall for the months that follow. The same will be true if there are temporary reductions or increases in prices, advertising campaigns, or many other special actions taken by the company. Similarly, there are several events outside the control of the company which can also influence sales. For instance, changes in taxes, changes in monetary policies, possible strikes, actions or reactions of competitors and so forth do often influence the sales significantly.

Special events or actions, and their influence on sales exist, as well as influences determined by trend and seasonality, those caused by calendar or climatic effects, or exogenous variables. A major challenge for forecasting is to be able to separate the individual influences of each one of the factors affecting sales. Even if this can only be done purely in an *ex-post* manner, it can be of great help to management who can then understand what is happening and how their sales are affected by the various factors involved. However, this is a topic which has received little attention in the forecasting literature. Such developments as intervention analysis (Box and Tiao, 1975) have little relevance for the practitioner who neither understands them nor can use them because of lack of long enough series. It is a major objective of FORSYS, therefore, to allow the decomposition of sales to be performed in a way that each one of the five components of equation (1) can be separated and identified. This historical analysis is an important characteristic of FORSYS which provides the basis of assessing the influence of each one of the factors affecting sales. However, before doing anything, the influence of special

TABLE 1

Type 4 .	Base :	Value of special action					
PERIOD ·	value ·	before each implementation					
		0	1	2	3	4	5

Type 3 .	Base :	Value of special action					
PERIOD ·	value ·	before each implementation					
		0	1	2	3	4	5

Type 2 .	Base :	Value of special action					
PERIOD ·	value ·	before each implementation					
		0	1	2	3	4	5
1	20.0	11.6	10.4	11.1	11.1	11.6	
2	10.0	-13.6	-13.0	-14.9	-14.3	-13.6	
3	-10.0	-12.5	-12.5	-12.1	-13.0	-12.5	
4	-10.0	-11.5	-13.1	12.8	-12.3	-11.5	

200 ·   TABLEAU DE L'ANALYSE DES ACTIONS SPECIALES   · 200

Type 1 .	Base :	Value of special action					
PERIODE ·	value ·	before each implementation					
		0	1	2	3	4	5
1	10.0	8.5	6.7	8.5		
2	5.0	7.7	8.1	7.7		
3	-5.0	-4.3	-4.2	-4.5		
4	0.0	0.0	0.0	0.0		
5	0.0	0.0	0.0	0.0		
6	0.0	0.0	0.0	0.0		

Sales promotion aimed at retailers
Sept. - Oct. 1971

events must be identified and removed from X_t. This, in the opinion of this author, is one of the most important advantages of FORSYS, as far as practitioners are concerned.

Special events, or actions, are analysed each time that they occur. Such an analysis is not thrown away but kept, allowing management to build up a history of special events and how they have affected the sales in the past. Such a catalogue of influences can be used successfully for forecasting purposes by providing management with an easily accessible and easy to interpret history of similar situations. Although the principle may seem simple, its practical implications are enormous. The reader is asked to think how long a manager, or staff member, stays in a job. Obviously, nobody remembers what has happened after the person responsible has left, and, since a record is rarely kept, the new manager does not understand why sales have been erratic in some specific periods. Since no knowledge of special events is remembered, their influence is attributed to random reasons. No history of similar events can be built beyond the memory of those on the job. This is an obvious shortcoming. FORSYS provides for organizational memory by printing the sales history during and after special actions (see Table 1). Such print-outs are classified in a dossier of similar actions and they form an historical record that provides organizational memory.

More specifically, the analysis and quantification of special events permits the following:

(a)   a better historical analysis of sales, eliminating variations due to special actions and permitting the more accurate determination of the series' own characteristics such as the trend, the seasonality and the influence of exogenous factors;

(b) more reliable forecasting which is a logical result of a more accurate analysis of the past characteristics of the series;

(c) a quantification of the reactions of the market to special events in the company, thus facilitating the understanding of consumer behaviour and competitive reaction;

(d) simulation and sensitivity analysis of the financial viability of short-term marketing policies;

(e) an efficient and continuous monitoring policy of special events and their influence on sales.

Table 1 shows an illustration of the FORSYS output concerning some special actions. As can be seen, each special action and its net effect is isolated and shown under base value. The number on the right of base value shows the value of special action before each implementation. For the particular special action in Table 1, this base value has been $+10 + 5 - 5$, which means a gain of 10 for the period during which the special action was introduced, another gain of 5 during the next period and a loss of 5 during the subsequent period. This makes a net gain of $+10$ which is the result attributed to this special event which was a sales promotion aimed at retailers and carried out in September–October 1971.

FORECASTING

FORSYS does not assume that a series will increase monotonically. It rather searches among several alternative estimates of the trend. The choice of a particular alternative is made dependent on the time horizon of forecasting. In short term, the forecasts become a straightforward extrapolation of the most recent values. However, as the time horizon of forecasting increases, then the trend estimate based on the most recent periods is taken less into account and that based on the whole series is assumed to influence future sales to a greater extent.

There are three alternative ways by which forecasts for longer time horizons are computed by FORSYS. The first extrapolates the average trend of *all* available data to obtain the predictions for the future. The second still extrapolates the average trend, but, because of the uncertainty involved in future predictions, it extrapolates a diminishing percentage of the average trend. The third alternative uses an exponential declining trend. This trend is based on the average trend of all data but is reduced by a constant percentage as the time horizon of forecasting increases. This constantly declining percentage is calculated by examining the stability of the series. Thus, the more stable the series, the smaller the exponential reduction of the trend being used. On the other hand, for unstable series the trend is assumed to decline very fast until it reaches zero.

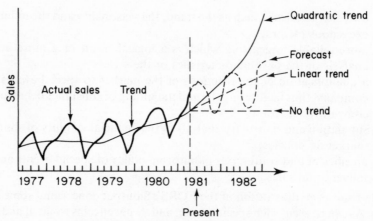

FIGURE 2   Possibilities of trend extrapolation

Figure 2 shows the principle of the possibility of several trend extrapolations. The choice is between a quadratic trend and a no trend (see Figure 2). Obviously, deciding on the best alternative trend is extremely important. Since the success of a forecasting system greatly depends upon such a choice of trend, FORSYS searches among many alternatives until it finds one which minimizes the forecasting errors. The details of establishing the optimal trend are beyond the scope of this paper. However, the interested reader is referred to Lewandowski (1979, pp. 245–249). Finally, it must be emphasized that the principle of searching for alternative trend extrapolations is unique in FORSYS and, as it turns out, most successful since the forecasts from FORSYS were more accurate than those of other methods (see Makridakis *et al.*, 1982) for longer forecasting horizons.

FORECASTING: THE CHALLENGE BEFORE US

Forecasting is the first step in any planning activity and the many decisions which need to be based on a more certain knowledge of the near or distant future. Without this knowledge of what the future will be like, it is impossible to plan production, the purchasing of raw materials, the personnel selection, to manage inventories or consider the financial implications involved. It is difficult to envisage an effective corporate management that does not expend considerable effort on some form of thinking about the future.

Nevertheless, forecasting, and above all short-term forecasting, has short-comings and difficulties. Whereas market forecasting for a period of 2 to 3 years can be done relatively accurately, forecasting for the next 3 to 6 months could be more uncertain because the reaction of consumers and competitors in the short term is influenced by random events and cannot, therefore, be anticipated easily.

It is important to remember that forecasting bears no relation to fortune-telling. Short-term forecasting methods attempt to detect, from an analysis of past data, signs of future developments and nothing more. There can be surprises, of course, but, apart from systematic changes, random fluctuations tend to iron out and longer term trends can be recognized. This author believes that this is very important.

Many forecasting users believe that all short-term fluctuations have permanent effects. This belief makes one over-react to short-term changes and take certain actions which are reversed at a later time. The challenge to the forecaster is, in the opinion of this author, to be able to operate his or her short-term forecasting system in such a way so as not to over-react to random changes while, at the same time, not ignoring systematic modifications in trends or other permanent changes in the data. How this can be done is a challenge that requires good experience and an intuitive feel of the market and the environment in which the organization operates. A forecasting system can only provide help and cannot replace good human judgement.

In summary, practitioners need to be aware of the strengths and weaknesses of the available short-term forecasting methods. In developing FORSYS, I attempted to overcome some of the major weaknesses in existing systems. However, any good forecasting system should include the following characteristics (which are part of FORSYS):

(a) Modern methods of short-term forecasting should permit the analysis of sales by quantifying the various components. It must be emphasized that this can be done much better than it would have been possible to do manually. Sales analysis permits the acquisition of extremely valuable information on the dynamics of the markets and the behaviour of the products which could not have been found judgementally.

(b) The use of modern forecasting methods opens up considerable possibilities since the actual impact of sales promotion campaigns (or special actions) on sales can be analysed and measured. Thus, short-term forecasting becomes an indispensable instrument for planning short and medium-term marketing activities and special events.

(c) Short-term sales forecasting can indicate when significant changes in sales are taking place. This monitoring ability of forecasting is indispensable and one of the least appreciated of forecasting uses. It works as an early warning system which allows the detection of significant changes in sales several months earlier than when this task is carried out manually. Monitoring of sales opens up a new area of applications in terms of management by exception. This can be done by regular reporting of significant deviations from plans and forecasts.

(d) Management using an integrated short-term forecasting system is provided with objective information on which to base its decisions. Short-term forecasting generated by a well-defined method is not influenced by personal motives, filtering of information, or simple distortion of information. This objectivity of formal forecasting, coupled with the management by exception reporting described above, can become an extremely effective management tool.

and finally,

(e)   A sales forecasting system allows the manager to experiment with alternative policies and to consider the effects of various possible levels of future sales. This simulation of alternative futures helps management to examine the 'sensitivity' of various management plans and reconsider their effect on the financial and competitive viability of the organization.

APPENDIX 1

A brief presentation of FORSYS

X_t, the time series, is decomposed as follows:

$$X_t = M_t S_t + e_t \tag{1}$$

The mean, M_t, is defined by a moving average process which is basically of exponential smoothing type. For instance, for a linear model, M_t is defined as:

$$M_t = 2(M1_t) - M2_t \tag{2}$$

where

$$M1_t = \sum^{\theta} \frac{X_{t-\theta}}{S_{t-\theta}} \alpha_{t-\theta} \prod^{\theta} (1 - \alpha_{t-\theta})^{\theta} \tag{3}$$

$$M2_t = \sum^{\theta} M1_{t-\theta} \alpha_{t-\theta} \prod^{\theta} (1 - \alpha_{t-\theta})^{\theta} \tag{4}$$

The smoothing constant α_t is given by:

$$\alpha_t = \alpha_{0_t} + \Delta \alpha_t.$$

The values of α_t vary as follows:

$$\alpha_{0_t} = \alpha_0 \rho^{f_1 [\sigma_t^{(1)}]}$$

$$\Delta \alpha_t = \kappa_0 \rho^{f_2 [\sigma_t^{(2)}]} - \kappa_1 \rho^{f_3 (\Sigma_t^*)}$$

where $\sigma_t^{(1)}$ is a measure of the stability of the series and is defined as:

$$\sigma_t^{(1)} = \left| \frac{\text{MAD}_t}{M_t} \right|$$

and where

$$\text{MAD}_t = |\varepsilon_t| \gamma + (1 - \gamma) \text{MAD}_{t-1}$$

$\sigma_t^{(2)}$ is a normalized measure of the randomness of the series. It is defined as:

$$\sigma_t^{(2)} = \left| \frac{\varepsilon_t}{\text{MAD}_t} \right|$$

and finally, Σ_t^* is a tracking signal defined as follows:

$$\Sigma_t^* = \frac{\Sigma_t}{\text{MAD}_t}$$

where

$$\Sigma_t = \Sigma_{t-1}(1 - \gamma_{S_t}) + \varepsilon_t$$

where γ_{S_t} can be thought of as the coefficient of decay, that is:

$$\gamma_{S_t} = \gamma_{S_0}[1 - \rho^{f_4(\sigma_t^{(2)})}]$$

The seasonal coefficients are found by an exponential smoothing process similar to that of (3) and (4) which is:

$$S_t = \sum^{\tau} \frac{X_{t-\tau}}{M_{t-\tau}} \beta_{t-\tau} \prod^{\tau} (1 - \beta_{t-\tau})^{\tau},$$

where

$$\beta_t = \beta_0 \bar{\rho}^{f_5(\Sigma_t^*)}.$$

The forecasting of the series is given by combining the components of (1), that is M_t and S_t. This results in the following projections:

$$\hat{X}_{t+\kappa}^{(1)} = M(\alpha)_t + \kappa T(\alpha)_t + \kappa^2 Q(\alpha)_t$$

$$\hat{X}_{t+k}^{(s)} = M(a^\delta)_t + \kappa T(a^\delta)_t + \kappa^2 Q(a^\delta)_t$$

$$\hat{X}_{t+\kappa}^{(2)} = M(\alpha^*)_t + \kappa T(\alpha^*)_t$$

Finally, the forecasts are found by

$$\hat{X}_{t+\kappa} = \{\hat{X}_{t+\kappa}^{(\delta)} \delta_{t+\kappa}\} S_{t+\kappa}$$

For more details, see Lewandowski (1979).

REFERENCES

Box, G. E. P. and Tiao, G. C. (1975). 'Intervention Analysis with Applications to Economic and Environmental Problems', *Journal of American Statistical Association*, **70**, 70–79.

Lewandowski, R. (1969). 'Ein voll adaptationsfähiges Modell zur kurzfristigen Prognose'. *AKOR-Tagung*, Aachen.

Lewandowski, R. (1979). *La Prévision à Court Terme*, Paris: Dunod.

Makridakis, S. *et al.* (1982). 'The Accuracy of Extrapolation (time series) Methods: Results of a Forecasting Competition', *Journal of Forecasting*, **1**, 111–153.

Wheelwright, S. C. and Clarke, D. G. (1976). 'Corporate Forecasting: Promise and Reality'. *Harvard Business Review*, November–December.

CHAPTER 11

Forecasting and Time Series Model Types of 111 Economic Time Series

H. Joseph Newton and Emanuel Parzen
Texas A&M University, U.S.A.

1. INTRODUCTION

'Is it possible to put an end to the argument of what forecasting methods are better and under what circumstances?', is the question raised by Professor Spyros Makridakis in several stimulating papers (1976), (1978), (1979). He has organized a 'forecasting competition' to which various forecasting experts would contribute forecasts of 111 economic and business time series which he has collected. This paper reports the results of our analysis of these series, based on the general approach to time series modelling, spectral analysis, and forecasting developed by Parzen, with the collaboration of Newton.

An appendix describes the theory of univariate time series modelling and forecasting used in this study. The main text summarizes the diverse models which are encompassed by our approach, and which arise in the study of the 111 time series being forecasted.

The methods of time series modelling and forecasting applied in this paper can be applied automatically but they are not rote formulae, since they are based on a flexible philosophy which provides several models for consideration and diverse diagnostics for qualitatively and quantitatively checking the fit of a model (see Parzen, 1979, 1980, 1981). The models considered are called ARARMA models because the model computed adaptively for a time series is based on sophisticated time series analysis of ARMA schemes (a short-memory model) fitted to residuals of simple extrapolation (a long-memory model obtained by parsimonious 'best lag' non-stationary autoregression).

A consumer of time series forecasting and/or modelling methods must evaluate the value of a proposed procedure in the context of the actual time series with which he, or she, is concerned. Our approach aims to be applicable in all the diverse fields to which time series analysis is being applied.

A major problem of time series forecasting is whether long-range forecasting and short-range forecasting require different methods to obtain satisfactory forecasts. This paper describes *iterated models* which provide qualitative

267

diagnostics as to the possibility of long-range forecasts (by diagnosing whether the time series is long-memory). Both long-range and short-range forecasts are provided by a model obtained by fitting a parsimonious non-stationary autoregression whose residuals $\tilde{Y}(t)$ are modelled by a stationary autoregression.

The modelling procedure is both automatic and flexible. In particular, two model orders are determined for $\tilde{Y}(t)$ and we would recommend computing and comparing forecasts from both models.

This paper aims to illustrate the results one obtains by typical graphs, and to describe the time series model types that one should expect to encounter when dealing with many economic time series.

2.　ITERATED MODELS APPROACH TO TIME SERIES ANALYSIS

The problem of forecasting future values of a time series from observations of its past values has an extensive literature which proposes many different approaches. The approach adopted here aims to fit automatically to a time series sample not one but several models. The class of models considered is suitable for time series modelling, spectral analysis, and forecasting and for time series encountered by researchers in the physical sciences, engineering sciences, biological sciences, and medicine, as well as to the social sciences, economics, and management sciences.

A time series may be predictable for a long time in the future or only over a limited future. We say the former has 'long memory' and the latter 'short memory'. A time series with long memory requires a 'non-stationary' model with periodic, cycle, and trend components. A time series with short memory requires a 'stationary' model which is a linear filter relating the time series to its innovations or random shocks. The linear filter is an AR, MA, or ARMA filter (autoregressive, moving average, or mixed autoregressive-moving average).

The model we fit to a time series $Y(.)$ is an iterated model

$$Y(t)\ \boxed{}\!\!\rightarrow \tilde{Y}(t)\ \boxed{}\!\!\rightarrow \varepsilon(t).$$

If needed to transform a long-memory series Y to a short-memory series $\tilde{Y}$, $\tilde{Y}(t)$ is chosen to satisfy one of the three forms

$$\tilde{Y}(t) = Y(t) - \hat{\phi}(\hat{\tau})Y(t - \hat{\tau}), \tag{1}$$

$$\tilde{Y}(t) = Y(t) - \hat{\phi}_1 Y(t - 1) - \hat{\phi}_2 Y(t - 2). \tag{2}$$

$$\tilde{Y}(t) = Y(t) - \hat{\phi}_1 Y(t - \tau - 1) - \hat{\phi}_2 Y(t - \tau) \tag{3}$$

Usually $\tilde{Y}(t)$ is short-memory; then it is transformed to a white noise, or no-memory, time series $\varepsilon(t)$ by an approximating autoregressive scheme AR(m) whose order m is chosen by an order determining criterion (we use CAT, introduced by Parzen (1974), (1977)).

In the present study, $\tilde{Y}(t)$ was found to be always short-memory. It is then modelled by a stationary autoregressive scheme. It is argued by Parzen that approximating AR schemes suffice for spectral analysis and forecasting. Only for model interpretation is it desirable to fit an ARMA scheme. In the present study not more than 15 per cent of the time series could be regarded as requiring an ARMA scheme.

To determine the best lag $\hat{\tau}$, we use non-stationary autoregression; either fix a maximum lag M and choose $\hat{\tau}$ as the lag minimizing over all τ

$$\sum_{t=M+1}^{T} \{Y(t) - \phi(\tau)Y(t-\tau)\}^2$$

or choose $\hat{\tau}$ as the lag minimizing over all τ

$$\sum_{t=\tau+1}^{T} \{Y(t) - \phi(\tau)Y(t-\tau)\}^2 \div \sum_{t=\tau+1}^{T} Y^2(t).$$

For each τ, one determines $\phi(\tau)$, and then one determines $\hat{\tau}$ (the optimal value of τ) as the value minimizing

$$\text{Err}(\tau) = \sum_{t=M+1}^{T} \{Y(t) - \phi(\tau)Y(t-\tau)\}^2 \div \sum_{t=M+1}^{T} Y^2(t)$$

or

$$\text{Err}(\tau) = \sum_{t=\tau+1}^{T} \{Y(t) - \phi(\tau)Y(t-\tau)\}^2 \div \sum_{t=\tau+1}^{T} Y^2(t).$$

The decision as to whether the time series is long-memory or not is based on the value of $\text{Err}(\hat{\tau})$. An *ad hoc* rule we use is if $\text{Err}(\hat{\tau}) < 8/T$, the time series is considered long-memory. In the present study all time series were judged to be long-memory by this criterion. When this criterion fails one often seeks transformations of the form of (2) or (3), using semi-automatic rules described in the appendix.

For the maximum lag M of non-stationary autoregression, the following rules were adopted in this study: $M = 2$ for yearly series, $M = 5$ for quarterly series, $M = 15$ for monthly series.

3.   FORECASTING FORMULAE

For forecasting purposes it suffices to adopt for $\tilde{Y}(t)$ a stationary autoregressive model of suitable order m whose coefficients $\alpha_1, \ldots, \alpha_m$ are estimated by Yule–Walker equations in the correlation function $\hat{\rho}(v)$ of $\tilde{Y}(t)$. In this paper the model adopted for all time series was of the form

$$\tilde{Y}(t) = Y(t) - \phi(\tau)Y(t-\tau)$$

$$\tilde{Y}(t) + \alpha_1 \tilde{Y}(t-1) + \cdots + \alpha_m \tilde{Y}(t-m) = \varepsilon(t)$$

The residual variances are denoted

$$\text{RVY} = \sum_{t=M+1}^{T} \tilde{Y}^2(t) \div \sum_{t=M+1}^{T} Y^2(t)$$

$$\text{RVYT} = \sum_{t=\tau+1}^{T} \varepsilon^2(t) \div \sum_{t=\tau+1}^{T} \tilde{Y}^2(t).$$

The last 18 points of the graphs of Y and $\tilde{Y}$ do not represent observed values of these series but forecasted values of horizons $h = 1$ to 18. The mathematical procedure by which they are derived is as follows.

Let

$$Y^\mu(t+h|t) = E\{Y(t+h)|Y(t), Y(t-1), \ldots\}$$

denote the predictor of $Y(t+h)$ given values $Y(t), Y(t-1), \ldots$. From the equation

$$Y(t+h) = \phi(\tau)Y(t-\tau+h) + \tilde{Y}(t+h)$$

one obtains, by conditioning with respect to $Y(t), Y(t-1), \ldots$

$$Y^\mu(t+h|t) = \phi(\tau)Y^\mu(t-\tau+h|t) + \tilde{Y}^\mu(t+h|t).$$

To obtain a formula for forecasts of $\tilde{Y}$ when we have fitted an AR(m) to $\tilde{Y}$:

$$\tilde{Y}(t) + \alpha_1 \tilde{Y}(t-1) + \cdots + \alpha_m \tilde{Y}(t-m) = \varepsilon(t)$$

write

$$\tilde{Y}(t+h) + \alpha_1 \tilde{Y}(t+h-1) + \cdots + \alpha_m \tilde{Y}(t+h-m) = \varepsilon(t+h)$$

$$\tilde{Y}^\mu(t+h|t) + \alpha_1 Y^\mu(t+h-1|t) + \cdots + \alpha_m \tilde{Y}^\mu(t+h-m|t) = 0.$$

One can now compute $Y^\mu(t+h/t)$ recursively for $h = 1, 2, \ldots$, using the fact that

$$\tilde{Y}^\mu(t+j|t) = \tilde{Y}(t+j) \qquad \text{if } j \le 0.$$

For example,

$$-\tilde{Y}^\mu(t+1|t) = \alpha_1 Y(t) + \cdots + \alpha_m Y(t-m+1).$$

Then one can compute $Y^\mu(t+h/t)$ recursively for $h = 1, 2, \ldots$ using the fact that

$$Y^\mu(t+j|t) = Y(t+j) \qquad \text{if } j \le 0.$$

For large values of h, one expects $\tilde{Y}^\mu(t+h/t) = 0$. Then

$$Y^\mu(t+j|t) = \phi(\tau)Y^\mu(t+h-|t).$$

When $\phi(\tau) \ge 1$, this does not damp down to zero, and provides the long-term predictability apparent in many of the series.

4.   SUMMARY OF ITERATED MODELS FITTED TO 111 TIME SERIES

Table 1 describes the lags of the most significant lag non-stationary scheme for $Y(t)$. For 60 per cent of the monthly series, the annual period ($\hat{\tau} = 12$) was most important; only 26 per cent of the quarterly series had an annual period ($\tau = 4$).

TABLE 1   Lag of Non-stationary AR

τ	1	2	3	4	5	12
20 Yearly	18	2	0	0	0	0
23 Quarterly	15	1	0	6	0	1
68 Monthly	24	0	3	0	0	41

Range of 'non-stationary' coefficients $\phi(\tau)$

	$0.99 \le \phi(\tau) \le 1.01$	$\phi(\tau) < 1.01$	$\phi(\tau) > 0.99$
20 Yearly	4	15	1
23 Quarterly	11	9	3
68 Monthly	27	27	14

TABLE 2   AR Orders Determined by CAT and Innovation Variances

m	0	1	2	3	4	5	6	7	8	9	10	11	12	13	14	15	16
20 Yearly	12	4	2	1	0	1	0										
23 Quarterly	8	5	2	1	1	4	2										
68 Monthly	12	18	13	6	3	1	2	1	0	1	1	0	4	2	3	0	1

σ_m^2	0.1	0.2	0.3	0.4	0.5	0.6	0.7	0.8	0.9	0.10
20 Yearly	0	0	1	0	1	2	2	1	1	12
23 Quarterly	0	2	2	1	2	4	1	3	0	8
68 Monthly	2	1	2	5	12	9	6	12	7	12

0.1 means range $0.1 \le \sigma_m^2 > 0.2$; similarly for 0.2 to 0.9.

The AR character of the residual series $\tilde{Y}(t)$ are described in Table 2. Order $m = 0$ indicates white noise (or no memory); 60 per cent of the yearly series obey the 'naive' model $\tilde{Y}(t) = \varepsilon(t)$, white noise.

Table 3 lists the names of 33 series arbitrarily chosen from the set of 111 series to represent typical series. We select this small number of series to discuss in detail. The different types of time series which can be diagnosed by our approach to time series modelling and forecasting are illustrated by the results in Table 4 and the graphs of Y and $\tilde{Y}$ for the series listed in Table 3.

TABLE 3   Typical Series for Detailed Discussion
(Y, Q, M are the the prefixes of Yearly, Quarterly and Monthly Series Respectively)

YA	Machinery and Equipment (YAC 17)
YB	National Product and Expenditure-Residential Construction (YAC 26)
YC	Population Movement Male Death (YAD 6)
YD	Crude Birth Rates (YAD 15)
YE	Deaths, Analysis by Age and Sex, All Ages, United Kingdom (YAD 24)
QA	Industrial Production: Textiles (QNI 1)
QB	Industry Germany (QNI 10)
QC	Company Data Germany (QNM 15)
QD	Company Data (QNM 6)
QE	Industrial Production: Durable Manufactures (QRC 13)
QF	Industrial Production: Total Austria (QRC 22)
QG	Value of Manufacturer's New Orders for Consumer Goods (QRC 4)
QH	Per Capita GNP in Current Dollars (QRG 13)
QI	Total Industrial Production (QRG 4)
MA	Company Data (MNB 11)
MB	Company Data (MNB 2)
MC	Company Data (MNB 20)
MD	Company Data (MNB 29)
ME	Company Data USA (MNB 38)
MF	Company Data UK (MNB 47)
MG	Company Data (MNB 56)
MH	Company Data (MNB 65)
MI	Textiles – Quoted at Paris Stock Exchange (MNC 17)
MJ	General Index of the Industrial Production (MNC 26)
MK	Reserves – Denmark (MNC 35)
ML	New Private Housing Units Started Total USA (MNC 44)
MM	Industrial Production Spain (MNG 28)
MN	Industrial Production: Finished Investment Goods Austria (MNG 37)
MO	Aluminium Production Netherlands (MNI 103)
MP	Lead Production Canada (MNI 122)
MQ	Production Tin Thailand (MN 122)
MR	Industry France (MNI 13)
MS	Motor Vehicles Production Canada (MNI 131)

Table 4 summarizes the basic model diagnostics of a time series $Y(t)$. These are length; most significant non-stationary autoregressive lag τ, and coefficients $\phi(\tau)$; the residual variance RVY of this non-stationary AR scheme; the best orders (denoted CAT 1 and CAT 2) of approximating AR schemes for $\tilde{Y}(t)$, their horizons HOR 1 and HOR 2, and the residual variance RVYT of the best approximating AR Scheme.

Some ARMA models for quarterly time series were:

QA $\tilde{Y} = (I - 1.04L^4)Y, (I - 0.74L)\tilde{Y} = (I - 0.85L^4)\varepsilon$

QH $\tilde{Y} = (I - 1.02L)Y, (I - 0.29L^4)\tilde{Y} = (I - 0.38L^3)\varepsilon$

TABLE 4   Diagnostics of Model Types of Typical Time Series

Series length	τ	$\phi(\tau)$	RVY	CAT 1	CAT 2	HOR 1	HOR 2	RVYT
YA	13	1	1.040	0.007	2	1	4	2	0.66
TB	13	1	1.066	0.007	1	2	2	2	1.00
YC	39	1	1.003	0.009	5	4	6	6	0.58
YD	9	1	0.990	0.000	1	2	1	1	1.00
YE	13	2	1.010	0.001	1	2	1	3	1.00
QA	32	4	1.038	0.008	6	3	11	7	0.32
QB	57	4	1.023	0.003	5	4	5	11	0.27
QC	24	1	1.004	0.001	1	2	1	2	1.00
QD	20	1	0.994	0.037	1	7	1	13	1.00
QE	36	1	1.012	0.001	1	2	2	2	0.86
QF	56	2	1.030	0.001	3	4	7	7	0.62
QG	52	1	1.011	0.009	1	2	1	1	1.00
QH	52	1	1.018	0.000	5	3	10	4	0.68
QI	56	1	1.013	0.000	5	2	14	3	0.65
MA	79	12	1.024	0.016	2	10	5	24	0.66
MB	30	12	1.065	0.103	1	2	1	3	1.00
MC	54	1	0.973	0.057	2	24	4	31	0.76
MD	57	1	1.004	0.001	2	4	3	5	1.00
ME	32	12	1.261	0.050	1	2	2	2	1.00
MF	80	1	0.968	0.063	9	5	13	5	0.65
MG	64	12	1.154	0.010	2	10	6	10	0.55
MH	52	1	0.999	0.000	1	3	1		1.00
MI	42	1	0.993	0.002	1	14	2	16	0.95
MJ	90	12	1.076	0.003	13	19	12	28	0.17
MK	5	1	0.979	0.055	2	7	2	6	0.89
ML	126	1	1.003	0.005	6	2	4	2	0.85
MM	43	12	0.995	0.010	1	2	4	7	0.55
MN	102	12	1.037	0.002	14	3	24	4	0.60
MO	65	1	1.006	0.003	1	5	2	7	0.79
MP	66	1	0.949	0.100	1	12	2	13	0.95
MQ	62	3	0.973	0.029	6	3	11	7	0.57
MR	120	12	1.053	0.004	16	13	19	19	0.63
MS	66	12	0.989	0.029	3	4	4	5	0.81

Some ARMA models for monthly time series were:

MA $\tilde{Y} = (I - 1.02L^{12})Y, (I - 0.41L + 0.32L^{12})\tilde{Y} = (I - 0.42L + 0.31L^5)\varepsilon$

MF $\tilde{Y} = (I - 0.97L)Y, (I + 0.31L^{10})\tilde{Y} = (I - 0.49)\varepsilon$

MJ $\tilde{Y} = (I - 1.08L)Y, (I - 0.75L - 0.21L^3)\tilde{Y} = (I - 0.54L^{12})\varepsilon$

MN $\tilde{Y} = (I - 1.04L^{12})Y, (I - 0.29L^2 - 0.28L^3 - 0.27L^{11} + 0.30L^{13})\tilde{Y}$
$\quad = (I - 0.42L^{12})\varepsilon$

MR $\tilde{Y} = (I - 1.05L^{12})Y, (I - 0.21L^5 - 0.41L^6)\tilde{Y} = (I - 0.55L^{12})\varepsilon$

REFERENCES

Box, G. E. P. and Jenkins, G. M. (1970). *Time Series Analysis: Forecasting and Control*, San Francisco: Holden Day.

Makridakis, S. (1976). 'A Survey of Time Series', *Int. Statist. Review*, **44**, 29–70.

Makridakis, S. (1978). 'Time Series Analysis and Forecasting: an Update and Evaluation', *Int. Statist. Rev.*, **46**, 255–278.

Makridakis, S. and Hibon, M. (1979). 'Accuracy of Forecasting: An Empirical Investigation', *Journal of the Royal Statistical Society*, Series A, **142**, 97–145.

Parzen, E. (1964). 'An Approach to Empirical Time Series Analysis', *J. Res. Nat. Bur. Standards*, **68D**, 937–951. Reprinted in Parzen (1967).

Parzen, E. (1967). *Time Series Analysis Papers*, San Francisco: Holden Day.

Parzen, E. (1967). 'The Role of Spectral Analysis in Time Series Analysis', *Review of the International Statistical Institute*, **35**, 125–141.

Parzen, E. (1974). 'Some Recent Advances in Time Series Modeling', *IEEE Transactions on Automatic Control*, **AC-19**, No. 6, December, 723–730.

Parzen, E. (1976). 'An Approach to Time Series Modeling and Forecasting Illustrated by Hourly Electricity Demands', Technical Report.

Parzen, E. (1977). 'Multiple Time Series: Determining the Order of Approximating Autoregressive Schemes', *Multivariate Analysis – IV*, P. Krishnaiah (ed.), Amsterdam: North-Holland, pp. 283–295.

Parzen, E. and Pagano, M. (1979). 'An Approach to Modeling Seasonally Stationary Time Series', *Journal of Econometrics*, **9**, 137–153.

Parzen, E. (1979). 'Forecasting and Whitening Filter Estimation', *TIMS Studies in the Management Sciences*, **12**, 149–165.

Parzen, E. (1980). 'Time Series Modeling, Spectral Analysis and Forecasting', *Directions in Time Series Analysis*, D. R. Brillinger and G. C. Tiao (eds), Institute of Mathematical Statistics, pp. 80–111.

Parzen, E. (1981). 'Time Series Model Identification and Prediction Variance Horizon', *Second Tulsa Symposium on Applied Time Series Analysis*, David Findley (ed.), New York: Academic Press, pp. 415–447.

Parzen, E. (1982). 'ARARMA Models for Time Series Analysis and Forecasting', *Journal of Forecasting*, **1**, 67–82.

APPENDIX: UNIVARIATE TIME SERIES MODELLING AND FORECASTING AUTOMATIC APPROACHES USING ARARMA MODELS

The model we propose fitting in general to a time series $Y(t)$ is an iterated model (with symbolic transfer functions G and g_α)

$$Y(t)\!-\!\boxed{G}\!\to\!\tilde{Y}(t)\!-\!\boxed{g_\alpha}\!\to\!\varepsilon(t)\ \text{white noise}$$

where $\tilde{Y}(t)$ is the result of a 'memory-shortening' transformation chosen to transform a long-memory time series to a short-memory one, and g_α is an innovation filter which is either an approximating **AR** filter or an **ARMA** filter. Parzen (1982) introduces the terminology **ARARMA** scheme for the iterated time series model with G determined by a non-stationary autoregressive estimation procedure; an **ARIMA** scheme, introduced by Box and Jenkins

(1970), corresponds to a pure differencing operator for G. Autoregressive analysis by Yule–Walker equations yields a stationary autoregressive scheme; a non-stationary autoregressive scheme is one which is fit by estimating its coefficients by ordinary least squares.

To identify the final model, or 'overall whitening filter', of a time series, one should determine its model memory type, and identify an iterative model for the time series as shown in the diagram.

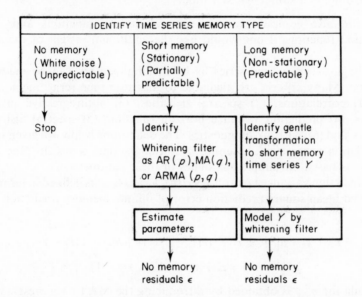

A confirmatory theory of statistical inference is available only for short-memory time series (which are ergodic). The modelling of a short-memory time series by a whitening filter can be regarded as a *science*, and it can be made semi-automatic. Given a sample of short-memory stationary time series $\hat{Y}(t)$, our modelling procedure in the time domain is to compute approximating autoregressive schemes.

(1)   Form the sample correlation function

$$\hat{\rho}(v) = \sum_{t=1}^{T-v} \tilde{Y}(t)\tilde{Y}(t+v) \div \sum_{t=1}^{T} \tilde{Y}^2(t)$$

but do not base any decision upon it, or upon the partial correlations. Rather, compute approximating autoregressive schemes.

(2)   Solve successive order $m = 1, 2, \ldots$ Yule–Walker equations for auto-regressive coefficients $\alpha_{1m}, \ldots, \alpha_{mm}$ and residual variance σ_m^2.

(3)   Use an autoregressive order-determining criterion (either CAT or AIC)

to determine $\hat{m}(1)$ and $\hat{m}(2)$, the best and second best orders of approximating autoregressive schemes.

(4)   Compute PVH(h), the prediction variance horizon function for the insight it provides on the memory type and ARMA type of the time series. Compute horizons HOR 1, HOR 2 using approximating AR schemes of orders $\hat{m}(1)$ and $\hat{m}(2)$.

(5)   Compute a subset AR model.

(6)   Compute a subset ARMA model.

One can also compute various spectral density functions and spectral distribution functions if one would like the additional insight of the *spectral domain*.

The diagnosis of a time series as being long-memory can be made semi-automatic. Many criteria are available to diagnose time series memory type, using (1) correlations, (2) spectral densities, (3) autoregressive prediction variances, (4) prediction variance horizon function, (5) spectral distribution functions, and (6) *S*-PLAY diagnostics. The definitions below are given in terms of population parameters, assuming a stationary time series. In practice, the diagnosis is based on sample analogues of these parameters.

The prediction variance horizon PVH(h), $h = 1, 2, \ldots$, is defined in terms of the normalized mean square prediction error of infinite memory prediction h steps ahead:

$$\sigma_{h,\infty}^2 = E\{|Y^v(t+h|t)|^2 \div E\{Y^2(t)\}\}, \qquad Y^v(t+h|t) = Y(t) - Y^\mu(t+h|t),$$

$$Y^\mu(t+h|t) = E\{Y(t+h)|Y(t), Y(t-1), \ldots\}.$$

A formula for $\sigma_{h,\infty}^2$ is obtained by introducing the MA (∞) representation of

$$Y(t) = \varepsilon(t) + \beta_1 \varepsilon(t-1) + \cdots.$$

Then

$$\sigma_{h,\infty}^2 = \sigma_\infty^2 \{1 + \beta_1^2 + \cdots + \beta_{h-1}^2\}.$$

The graph of $\sigma_{h,\infty}^2$ increases monotonically from σ_∞^2 at $h = 1$ to 1 as h tends to ∞. We define

$$\text{PVH}(h) = 1 - \sigma_{h,\infty}^2, \qquad h = 1, 2, \ldots$$

and define horizon HOR to be the smallest value of h for which PVH(h) ≤ 0.05 (whence $\sigma_{h,\infty}^2 \geq 0.95$).

The infinite moving average coefficients β_k are estimated by inverting the transfer function $g_m(z)$ of an approximating autoregressive scheme to obtain, for $k = 1, 2, \ldots$

$$\alpha_0 \beta_k + \alpha_1 \beta_{k-1} + \cdots + \alpha_k \beta_0 = 0.$$

The classification of memory type by prediction horizon HOR is:

No Memory	Short Memory	Long Memory
HOR $= 0$	$0 <$ HOR $< \infty$	HOR $= \infty$

By HOR $= \infty$, we mean HOR is comparatively large: experiments lead us to conclude that one should compare HOR with the order ORD of the approximating autoregressive scheme. Let HOR/ORD denote the ratio of HOR to ORD; identify time series as follows: If HOR/ORD ≤ 1, then MA(q), with $q \leq$ HOR $- 1$. If HOR/ORD ≥ 4 (say) and PVH decays slowly, then long memory. If PVH declines smoothly and exponentially, then an AR(p) is indicated. If PVH has 'bends', then ARMA. If PVH has many level stretches with period τ, then an ARMA model is indicated of the form

$$Y(t) = \frac{I + \beta_1 L + \beta_2 L^2 + \cdots + \beta_q L^q}{I - \alpha_\tau L^\tau} \, \varepsilon(t).$$

The final identification of the orders p and q should be by parameter estimation or by use of S-arrays.

The determination of most appropriate 'gentle' transformation of Y to $\tilde{Y}$, where Y is long-memory and $\tilde{Y}$ is short-memory must inevitably involve the physical nature of the observed time series. A semi-automatic approach can be developed by considering the following examples of long-memory time series.

A time series $Y(t)$, $t = 0, +1, \ldots$, is called *periodic* with period τ, if

$$Y(t + \tau) - Y(t) = 0, \qquad \text{all } t.$$

It follows a linear trend $Y(t) = a + bt$, if for all t

$$Y(t + 1) - Y(t) = b, \qquad \text{a constant.}$$

It is a pure harmonic of period τ if for all t

$$Y(t) - \phi Y(t - 1) + Y(t - 2) = 0, \qquad \phi = 2 \cos \frac{2\pi}{\tau}.$$

Then

$$Y(t) = A \cos \frac{2\pi}{\tau} t + B \sin \frac{2\pi}{\tau} t.$$

As gentle memory shortening transformations, it is natural to consider

$$\tilde{Y}(t) = Y(t) - \hat{\phi}(\hat{\tau}) Y(t - \hat{\tau}), \tag{1}$$

$$\tilde{Y}(t) = Y(t) - \hat{\phi}_1 Y(t - 1) - \hat{\phi}_2 Y(t - 2) \tag{2}$$

$$\tilde{Y}(t) = Y(t) - \hat{\phi}_1 Y(t - (m - 1)) - \hat{\phi}_2 Y(t - m) \tag{3}$$

whose coefficients τ, $\hat{\phi}(\tau)$, $\hat{\phi}_1$, $\hat{\phi}_2$ are determined adaptively from the data. Our first choice is (1); the lag $\hat{\tau}$ is chosen to minimize over τ

$$\text{Err}(\tau) = \sum_{t=\tau+1}^{T} \{Y(t) - \hat{\phi}(\tau)Y(t-\tau)\}^2 \div \sum_{t=\tau+1}^{T} Y^2(t)$$

and $\hat{\phi}(\tau)$ is chosen to minimize over $\phi(\tau)$

$$\sum_{t=\tau+1}^{T} \{Y(t) - \phi(\tau)Y(t-\tau)\}^2.$$

The stationary correlation function $\rho(\tau)$ of $(Y(t), t = 1, 2, \ldots, T)$ is defined by

$$\hat{\rho}(\tau) = \sum_{t=1}^{T-\tau} Y(t)Y(t+\tau) \div \sum_{t=1}^{T} Y^2(t).$$

Define

$$\text{SSQ}(v) = \sum_{t=1}^{v} Y^2(t).$$

One can show that

$$\hat{\phi}(\tau) = \hat{\rho}(\tau) \frac{\text{SSQ}(T)}{\text{SSQ}(T-\tau)}$$

$$\text{Err}(\tau) = 1 - |\hat{\phi}(\tau)|^2 \frac{\text{SSQ}(T-\tau)}{\text{SSQ}(T) - \text{SSQ}(\tau)}.$$

The most significant lag $\hat{\tau}$ is defined as the value minimizing $\text{Err}(\tau)$.

We propose three possible actions at the initial stage of analysis of a time series $(Y(t), t = 1, \ldots, T)$:

L. Declare time series to be long-memory, and form $\tilde{Y}(t)$ by (1).
M. Declare time series to be moderately long-memory, and form $\tilde{Y}(t)$ by (2).
S. Declare time series to be short-memory, and form $\tilde{Y}(t) = Y(t)$, or $\tilde{Y}(t) = Y(t) - \bar{Y}$

where $\bar{Y}$ is the sample mean. After computing $\bar{Y}$, one performs a naive test to decide if it should be set equal to 0; a naive test is $|\bar{Y}| \le 2\sigma/\sqrt{T}$ where σ is the sample standard deviation.

(1) Compute and print $\hat{\phi}(\tau)$ and $\text{Err}(\tau)$ for $\tau = 1, 2, \ldots, M$, where M is suitably chosen (15 for yearly, quarterly, or monthly data):
(2) Determine $\hat{\tau}$. If $\text{Err}(\hat{\tau}) \le 8/T$, go to L.
(3) If $\phi(\hat{\tau}) \ge 0.9$, and $\hat{\tau} > 2$, go to L.

(4)   If $\phi(\hat{\tau}) \geq 0.9$ and $\hat{\tau} = 1$ or 2 determine the best fitting non-stationary AR(2) scheme minimizing

$$\sum_{t=3}^{T} \{Y(t) - \phi_1 Y(t-1) - \phi_2 Y(t-2)\}^2.$$

Let $\hat{\phi}_1, \hat{\phi}_2$ denote the minimizing values of ϕ_1 and ϕ_2. Then go to M.

(5)   If $\hat{\phi}(\hat{\tau}) \leq 0.9$ go to S.

(6)   If $\hat{\phi}(\tau)$ is approximately 1 for some τ, one may set this value of τ equal to $\hat{\tau}$ and go to L. One compares the stationary analysis of this choice of memory-shortening transformation with that determined by the value of τ minimizing Err(t).

(7)   Non-stationary prediction analysis of a time series in general finds coefficients $\phi_1, \ldots, \phi_m$ minimizing (for a specified memory m)

$$\sum_{t=m+1}^{T} \{Y(t) - \phi_1 Y(t-1) - \cdots - \phi_m Y(t-m)\}^2.$$

We recommend a subset regression solution which attempts to determine the most significant lags $j_1, \ldots, j_m$ minimizing

$$\sum_{t=m+1}^{T} \{Y(t) - \phi_{j_1} Y(t-j_1) - \cdots - \phi_{j_n} Y(t-j_n)\}^2$$

and determines the solution for a specified set of lags $j_1, \ldots, j_n$. One may take $n = 2$, and j_1 and j_2 are two adjacent lags ($m-1$ and m) for which $\phi(\tau)$ is approximately 1; one then obtains the transformation of type (3).

A model frequently fitted to monthly economic time series is the so-called 'airline' model (see Parzen, 1979):

$$(I - L)(I - L^{12})Y(t) = (I - \theta_1 L)(I - \theta_{12}L^{12})\varepsilon(t).$$

It seems doubtful that this model would be judged adequate by our criteria, which proposes

$$\tilde{Y}(t) = (I - \phi(12)L^{12})Y(t)$$

$$g_{13}(L)\tilde{Y}(t) = \varepsilon(t).$$

If one desires a parsimonious ARMA model for $\tilde{Y}(t)$ it may be given by

$$\tilde{Y}(t) + \alpha_1 Y(t-1) + \alpha_{12}\tilde{Y}(t-12) + \alpha_{13}\tilde{Y}(t-13) = \varepsilon(t)$$

or

$$\tilde{Y}(t) + \alpha_1\tilde{Y}(t-1) + \alpha_2\tilde{Y}(t-2) = \varepsilon(t) + \beta_{12}\varepsilon(t-12).$$

It should be noted that double differencing is not recommended by us as a memory-shortening transformation. When the need for double differencing arises, it appears as a situation in which long-memory components continue to be present even after several iterations; then the final iterated model is of the form

$$Y(t) \overline{}\!\rightarrow \tilde{Y}^{(1)}(t) \overline{}\!\rightarrow \tilde{Y}^{(2)}(t) \overline{}\!\rightarrow \varepsilon(t).$$

An iterated filter model provides not only forecasts and spectral analysis, but also model interpretation.

Classification of a Time Series into Memory Types

		No-memory	Short-memory	Long-memory
Correlation function	$\bar{\rho} = \sum_{v=1}^{\infty} \lvert \rho(v) \rvert$	$\bar{\rho} = 0$	$0 > \bar{\rho} > \infty$	$\bar{\rho} = \infty$
Spectral density	$M = \dfrac{\max f(\lambda)}{\min f(\lambda)}$	$\log M = 0$	$0 > \log M > \infty$	$\log M = \infty$
Residual variance	σ_α^2	$\sigma_\alpha^2 = 1$	$0 > \sigma_\alpha^2 > 1$	$\sigma_\alpha^2 = 0$
Prediction horizon	HOR	HOR $= 0$	$0 >$ HOR $> \infty$	HOR $= \infty$
Spectral distribution	$F(\lambda)$	$F(\lambda)$ uniform	$F(\lambda)$ continuous	$F(\lambda)$ has sharp jumps
S-play	S-array of Gray	Infinities in column 1	ARMA(p, q) if Constant row p, Alternating constant column q, Infinity column $-(q+1)$	Constant row 1 (trend) Constant row 2 (seasonal)

Graphs of Y and $\tilde{Y}$ (denoted YT) for the 33 times series listed in Table 3. The break in the graphs indicates the end of the observed values of the time series and the beginning of predictions of the next 18 values.

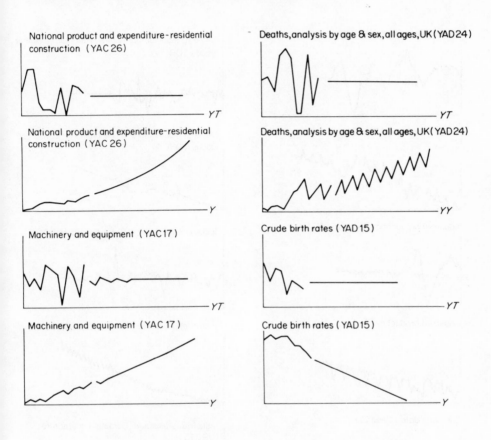

National product and expenditure - residential construction (YAC 26)

National product and expenditure-residential construction (YAC 26)

Machinery and equipment (YAC 17)

Machinery and equipment (YAC 17)

Deaths, analysis by age & sex, all ages, UK (YAD 24)

Deaths, analysis by age & sex, all ages, UK (YAD 24)

Crude birth rates (YAD 15)

Crude birth rates (YAD 15)

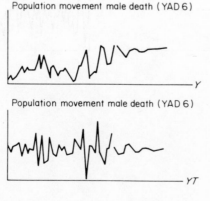

Population movement male death (YAD 6)

Population movement male death (YAD 6)

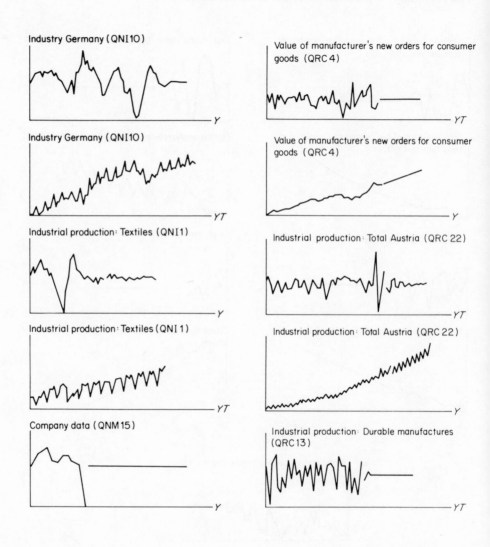

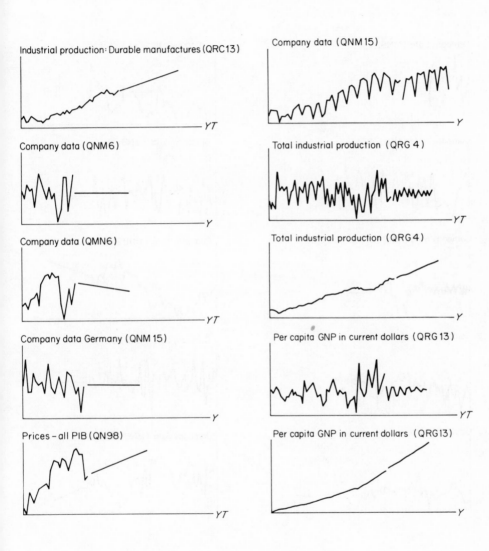

Industrial production: Durable manufactures (QRC13)

Company data (QNM15)

Company data (QNM6)

Total industrial production (QRG 4)

Company data (QMN6)

Total industrial production (QRG4)

Company data Germany (QNM 15)

Per capita GNP in current dollars (QRG 13)

Prices – all PIB (QN98)

Per capita GNP in current dollars (QRG13)

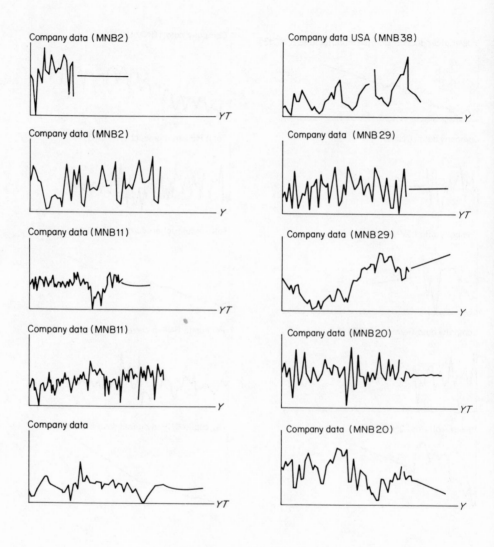

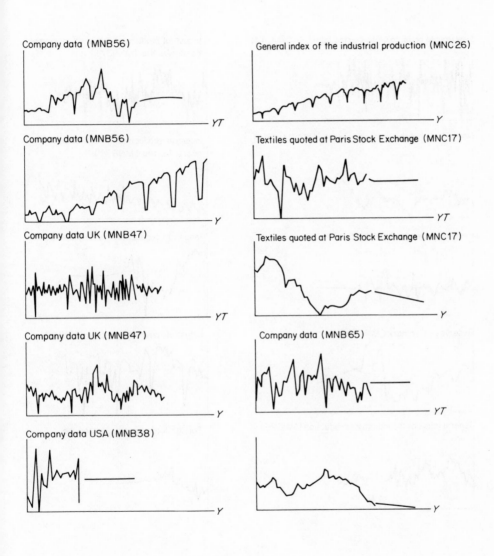

Company data (MNB56)

Company data (MNB56)

Company data UK (MNB47)

Company data UK (MNB47)

Company data USA (MNB38)

General index of the industrial production (MNC26)

Textiles quoted at Paris Stock Exchange (MNC17)

Textiles quoted at Paris Stock Exchange (MNC17)

Company data (MNB65)

New private housing units started total USA (MNC44)

New private housing units started total USA (MNC44)

Reserves – Denmark (MNC35)

Reserves – Denmark (MNC35)

General index of the industrial production (MNC26)

Industrial production: Finished investment goods Austria (MNG37)

Industrial production: Finished investment goods Austria (MNG37)

Industrial production: Spain (MNG28)

Industrial production: Spain (MNG28)

Foreign trade exports Switzerland

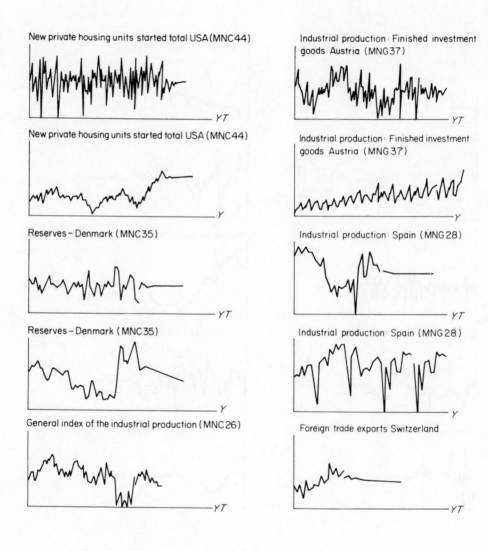

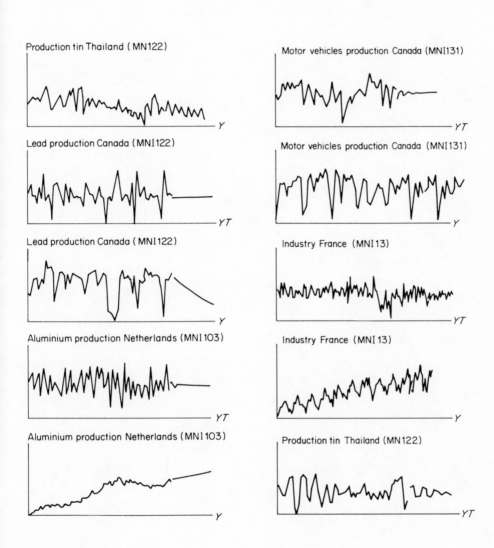

Production tin Thailand (MN122)

Lead production Canada (MNI122)

Lead production Canada (MNI122)

Aluminium production Netherlands (MNI103)

Aluminium production Netherlands (MNI103)

Motor vehicles production Canada (MNI131)

Motor vehicles production Canada (MNI131)

Industry France (MNI13)

Industry France (MNI13)

Production tin Thailand (MN122)

CHAPTER 12

Combining Forecasts

Robert L. Winkler
Indiana University, U.S.A.

1. INTRODUCTION

Different forecasting methods obviously can and do tend to produce different forecasts for the same series. Otherwise, there would be no point in trying to compare the methods. In comparing the individual forecasting methods considered in this study, we see that their properties and performance differ somewhat. In this sense, a forecast from any given method may provide some useful information that is not conveyed in forecasts from other methods. To the extent that we would like to base forecasts (and decisions which use such forecasts as inputs) on all available information, it seems reasonable to aggregate information from different forecasting methods by combining forecasts.

The combination of forecasts has been tried before, with some degree of success. Two noteworthy papers in this regard are Bates and Granger (1969) and Newbold and Granger (1974). In the latter paper, the authors arrive at the following conclusions (p. 143): 'It does appear . . . that Box–Jenkins forecasts can frequently be improved upon by combination with either Holt–Winters or stepwise autoregressive forecasts, and we feel that our results indicate that in any particular forecasting situation combining is well worth trying, as it requires very little effort. Further improvement is frequently obtained by considering a combination of all three types of forecast'.

Two types of combining rules were considered in this study. The first is perhaps the most simple combination rule that could be used: an average of the forecasts provided by different methods. The second is a generalization of a simple average to a weighted average of forecasts, where the weights depend on the relative accuracy of the individual methods and on covariances of forecast errors among the methods.

Details concerning the simple averages and weighted averages are provided in Sections 2 and 3, respectively. Both approaches performed quite well, and an

Supported in part by the US National Science Foundation under Grant IST80 18578 and in part by INSEAD.

evaluation of their performance is given in Section 4. Some implications of these results are discussed in Section 5.

2. SIMPLE AVERAGES OF FORECASTS

The simplest type of combination rule is to take an average of forecasts from different methods. If p methods are used, the combined forecast for period t is of the form

$$\hat{x}_t = \frac{\hat{x}_t^{(1)} + \hat{x}_t^{(2)} + \cdots + \hat{x}_t^{(p)}}{p}, \tag{1}$$

where $\hat{x}_t^{(i)}$ is the forecast from method i. The individual methods are treated in a symmetric manner in (1), with each forecast receiving a weight of $1/p$. This means that information concerning the relative accuracy of the individual methods is not taken into consideration. However, it also means that no model fitting is necessary for the combination rule.

In this study, $p = 6$ methods were chosen to be combined. The six methods are

(1)   deseasonalized exponential smoothing;
(2)   deseasonalized adaptive response rate exponential smoothing;
(3)   deseasonalized Holt's exponential smoothing;
(4)   deseasonalized Brown's linear exponential smoothing;
(5)   Holt–Winters linear and seasonal exponential smoothing; and
(6)   automatic filtering.

The simple average of the forecasts provided by these six methods is labelled 'Combining A'. The six methods are relatively easy and cheap to implement, which means that Combining A is not difficult to apply.

3. WEIGHTED AVERAGES OF FORECASTS

Why might we want to consider a weighted average instead of a simple average? Perhaps one method tends to be more accurate than another, and thus it might seem reasonable to give the first method more emphasis than the second when the methods are combined. Also, potential dependence among the methods (in terms of correlations among their forecast errors) may influence the relative emphasis that should be given to different methods. For an extreme example, suppose that we have three methods and that the correlations among their forecast errors are $\rho_{12} = 0$, $\rho_{13} = 0$, and $\rho_{23} = 1$. In this case, the forecasts provided by the second and third methods are redundant and are thus given too much weight in a simple average.

Dependence among methods and differences in the accuracy of methods can be taken into account by using a weighted average. If p methods are used, the combined forecast for period t is of the form

$$\hat{x}_t = \sum_{i=1}^{p} w_i \hat{x}_t^{(i)} \tag{2}$$

with

$$w_i = \sum_{j=1}^{p} \alpha_{ij} \bigg/ \sum_{h=1}^{p} \sum_{j=1}^{p} \alpha_{hj}, \tag{3}$$

where $\hat{x}_t^{(i)}$ is the forecast from method i and the α_{ij} terms are elements of the inverse of the covariance matrix of percentage errors. That is, if

$$u_t^{(i)} = e_t^{(i)}/x_t = \frac{x_t - \hat{x}_t^{(i)}}{x_t} \qquad \text{for } i = 1, \ldots, p \tag{4}$$

and

$$\mathbf{S} = (\beta_{ij}), \tag{5}$$

where

$$\beta_{ij} = \text{cov}\{u_t^{(i)}, u_t^{(j)}\} \qquad \text{for } i \neq j \tag{6}$$

and

$$\beta_{ii} = V\{u_t^{(i)}\}, \tag{7}$$

then α_{ij} is the element in row i and column j of $\mathbf{S}^{-1}$. Newbold and Granger (1974) show that this type of procedure for determining weights minimizes the variance of the combined forecast error, and Winkler (1981) shows that (2) can be developed as a posterior mean in a Bayesian model.

Unlike the simple average given by (1), the weighted average given by (2) necessitates some model fitting. The covariance matrix $\mathbf{S}$, and hence its inverse $\mathbf{S}^{-1}$, are not known in advance. The sample covariance matrix from the forecasts developed for the model-fitting data ($t = 1, \ldots, n - m$) was used as an estimate of $\mathbf{S}$:

$$\hat{\mathbf{S}} = (\hat{\beta}_{ij})$$

where

$$\hat{\beta}_{ij} = \sum_{t=1}^{n-m} \{u_t^{(i)} - \bar{u}^{(i)}\}\{u_t^{(j)} - \bar{u}^{(j)}\}/(n - m) \tag{9}$$

for $i, j = 1, \ldots, p$. The weights in (3) were then based on $\hat{\alpha}_{ij}$ for $i, j = 1, \ldots, p$, the elements of $\hat{\mathbf{S}}^{-1}$. These weights were computed separately for each series.

The methods combined via (2) in this study were the same six methods listed in Section 2, and the weighted average is labelled 'Combining B'. Combining B is still not too difficult to apply, but because of the model-fitting procedure it requires more time and effort to implement than does Combining A, the simple average.

4. THE PERFORMANCE OF COMBINING A AND COMBINING B

The results indicate that Combining A performed very well indeed in terms of accuracy. It has the best average ranking for both 1001 series and 111 series, and no other method is even close in terms of average ranking. As for MAPE, Combining A has the lowest MAPE for several time horizons and is beaten in overall MAPE only slightly by a few methods.

In terms of pairwise comparisons, Combining A outperformed each of the individual methods over half of the time for 1001 series. The percentage of the time that Combining A was better than a competing method ranged from 53.76 per cent to 69.91 per cent, averaging 59.13 per cent. For 111 series, only Lewandowski's FORSYS system was better than Combining A over 50 per cent of the time. Combining A beat Lewandowski only 48.76 per cent of the time but beat other methods from 52.75 per cent to 69.11 per cent of the time.

The performance of Combining A is quite robust, with good results for all of the different types of series. In general, it seems to have the greatest advantage over other methods for short time horizons. Of course, this could well be due to the particular six methods that are being averaged.

Combining B also performed quite well in the study, better than most of the other methods. The only individual method with a better average ranking for 1001 series is deseasonalized exponential smoothing, and for 111 series only Lewandowski's FORSYS system and Holt–Winters linear and seasonal exponential smoothing had better average rankings. Only a few other methods had lower values of MAPE.

Combining B beat all individual methods in pairwise comparisons for 1001 series. The percentage of time that Combining B was better than a competing method ranged from 51.21 per cent to 66.04 per cent. For 111 series, four individual methods beat Combining B. As is the case with Combining A, Combining B appears to be quite robust.

Both the simple average (Combining A) and the weighted average (Combining B) look very good in this study. The notion that combining forecasts provides a more informative forecast which tends to yield smaller forecast errors is certainly borne out by the empirical results. Both Combining A and Combining B tend to perform better than virtually all or perhaps even all of the individual methods, including the six methods that are being combined as well as the other individual methods considered in the study. The next question of interest is how the two methods for combining forecasts compare with each other.

In comparisons of Combining A versus Combining B, the former comes out slightly ahead on most measures. It has a better average ranking and beats Combining B 53.74 per cent of the time for 1001 series and 55.10 per cent of the time for 111 series. Combining B has a slightly smaller MAPE than Combining A, however. If the weighted average has a slight advantage over the simple average, it might be for longer time horizons and monthly data, but these results are not clear-cut. In general, the differences between Combining A and Combining B are not large, but the former appears to have an edge on an overall basis despite the fact that the latter is more 'sophisticated' and more difficult to apply because it necessitates some model fitting.

5. IMPLICATIONS FOR FORECASTING PRACTICE AND FUTURE WORK

When expert advice is sought, it is sometimes thought useful to consult more than one expert in order to obtain additional information. The performance of Combining A and Combining B in this study suggests that when forecasting, it can be useful to generate forecasts from more than one forecasting method in order to obtain additional information. A combination of forecasts from two or more methods may well be more accurate than forecasts from individual methods. In general, the idea of combining forecasts has much to recommend it for use in actual forecasting practice.

What about the choice between simple averages and weighted averages of forecasts? If the process generating vectors of forecast errors is stationary, the average forecast error is zero (or close to zero), and a reliable estimate of the covariance matrix for forecast errors can be obtained, then a weighted average could take advantage of this information and be more accurate than a simple average. In this study, however, the simple average tended to be slightly more accurate, on average, than the weighted average. Adding to these empirical results the fact that simple averages are easier and cheaper to generate than weighted averages, the nod would have to go to the simple average at the present time. Unless differential weighting can be shown to lead to improved forecasts in practice, it is not worth the extra cost and effort. In a follow-up study (Winkler and Makridakis, 1983), some alternative procedures for estimating weights are shown to provide weighted averages that outperform Combining A.

Combining forecasts, then, seems to be a very promising approach that has not been studied extensively. Thus, it deserves further study. More evidence is needed about the relative merits of different combination rules and on choices of the parameters (e.g. the methods to be combined, the choice of weights) for a given combination rule.

First, consider the question of how many methods should be combined. Six methods were combined in this study, but that is certainly not a 'magic number' in any sense. In fact, a simple average of four methods (the first four of the six in Combining A) was also considered, and the results are very similar to those for Combining A. A follow-up study of the effect of the number of methods in the average (Makridakis and Winkler, 1983) indicates that increasing the number of methods tends to increase accuracy and reduce the effect of the specific methods being averaged, with the impact of adding an extra method diminishing as the number of methods already in the average increases. The choice of the number of methods should involve a balancing of anticipated increases in accuracy with increased costs in terms of money and effort.

Next, how about the question of which methods to include in a combination? This question is intertwined with the choice of the number of methods, of course. The six methods used in Combining A and Combining B were chosen because they were relatively easy to use, making the resulting averages feasible to consider in practice. It is easier to generate a forecast via Combining A than via some of the more complex individual forecasting methods. But certainly other combinations, some of which would share Combining A's trait of being easy to use and some of which would not share this trait, should be investigated. As for speculation in advance of concrete results, it should seem that it would be advantageous to use methods that are dissimilar to reduce the dependence among methods and increase the amount of 'new' information provided by additional methods. The six methods used in Combining A are, in general, quite similar; as noted above, they were chosen with an eye towards ease of application, not towards maximization of accuracy in any sense. Also, it might be useful to attempt to tailor the choice of methods to particular forecasting situations. For example, Lewandowski's FORSYS system appears to be particularly valuable for long forecast horizons. Thus, it might be a prime candidate for inclusion in situations with long horizons but not necessarily in situations with short horizons.

The joint behaviour of forecast errors from different forecasting methods as well as forecast errors from various combining rules deserves attention. Do joint distributions of forecast errors demonstrate enough stability to make it possible to take advantage of knowledge about these distributions via the use of weighted averages? Is modelling of forecast methods and errors worthwhile, or should we keep matters simple and just stick with simple averages?

Finally, it is important to note that combining forecasts is not confined to combinations utilizing time series methods as was the case in this study. The desire to consider any and all available information means that forecasts from different types of sources should be considered. For example, we could combine forecasts from time series methods with forecasts from econometric models and with subjective forecasts from experts. Apparently combinations involving only subjective forecasts are currently used by some companies for major decisions (Koten, 1981). The combination of so-called 'objective' forecasts with subjective forecasts is a very promising notion.

REFERENCES

Bates, J. M. and Granger, C. W. J. (1969). 'The Combination of Forecasts', *Operational Research Quarterly*, **20**, 451–468.

Koten, J. (1981). 'They Say No Two Economists Ever Agree, So Chrysler Tries Averaging Their Opinions', *Wall Street Journal*, 3 November.

Makridakis, S. and Winkler, R. L. (1983). 'Averages of Forecasts: Some Empirical Results', *Management Science*, **29**, 983–996.

Newbold, P. and Granger, C. W. J. (1974). 'Experience with Forecasting Univariate Time Series and the Combination of Forecasts', *Journal of the Royal Statistical Society*, Series A, **137**, 131–165.

Winkler, R. L. (1981). 'Combining Probability Distributions from Dependent Information Sources', *Management Science*, **27**, 479–488.

Winkler, R. L. and Makridakis, S. (1983). 'The Combination of Forecasts', *Journal of the Royal Statistical Society*, Series A, **146**, 150–157.

Index